A Practical Course on Quantum Monte Carlo

Online at: https://doi.org/10.1088/978-0-7503-6310-5

IOP Series in Quantum Technology

Series Editor: **Barry Garraway** (School of Mathematical and Physical Sciences, University of Sussex, UK)

About the series

The IOP Series in Quantum Technology is dedicated to bringing together the most up to date texts and reference books from across the emerging field of quantum science and its technological applications. Prepared by leading experts, the series is intended for graduate students and researchers either already working in or intending to enter the field. The series seeks (but is not restricted to) publications in the following topics:

- Quantum biology
- Quantum communication
- Quantum computation
- Quantum control
- Quantum cryptography
- Quantum engineering
- Quantum machine learning and intelligence
- Quantum materials
- Quantum metrology
- Quantum optics
- Quantum sensing
- Quantum simulation
- Quantum software, algorithms and code
- Quantum thermodynamics
- Hybrid quantum systems

A list of titles published in this series can be found here: https://iopscience.iop.org/bookListInfo/iop-series-in-quantum-technology.

A Practical Course on Quantum Monte Carlo

Vesa Apaja
Department of Physics, Nanoscience Center, University of Jyväskylä, Jyväskylä, Finland

IOP Publishing, Bristol, UK

ISBN   978-0-7503-6310-5 (ebook)
ISBN   978-0-7503-6308-2 (print)
ISBN   978-0-7503-6311-2 (myPrint)
ISBN   978-0-7503-6309-9 (mobi)

DOI   10.1088/978-0-7503-6310-5

Version: 20250801

IOP ebooks

British Library Cataloguing-in-Publication Data: A catalogue record for this book is available from the British Library.

Published by IOP Publishing, wholly owned by The Institute of Physics, London

IOP Publishing, No.2 The Distillery, Glassfields, Avon Street, Bristol, BS2 0GR, UK

US Office: IOP Publishing, Inc., 190 North Independence Mall West, Suite 601, Philadelphia, PA 19106, USA

Contents

5   Path integral Monte Carlo   5-1

Preface

Learning quantum mechanics traditionally starts with the historical nexus points at the dawn of quantum ideas: Max Planck's struggles to find what lies behind the black-body spectrum, Albert Einstein's photoelectric effect, the double-slit experiment, and so on. That is the easy part; then you will get deep into hydrogen atom and harmonic oscillator calculations, along with Legendre and Hermite polynomials and series solutions. Most students and lecturers are in a hurry to go through all the topics considered important, which gives the impression that quantum mechanics is a collection of recipes. It is not, although I am not suggesting you should ponder the philosophy of quantum mechanics or what it 'means,' just that quantum mechanics has something to say about the world around us and how we perceive it.

I would like to think that the essence of quantum mechanics is the Heisenberg uncertainty principle. We simply cannot know everything about a quantum system, that is, *any* system, not even in principle. This lack of knowledge is where quantum Monte Carlo (QMC) has an advantage, because there are often far too many hidden factors to account for with pen and paper. QMC tries out a few—often a few million—options at random and tells us roughly what is going on. To me, QMC makes the wave function and all the what-ifs of quantum mechanics tangible. No mysteries, just possibilities the world has hidden from our direct view, but which are nevertheless there and make up what we call reality.

The explanations in this book are sometimes lengthy because they follow my way of thinking, which is not always the shortest route. If you go on reading, which I hope you do, you will find me revisiting topics such as 'What does diffusion have to do with quantum mechanics?' (because it is a natural way to express what happens if we know less) and 'Why is free-space diffusion so popular in QMC?' (and why it causes so much trouble at boundaries).

Most equations in this book are written in coordinate space, mainly because potential energy has particle coordinates. Quantities expressed in coordinates are sometimes easier to grasp than abstract quantum states. For example, the finite temperature partition function Z is a sum over all states, where one sums a certain quantity. In coordinate space, Z is a sum (an integral) of all positions. Later in this book, we will write Z as a path integral over all possible paths that start from any place x and end in the same place x. Whatever was coded as quantum states is now represented in spatial coordinates x, and we all know what coordinates 'are' and how to write computer programs that use them.

The programs referred to in the book are available on GitHub, and you can clone them using `git clone https://github.com/VesaApaja/Book_Codes.git`. I wrote the programs in Python and Julia using old-school scripts to keep them human-readable, regardless of changes made to the Jupyter notebooks or other graphical environments' formats. Julia compilation takes place during the first pass, but the slightly annoying delay is rewarded with faster execution. I also appreciate the way math formulas look almost the same in Julia programs. Even Greek letters are there, so it is a simple matter to reverse engineer what a Julia function is computing. In the path integral Monte Carlo Julia program, I tried to pay attention to fast execution, avoiding the allocation of temporary arrays and using explicit loops that can be optimized by the compiler.

This book is based on the lecture notes I wrote during my years in the Department of Physics at Johannes Kepler University in Linz, Austria, and for the QMC lectures given at the University of Jyväskylä, Finland. I would like to express my deep thanks to two colleagues whose support has been invaluable, Dr. Ferran Mazzanti at Universitat Politècnica de Catalunya, Spain, and Dr. Robert Zillich at Johannes Kepler University, and to all my students, whose thoughtful questions and suggestions helped shape this work.

Last but not least, I would like to thank my wife Karoliina and my son Otso for being there for me.

Author biography

Vesa Apaja

I am a seasoned theoretical physicist with a strong focus on quantum many-body physics and computational methods. My research career began with quantum fluids as a research assistant at the Department of Physics, University of Oulu, Finland (1990–99), where I became increasingly involved in computational many-body theory and also wrote my first QMC code.

In 1999, I joined the Many-Body Theory Group at the Institut für Theoretische Physik, Johannes Kepler Universität, Linz, Austria. As a Universitätsassistent, I broadened my research interests to include excitations in quantum fluids, diagrammatic many-body theory, and statistical physics.

Since 2006, I have been a senior researcher at the Department of Physics, Nanoscience Center, University of Jyväskylä, Finland. My work spans a diverse range of condensed-matter topics, including flat band lattice dynamics and surface electrochemistry using kinetic Monte Carlo techniques.

Over the years, I have authored lecture notes on statistical physics, efficient numerical programming, numerical applications in physics, and QMC. Teaching these subjects has become a major focus of my recent academic activity.

I am also an experienced programmer, proficient in Python, Julia, Fortran, and C++, which I use to develop advanced simulations and computational models for analyzing complex physical systems.

IOP Publishing

A Practical Course on Quantum Monte Carlo

Vesa Apaja

Chapter 1

Introduction

Quantum Monte Carlo (QMC) is about solving nonrelativistic quantum mechanics problems with the aid of random numbers. In the vast majority of scenarios in solid-state physics and quantum chemistry, the energies are so low, $E \ll mc^2$, that we can safely ignore the possibility that new particles with mass m emerge or are annihilated. Electron velocities are well below the speed of light; albeit in gold atoms, the 1s electrons have relativistic speeds.

QMC has nothing to do with quantum computing, which may be a small disappointment to some readers. What QMC *can* offer is a down-to-earth view of quantum mechanics. With this method, quantum problems are solved by completing numerical tasks. What you write in computer code is very precise. Your algorithms may be faulty, but computers do not accept any hand-waving, and there is no room for vague expressions. Once you read QMC code, you see all there is to see. The wave function, a mysterious complex-valued object, is in code just that: a function. You throw in a few numbers and the code spits out one complex number, sometimes a real number. That is the definition of a function. Wave functions are central in QMC, and nonrelativistic quantum mechanics is all about solving them using the Schrödinger equation and computing the physical properties they imply.

1.1 The Schrödinger equation

In nonrelativistic quantum mechanics, the Hamiltonian of particles in potential $V(\mathbf{x})$ is the operator

$$\hat{\mathcal{H}} = -\sum_{i=1}^{N} \frac{\hbar^2}{2m_i} \nabla_i^2 + V(\mathbf{x}),$$
(1.1)

where m_i is the mass of the ith particle. To shorten the notation, I denote the positions of N particles as the single vector

$$\mathbf{x} := (\mathbf{r}_1, \mathbf{r}_2, \ldots, \mathbf{r}_N),$$
(1.2)

doi:10.1088/978-0-7503-6310-5ch1   1-1

where $\mathbf{r}_i$ is the d-dimensional position of the ith particle. As indicated, the potential $V(\mathbf{x})$ is usually known in coordinate space. The simplest case is the one with only a confining potential that acts separately on each particle. Such a single-particle potential can be written as a sum

$$V(\mathbf{x}) = \sum_{i=1}^{N} V_{\text{conf}}(\mathbf{r}_i). \qquad (1.3)$$

For a harmonic oscillator centered at the origin, one has $V_{\text{conf}}(\mathbf{r}_i) = k \, |\mathbf{r}_i|^2$, with spring constant k.

In atomic and molecular systems, the interparticle potential is the Coulomb potential. Customarily, one makes the Born–Oppenheimer (BO) approximation and assumes that in electronic timescales, the ions are immobile. Keeping A ions with charges Z_j fixed at positions $\mathbf{R}_j$, the N electrons move in the electron–nucleus potential

$$V_{\text{en}}(\mathbf{x}): = \sum_{j=1}^{A}\sum_{i=1}^{N} - \frac{Z_j}{|\mathbf{r}_i - \mathbf{R}_j|}, \qquad (1.4)$$

and the electron–electron potential is a sum over electron pairs,

$$V_{\text{ee}}(\mathbf{x}): = \sum_{i<j,1}^{N} \frac{1}{|\mathbf{r}_i - \mathbf{r}_j|}. \qquad (1.5)$$

The potentials are given in Hartree atomic units (a.u.), where four fundamental constants are set to unity:

$$\hbar = 1 \qquad \text{Planck's constant} \qquad (1.6)$$

$$e = 1 \qquad \text{elementary charge} \qquad (1.7)$$

$$m_e = 1 \qquad \text{electron mass} \qquad (1.8)$$

$$k_e = 1 \qquad \text{Coulomb constant.} \qquad (1.9)$$

Hartree atomic units are convenient in atomic physics, and the electron Hamiltonian is

$$\hat{\mathcal{H}} = -\frac{1}{2}\sum_{i=1}^{N} \nabla_i^2 + V_{\text{ee}}(\mathbf{x}) + V_{\text{en}}(\mathbf{x}) \quad \text{electron Hamiltonian.} \qquad (1.10)$$

In liquids and solids, the interparticle potential typically has a repulsive core and a long-range attraction, given roughly by the Lennard-Jones 6-12 potential,

$$V_{\text{LJ}}(r) = \epsilon_0 \left[\left(\frac{r_0}{r}\right)^{12} - \left(\frac{r_0}{r}\right)^{12} \right] \qquad \text{Lennard-Jones,} \qquad (1.11)$$

with constants ϵ_0 and r_0. The repulsive interaction may arise from filled electron orbitals that do not like to overlap, while the attraction comes from a long-range van der Waals interaction. On a more fundamental level, there is only the Coulomb potential, but adding up the contributions of Coulomb potentials between atoms is such a tremendous task that one takes a shortcut and uses a higher-level potential that captures the atom–atom interaction in liquid and solid densities and computes the potential energy as a sum over particle pairs of such pairwise interactions:

$$V(\mathbf{x}) = \sum_{i<j,1}^{N} V_{\mathrm{LJ}}(|\mathbf{r}_i - \mathbf{r}_j|). \tag{1.12}$$

The most convenient choice of units is less obvious, but if particles have the same mass, one can write the Hamiltonian in the form

$$\hat{\mathcal{H}} = -D\,\nabla_{\mathbf{x}}^2 + V(\mathbf{x}) \qquad \text{liquid Hamiltonian}, \tag{1.13}$$

where the N-particle Laplacian is

$$\nabla_{\mathbf{x}}^2 := \sum_{i=1}^{N} \nabla_i^2 . \tag{1.14}$$

The notation $D = \hbar^2/(2m)$ will be useful in the diffusion Monte Carlo (DMC) (chapter 4), where D takes the role of a diffusion constant. Nothing prevents us from also using the D-notation in atomic and molecular systems.

The Hamiltonian is what we assume about the system under investigation. We assume there is a certain fixed number of particles, and the interaction between them and the surrounding world is given. The obvious next question is: under these assumptions, what can we know about the system? That information is encoded in the abstract time-dependent Hilbert-space state vector $|\psi(t)\rangle$, represented in coordinate space by the wave function

$$\Psi(\mathbf{x}, t) := \langle \mathbf{x}|\Psi(t)\rangle. \tag{1.15}$$

The assumptions in $\hat{\mathcal{H}}$ and the knowledge about the system $|\Psi\rangle$ are related; in nonrelativistic quantum mechanics, the relationship between them is given by the time-dependent Schrödinger equation,

$$\hat{\mathcal{H}}|\Psi(t)\rangle = -i\hbar\frac{\partial}{\partial t}|\Psi(t)\rangle, \tag{1.16}$$

in coordinate space

$$\hat{\mathcal{H}}\Psi(\mathbf{x}, t) = -i\hbar\frac{\partial}{\partial t}\Psi(\mathbf{x}, t). \tag{1.17}$$

Why use the Hamiltonian and not the Lagrangian? Quantum field theory, a combination of special relativity and quantum mechanics, uses the Lagrangian as a starting point because Lorentz invariance is easier to write in a Lagrangian formulation from the outset. In nonrelativistic quantum mechanics, the Hamiltonian gives us direct access to the energy of the system. In the ith eigenstate $|\phi_i\rangle$ of the Hamiltonian, the energy is

$$\hat{\mathcal{H}}|\phi_i\rangle = E_i|\phi_i\rangle . \tag{1.18}$$

If the potential is time-independent, then the time evolution of an arbitrary state $|\Psi\rangle$ is given by

$$|\Psi(t)\rangle = e^{-i\hat{\mathcal{H}}/\hbar t}|\Psi(0)\rangle. \tag{1.19}$$

Every state can be written as a superposition of eigenstates $|\phi_i\rangle$, meaning all possible properties of the system are present in a combination of eigenstates. Therefore,

$$|\Psi(0)\rangle = \sum_i c_i|\phi_i\rangle, \tag{1.20}$$

where the sum is over all eigenstates (generalized to an integral if necessary), and the expansion coefficients c_i are complex numbers. The time evolution of any state follows from the time evolution of the eigenstates,

$$|\Psi(t)\rangle = e^{-i\hat{\mathcal{H}}t/\hbar}\sum_i c_i|\phi_i\rangle = \sum_i c_i e^{-iE_i t/\hbar}|\phi_i\rangle. \tag{1.21}$$

From the wave function, one obtains the probability density of finding particles at $\mathbf{x}$ at time t,

$$P(\mathbf{x}, t) = |\Psi(\mathbf{x}, t)|^2. \tag{1.22}$$

The identification of $|\Psi|^2$ as the probability was made by Max Born in 1926 [1]. Conservation of probability is a fundamental principle in nonrelativistic quantum mechanics. The probability density would be even more useful if only there were an equation to solve it from directly. Using the Schrödinger equation, one gets as far as

$$-i\hbar\frac{\partial|\Psi(\mathbf{x}, t)|^2}{\partial t} = -\frac{\hbar^2}{2m}(\Psi^*(\mathbf{x}, t)\,\nabla_{\mathbf{x}}^2\,\Psi(\mathbf{x}, t) + \Psi(\mathbf{x}, t)\,\nabla_{\mathbf{x}}^2\,\Psi^*(\mathbf{x}, t))$$
$$+ 2V(\mathbf{x})|\Psi(\mathbf{x}, t)|^2, \tag{1.23}$$

but there is no way to write it all just in terms of $|\Psi(\mathbf{x}, t)|^2$. The phase of the wave function cannot be completely eliminated.

The ground-state wave function represents the least amount of information we can have about the positions of particles in a system. Take a particle in a box in one dimension, extending from zero to L; these assumptions are written down in the Hamiltonian. The Hamiltonian eigenstates are

$$\phi_i(x) = \left(\frac{2}{L}\right)^{1/2} \sin(k_i x), \tag{1.24}$$

where the quantized values are $k_i := i\,\pi/L$ and the index i takes the values $i = 1, 2, 3, \dots$. A fact worth noticing is that the wave function has no reference to the particle mass m, and you can immediately calculate the probability density of particles in each eigenstate. The states have energies $\epsilon_i = \hbar^2 k_i^2/(2m)$, which do depend on m. These eigenstates list the possibilities of keeping the particle in the box, as $\hat{\mathcal{H}}$ stipulates. The indices start from $i = 1$, because a state with $i = 0$ would have $\phi_0 = 0$, which is impossible because, contrary to the assumptions, there would not be any particles. Classically, one would assume that the least information about the system corresponds to a state where the particle is at a random position in the box. However, this is not the quantum ground state, because steep gradients cost energy. The quantum ground state

$$\phi_1(x) = \left(\frac{2}{L}\right)^{1/2} \sin\left(\frac{\pi}{L}x\right) \tag{1.25}$$

is a compromise between achieving random spreading of the particle position while avoiding walls with large gradients. Random spreading is obviously related to diffusion, a classical idea of particles tending to move to regions of lower density. This analogy will be used in chapter 4, where we formulate the DMC method of finding the quantum ground state.

If we know the particle is in the first excited state,

$$\phi_2(x) = \left(\frac{2}{L}\right)^{1/2} \sin\left(2\frac{\pi}{L}x\right), \tag{1.26}$$

we have a little bit more information, namely that the particle will never be in the middle, i.e. at $x = L/2$. Adding nodes to the wave function increases the number of impossible locations, and consequently, we know more about the system. Information content in a state increases with increasing energy.

Solving the time evolution of a state is generally a very difficult task. According to equation (1.41), the number of Hamiltonian eigenstates one would need to know increases rapidly over time. The energy eigenstates rotate in the complex plane with a frequency proportional to the states energy, and numerically high-frequency oscillations resemble noise. Even solving just the ground state of a many-body system is a remarkable achievement, and we are generally quite satisfied with a good approximation. The computation of the ground-state wave function and the ground-state properties of many-body systems is the main topic of this book, but the excited states are not completely forgotten, as they play a role in finite-temperature calculations.

1.2 Indistinguishable particles and the ground state

Indistinguishable particles in three dimensions are either *bosons or fermions*. Bosons have symmetric wave functions and fermions have antisymmetric wave functions under particle exchange. The indistinguishability of identical particles goes as deep as one can get: there is no way to tell particles apart, not even in principle.[1] Every particle carries both position and *spin*.[2] The spin-statistics theorem [2–4] says that integral-spin (0, 1, 2,...) particles have symmetric wave functions, and half-integral-spin ($\frac{1}{2}, \frac{3}{2}, \frac{5}{2}$,...) particles have antisymmetric wave functions, something that cannot be proven without special relativity. Nonrelativistic quantum mechanics takes spin

---

[1] This hints that there is an underlying 'field' that spawns identical particles, but that is the concern of quantum field theory.
[2] Hypothetical Majorana fermions are particles with spatially separated position and spin degrees of freedom, but so far, all known particles have position and spin tied together.

and wave function (anti)symmetry into account as an axiomatic, extra requirement for the solution of the Schrödinger equation. In this book, I only consider spin-independent interactions $V(\mathbf{x})$, so the whole Hamiltonian is spin-independent.

Position and spin can be expressed compactly as a composite position–spin coordinate,

$$x_i: = (\mathbf{r}_i, \sigma_i) \qquad i\text{ th particle}, \tag{1.27}$$

but to simplify notation, one customarily writes just i instead of x_i. Particle exchange means the exchange of particle labels, both position and spin. Consider two particles. Their wave function satisfies

$$\psi(1, 2) = \psi(2, 1) \quad \text{bosons, symmetric wave function} \tag{1.28}$$

$$\psi(1, 2) = -\psi(2, 1) \quad \text{fermions, antisymmetric wave function}, \tag{1.29}$$

where

$$\psi(1, 2): = \psi(x_1, x_2): = \psi((\mathbf{r}_1, \sigma_1), (\mathbf{r}_2, \sigma_2)). \tag{1.30}$$

For convenience, the wave function is split into spatial and spin parts,

$$\psi(1, 2) = \Psi(\mathbf{r}_1, \mathbf{r}_2)\Xi(\sigma_1, \sigma_2): = \Psi(\mathbf{r}_1, \mathbf{r}_2)\Xi(1, 2), \tag{1.31}$$

where the last form is yet another shorthand notation.[3] We are then at liberty to assign symmetry or antisymmetry to either part and combine them to make a (total) wave function symmetric or antisymmetric. The spatial part of the wave function describes the probability distribution of finding the particle at a particular location, and the spin part describes the orientation of its spin.

For example, a state as simple as $\phi_0(\mathbf{x}) = 1$ in a finite region $\mathbf{x}$ would be a boson state because it is symmetric and normalizable. This would correspond to the classical particle-in-a-box probability $P(\mathbf{x}) = 1$, but the fact that the probability drops discontinuously to zero at the walls would make it a high-energy quantum state. The unity wave function also describes noninteracting bosons because their positions $\mathbf{x}$ have no effect. Potentials, confinement, and interparticle potential introduce correlations to particle positions, and much of this book is about how to add them to the wave function.

1.2.1 Introducing orbitals

The function $\Psi(\mathbf{r}_1, \mathbf{r}_2) = \mathbf{r}_1\mathbf{r}_2$ is symmetric; however, is not a good wave function, because it is not normalizable. Due to the Heisenberg uncertainty principle, localizing particles to points is not meaningful, so the coordinates $\mathbf{r}_1$ and $\mathbf{r}_2$ can only be the *possible locations* of the two particles. Hence, we need to tie the coordinates to some 'intermediate' functions that dilute their sharpness. The minimal choice is that both coordinates carry with them a function φ, such that

---

[3] For that matter, $\psi(1, 2) = \Psi(1, 2)\Xi(1, 2)$ is also understandable, since the function name uniquely determines whether '1' means $\mathbf{r}_1$, or σ_1, or a combination of both.

$$\Psi(\mathbf{r_1}, \mathbf{r_2}) = \varphi(\mathbf{r_1})\varphi(\mathbf{r_2}). \tag{1.32}$$

The function φ is called a *single-particle wave function* or a *single-particle orbital*. In other words, both particles are in the same single-particle state. If you apply this idea to very many bosons, you have the wave function of a *Bose–Einstein condensate*. Another choice is that the two particles carry different functions,

$$\Psi(\mathbf{r_1}, \mathbf{r_2}) = \varphi_a(\mathbf{r_1})\varphi_b(\mathbf{r_2}) + \varphi_b(\mathbf{r_1})\varphi_a(\mathbf{r_2}). \tag{1.33}$$

In other words, particles are in different single-particle states in a symmetric combination. The two particles mind their own business because there is nothing that depends on their mutual distance, so this is a wave function for *noninteracting particles*. It is still not a classical ideal gas of two particles; the symmetry of the wave function does make a difference.

Since we have not mentioned spin, the orbitals $\varphi(\mathbf{r})$ we have been talking about are called *spatial orbitals*. They hold information about the probability of finding a particle at a given location but say nothing about the spin.

What about antisymmetric spatial wave functions and the Pauli principle? The function $\Psi(\mathbf{r_1}, \mathbf{r_2}) = \mathbf{r_1} - \mathbf{r_2}$ is antisymmetric but not normalizable. Again, we introduce intermediate functions. The minimal choice is

$$\Psi(\mathbf{r_1}, \mathbf{r_2}) = \varphi_1(\mathbf{r_1})\varphi_2(\mathbf{r_2}) - \varphi_2(\mathbf{r_1})\varphi_1(\mathbf{r_2}) = \begin{vmatrix} \varphi_1(\mathbf{r_1}) & \varphi_2(\mathbf{r_1}) \\ \varphi_1(\mathbf{r_2}) & \varphi_2(\mathbf{r_2}) \end{vmatrix}. \tag{1.34}$$

A single function φ will not do, because it would leave $\Psi(\mathbf{r_1}, \mathbf{r_2}) = 0$. The determinant-form wave function is a *Slater determinant* [5, 6]. The extension to more particles is straightforward: just add more orbitals and make bigger determinants.

In physics, a position vector $\mathbf{r}$ is written in a coordinate representation, with a specific point of reference. In atoms, a natural choice for the origin is the location of the nucleus. In molecules, the physical reference point can be a center of mass or one of the nuclei. For large numbers of molecules in a system, the choice of the origin can be anything, even a point outside the system. This increasing freedom in the choice of the physical reference point reflects the fact that the concept of a single-particle orbital $\varphi(\mathbf{r})$ loses some of its descriptive potential. Nevertheless, we stubbornly use single-particle orbitals in models for large systems because they offer a convenient way to keep the total wave function antisymmetric.

Fixing the point of reference to a nucleus has the downside that while it can capture the nucleon–electron correlations caused by the Coulomb interaction, it is less efficient in capturing the electron–electron correlations. Ignoring the latter, the occupied orbitals—those whose coordinates represent the possible locations of an electron—would be the lowest-energy ones, and the total wave function is a single Slater determinant of these occupied orbitals. Taking into account electron–electron correlations breaks this simple picture, and if you work within the same nucleus-based orbitals as before, you need to add orbitals that were previously unoccupied and collect them to separate Slater determinants. The inefficiency of nucleus-based orbitals manifests itself as a need to add many Slater determinants to obtain accurate electron–electron correlations. We shall frequently take a shortcut and reduce the need to sum too many Slater determinants by using an approximate electron–electron correlation function factor in the many-body wave function.

Orbitals can also be 'detached from reality' and treated as a computational tool. In the Kohn–Sham density functional theory, the single-particle orbitals $\varphi_i^{KS}(\mathbf{r})$ are a set of fictitious functions that give the electron density as follows:

$$n(\mathbf{r}) = \sum_i |\varphi_i^{KS}(\mathbf{r})|^2, \tag{1.35}$$

where the sum is over all occupied orbitals. Now things become interesting: a Kohn–Sham orbital $\varphi_i^{KS}(\mathbf{r})$ is not occupied by a real electron, so the coordinate $\mathbf{r}$ does not represent a possible electron location. Instead, the system is made of fictitious, noninteracting spin-half fermions whose density $n(\mathbf{r})$ matches that of real, interacting electrons. If there are N electrons, the occupied orbitals are the lowest-energy ones that can be occupied by N fictitious fermions; the energies of the orbitals are solved self-consistently along with the orbitals themselves.

1.2.1.1 Electron spin state

In nonrelativistic quantum physics, spin is introduced as a postulate, and whatever is needed is added as an additional component to the formalism.

An electron is a spin-1/2 particle, meaning the eigenvalue of $\hat{S}^2$ is always $s(s+1) = 3/4$ with $s = 1/2$, and it is enough to specify the eigenvalue of the projected spin operator $\hat{S}_z$. Technically, the single-electron spin state is a two-component spinor, and all operators acting on it are 2x2 matrices. This matrix–vector notation becomes rather tedious in multielectron systems. In *quantum chemistry*, one customarily introduces the *spin functions* $\alpha(\sigma)$ and $\beta(\sigma)$. Here, σ is called the *spin variable* or the spin coordinate, $\sigma = \frac{1}{2}$ for spin up, and $\sigma = -\frac{1}{2}$ for spin down. The definitions of spin functions are as follows:

$$\alpha(\frac{1}{2}) = 1, \qquad \alpha(-\frac{1}{2}) = 0 \tag{1.36}$$

$$\beta\left(\frac{1}{2}\right) = 0, \qquad \beta\left(-\frac{1}{2}\right) = 1. \tag{1.37}$$

Here, α and β are called the spin-up function and the spin-down function, respectively, because they project out the corresponding spin values. The reason for introducing spin functions is that now it is possible to write, for example,

$$\psi(\mathbf{r}, \sigma) = \varphi_\alpha(\mathbf{r})\alpha(\sigma) + \varphi_\beta(\mathbf{r})\beta(\sigma), \tag{1.38}$$

instead of

$$\psi(\mathbf{r}, \sigma) = \begin{pmatrix} \varphi_\alpha(\mathbf{r}) \\ \varphi_\beta(\mathbf{r}) \end{pmatrix}. \tag{1.39}$$

1.2.1.2 Spin orbital: spatial orbital and spin
A *spin orbital* is a single-particle orbital that has both spin and spatial information:[4]

$$\chi(\boldsymbol{x}): =\chi(\mathbf{r}, \sigma) \qquad \text{spin orbital.} \tag{1.40}$$

The motivation is to tie together position and spin; however, the notation is somewhat artificial because position is a continuous variable, while spin is a discrete variable. In this formalism, an electron has three spatial coordinates and a spin coordinate. The possible spin orbitals are either $\chi(\boldsymbol{x}) = \varphi(\mathbf{r})\alpha(\sigma)$ or $\chi(\boldsymbol{x}) = \varphi(\mathbf{r})\beta(\sigma)$.

1.2.1.3 Hydrogen atom
In a hydrogen atom, the electron state is a spin orbital specified by four quantum numbers,

$$|\psi_{n,l,m_l,m_s}\rangle = |\Psi_{n,l,m_l}\rangle|\sigma_s^{m_s}\rangle. \tag{1.41}$$

An electron in a $2P_z$ spin-down state is in the spin orbital $|\Psi_{2,1,0}\rangle|\beta\rangle$. The quantum numbers give the eigenvalues of the corresponding operators:

$$\hat{\mathcal{L}}^2|\psi_{n,l,m_l,m_s}\rangle = l(l + 1)|\psi_{n,l,m_l,m_s}\rangle \quad \text{(orbital) angular momentum} \tag{1.42}$$

$$\hat{\mathcal{L}}_z|\psi_{n,l,m_l,m_s}\rangle = m_l|\psi_{n,l,m_l,m_s}\rangle \quad \text{angular momentum } z \text{ component} \tag{1.43}$$

$$\hat{S}^2|\psi_{n,l,m_l,m_s}\rangle = s(s + 1)|\psi_{n,l,m_l,m_s}\rangle \quad \text{spin (angular momentum)} \tag{1.44}$$

$$\hat{S}_z|\psi_{n, l, m_l, m_s}\rangle = m_s|\psi_{n, l, m_l, m_s}\rangle \quad \text{spin } z \text{ component.} \tag{1.45}$$

1.2.2 Exact multielectron wave function from spin orbitals

The process of constructing a multielectron wave function from spin orbitals is described in the classic book by Szabo and Ostlund [7]. Any single-variable function of a composite position and spin coordinate $\boldsymbol{x}_i$ can be expanded as follows:

$$\Psi(\boldsymbol{x}_1) = \sum_i a_i \chi_i(\boldsymbol{x}_1). \tag{1.46}$$

The weak point is that this is valid only for a complete set of spin orbitals, so consider the process as a formal proof. If another coordinate $\boldsymbol{x}_2$ is held fixed, then we have

$$\Psi(\boldsymbol{x}_1, \boldsymbol{x}_2) = \sum_i a_i(\boldsymbol{x}_2)\chi_i(\boldsymbol{x}_1), \tag{1.47}$$

---

[4] A spatial orbital has just spatial information, so a spin orbital has just spin information, right? Wrong.

where the expansion coefficients depend on x_2 as if it were a parameter. The notation already anticipates that parameters x_2 can be taken as function arguments. We also expand $a_i(x_2)$ in the complete set of spin orbitals, obtaining

$$a_i(x_2) = \sum_j b_{ij} \chi_j(x_2), \tag{1.48}$$

and we have a general two-particle wave function expanded in spin orbitals:

$$\Psi(x_1, x_2) = \sum_{i,j} b_{ij} \chi_i(x_1) \chi_j(x_2). \tag{1.49}$$

This wave function has no definite symmetry, but we can easily make an antisymmetric combination by demanding that $\Psi(x_1, x_2) = -\Psi(x_2, x_1)$. This condition gives $b_{ij} = -b_{ji}$ and $b_{ii} = 0$, and the sum can be written as

$$\Psi(x_1, x_2) = \sum_{i<j} b_{ij} [\chi_i(x_1)\chi_j(x_2) - \chi_j(x_1)\chi_i(x_2)] \tag{1.50}$$

$$= \sum_{i<j} b_{ij} \begin{vmatrix} \chi_i(x_1) & \chi_j(x_1) \\ \chi_i(x_2) & \chi_j(x_2) \end{vmatrix} :\,= \sum_{i<j} b_{ij} 2^{1/2} |\chi_i \chi_j\rangle \quad \text{antisymmetric.} \tag{1.51}$$

The last form is a chemists' shorthand notation, which lists the diagonal elements in the determinant. This is a valid two-electron wave function, and its extension to N electrons is obvious.

1.2.3 Solving single-particle orbitals

The electron Hamiltonian in equation (1.10) is independent of spin, so the minimization of the energy gives an equation for the spatial orbitals. The optimization of a state with a single Slater determinant is the so-called Hartree–Fock method [8, 9]. The solved single-particle orbitals, the *Hartree–Fock orbitals*, will be used later in fermion QMC calculations in chapter 4.

The Pauli exclusion principle is taken into account by stating that each spatial orbital can be occupied by a maximum of two electrons with opposite spins (spin up and spin down). The Hartree–Fock ground-state energy is the energy of the occupied lowest-energy Hartree–Fock orbitals. One can, however, pick any combination of orbitals and make an N-fermion Slater determinant state. It can be shown that a linear combination of these Slater determinants gets closer to the exact fermion ground state the more determinants one adds. It is possible to optimize the linear combination factors using the configuration interaction (CI) method, one of the many so-called post-Hartree–Fock methods. To be competitive with this method, QMC needs to be either more accurate or computationally less demanding.

1.2.4 Boson ground state

If the normalized boson ground-state wave function is $\phi_0(\mathbf{x})$, then the ground-state energy is

$$E_0 = \langle \phi_0 | \hat{\mathcal{H}} | \phi_0 \rangle = \int d\mathbf{x} \phi_0^*(\mathbf{x}) \left[-\frac{\hbar^2}{2m} \nabla_\mathbf{x}^2 \phi_0(\mathbf{x}) + V(\mathbf{x})\phi_0(\mathbf{x}) \right] \quad (1.52)$$

$$= \int d\mathbf{x} \left[\frac{\hbar^2}{2m} | \nabla_\mathbf{x} \phi_0(\mathbf{x})|^2 + V(\mathbf{x})|\phi_0(\mathbf{x})|^2 \right]. \quad (1.53)$$

Since a normalizable ground-state wave function and its gradients must vanish at infinity, integration by parts leaves no surface terms.

The boson ground-state wave function is a real function. To prove this, we can write $\phi_0(\mathbf{x}) = |\phi_0(\mathbf{x})|\exp(i\mathcal{X}(\mathbf{x}))$ with a real function $\mathcal{X}(\mathbf{x})$. The ground-state energy would be

$$E_0 = \int d\mathbf{x} \left[\frac{\hbar^2}{2m}[(\nabla_\mathbf{x} |\phi_0(\mathbf{x})|)^2 + |\phi_0(\mathbf{x})|^2(\nabla_\mathbf{x} \mathcal{X}(\mathbf{x}))^2 + V(\mathbf{x})|\phi_0(\mathbf{x})|^2] \right], \quad (1.54)$$

so the complex phase always increases the energy, and for the ground state, we must choose $\mathcal{X}(\mathbf{x}) = 0$.

Reverting the partial integration, one also finds that if the real function $\phi_0(\mathbf{x})$ is the boson ground-state wave function, then so is $|\phi_0(\mathbf{x})|$. Penrose and Onsager [10] argued that a sign change in $\phi_0(\mathbf{x})$ at any location where the potential is finite would lead to a discontinuous $\nabla_\mathbf{x}|\phi_0(\mathbf{x})|$, so the energy would increase. As a consequence:

> The boson ground state can be chosen to be a real, nonnegative function.

Penrose and Onsager also show that there can be only one boson ground state; that is, the ground state is nondegenerate. For if there were two nondegenerate and nonnegative ground states, $\phi_0^a(\mathbf{x})$ and $\phi_0^b(\mathbf{x})$, then $\phi_0^a(\mathbf{x}) - \phi_0^b(\mathbf{x})$ would also be a nondegenerate and nonnegative ground state; however, in this case, all three could not be simultaneously normalized.

1.2.5 Fermion ground state

In January 1925, Wolfgang Pauli announced the famous exclusion principle, stating that no two electrons in an atom can occupy a state with the same values for the four quantum numbers. In atoms, three quantum numbers describe how electrons are distributed in space, and he postulated a fourth quantum number, the spin. Soon, the principle was extended to apply to all fermions:

> No two fermions can be in the same quantum state.

In other words, no two fermions can have the same position and spin at the same time. If you have two fermions with the same spin, they cannot be at the same location in space. This implies that the fermion ground-state wave function has nodes. In other words, some combinations of particle positions $\mathbf{x}_{node}$ cannot be realized because $\Psi(\mathbf{x}_{node}) = 0$. We showed earlier that the energy increases if one takes the boson ground state and adds a node, and one can picture the fermion ground state as a very special kind of boson state with multiple nodes:

For the Hamiltonian

$$\hat{\mathcal{H}} = -\frac{\hbar^2}{2m} \nabla_{\mathbf{x}}^2 + V(\mathbf{x}),$$

the fermion ground state is always energetically above the boson ground state.

The fermion energy must be minimized in a restricted function space to prevent the system from falling into the boson ground state lurking at a lower energy. It turns out that this makes the fermion problem a lot more difficult to solve using QMC.

1.3 Solving quantum problems with Monte Carlo

Readers who have attended quantum mechanics lectures have solved the Schrödinger equation of the hydrogen atom and the particle-in-a-box problem with paper and pen. They can be solved numerically, too, with no need to resort to Monte Carlo methods. Such exactly solvable quantum systems are, however, scarce in the real world. Analytical solutions either cannot be found or are too complicated to be of use. The reason for this is that in the quantum world, particle positions are only possibilities. The more particles there are, the more possibilities exist, and their number increases dramatically with the particle count. This is known as the *curse of dimensionality*.

The curse of dimensionality defies the systematic evaluation of integrals. If you discretize coordinates with only ten points in each direction, and a computer can evaluate the integrand for 10^7 points per second, the integration for just twenty particles in three dimensions would take

$$\frac{10^{3 \times 20}}{10^7} \quad s = 10^{53} \quad s \approx 10^{45} \quad \text{years.} \tag{1.55}$$

After this many years, you would have a very crude result.[5] Something has to give, and it is the evaluation of the integrand at systematically chosen points. If we cannot pick points systematically, the obvious choice is to pick points at random. This is the Monte Carlo method.

---

[5] A quantum computer could do better, but we shall see.

The workhorse of QMC is the (pseudo) *random number generator* (RNG), an algorithm that produces sufficiently random x-values between zero and one, denoted by $x \in U[0, 1]$. There are many good RNGs for Monte Carlo, such as the Mersenne Twister MT19937 [11], PCG64 [12], and Xoshiro256 [13].[6] It should be emphasized that numbers produced by any deterministic algorithm are at best pseudo random numbers, because there is no real source of randomness. Inventing random number algorithms is not for the feeble-minded, and in this book RNGs are used as black-box routines in the hope that the inevitable non-randomness in their output does not show up as systematically wrong physical results.[7]

1.3.1 One-dimensional integrals

The integral

$$\int_0^1 dx f(x) \approx \frac{1}{N} \sum_{i=1}^N f(x_i) \qquad (1.56)$$

can be written as a Monte Carlo algorithm:

Pick N random points between zero and one, and calculate the average of f at these points.

If the integrand is a constant c at any point x, then one random point would be enough to find that $f(x_1) = c$, and the integral is c.

A constant or nearly constant integrand makes Monte Carlo integration accurate.

Add another function $P(x) \geqslant 0$ as a factor. As above, the integral can be interpreted as

$$\int_0^1 dx P(x) f(x) \approx \frac{1}{N} \sum_{i=1}^N P(x_i) f(x_i) \qquad (1.57)$$

and evaluated as shown below:

Pick N random points x_i between 0 and 1, and calculate the average of P times f at these points.

---

[6] Implemented, for example, in the Julia Random module, in NumPy [14], and in SciPy [15].

[7] If you abandon randomness and write an algorithm that produces points that fill the space more and more densely, you are entering the world of *quasi Monte Carlo*.

However, we can also read the integral as

$$\int_0^1 dx P(x) f(x) = \int_0^1 (dx P(x)) f(x) \approx \frac{1}{N}\sum_{i=1}^{N} f(x_i) \qquad (1.58)$$

and evaluate it using the following algorithm:

Sample N random points x_i from the distribution $P(x)$ and calculate the average of fs at these points.

Consider, for the moment, that sampling points from a distribution is just a technical detail to be discussed in depth later.

1.3.2 Why use Monte Carlo?

If the curse of dimensionality is the problem, we need to show that, unlike many other methods, Monte Carlo is not affected by it. Start from one dimension, where you can evaluate the integrals using, for example, the trapezoidal rule,[8]

$$\int_0^1 dx f(x) \approx h\left(f(x_1) + \frac{1}{2} f(x_2) + \cdots + \frac{1}{2} f(x_{N-1}) + f(x_N) \right) + \mathcal{O}(h^3), \quad (1.59)$$

where points x_i are evenly spaced between integration limits zero and one with point spacing h, and the error is of order h^3, marked $\mathcal{O}(h^3)$. The Simpson rule evaluates the sum using different weights, but the idea is the same. For a three-dimensional integral, you would evaluate f on a grid of evenly spaced points in a cube. Visualize the cube and replace the points with tiny spheres, and you will notice that you can see through the grid from many directions. If you systematically pick the points where the integrand is evaluated, you also systematically leave out points.

This argument can be made more rigorous. If you evaluate the integrand N times when calculating a d-dimensional integral, then you have about $N^{1/d}$ points for each direction. The distance between points is $h \approx N^{-1/d}$. Using the trapezoidal rule, the error for every integration cell of volume h^d is $h^{d+2} \approx N^{-\frac{d+2}{d}}$, so the total error is N times that, i.e. $N^{-2/d}$. In one dimension, the trapezoidal error diminishes rapidly as more points are added, as $1/N^2$, but the convergence rate gets smaller the higher the dimension.

Monte Carlo integration also approximates the integral as a sum,

$$\int_0^1 dx f(x) \approx \frac{1}{N}\sum_{i=1}^{N} f(x_i), \qquad (1.60)$$

which is the average of N points x_i sampled from the distribution f that fall between zero and one. The error estimate in Monte Carlo is based on the *central limit theorem*, derived in appendix A. The central limit theorem is valid for large N if f

---

[8] You can always scale the coordinates so that the integration takes place from zero to one.

has a well-defined mean μ and variance σ^2. The theorem shows that the error, *the standard error of the mean* (SEM), in the Monte Carlo integral reduces slowly but steadily:

$$\text{error} = \frac{\sigma}{\sqrt{N}} \qquad \text{standard error of the mean.} \qquad (1.61)$$

In other words, the Monte Carlo error estimate relies on the fact that the distribution of the mean (the result of the integral) is a normal distribution (a Gaussian), and the width of the distribution tells us how accurate the mean is. The distribution gets narrower as N increases, which is why the error diminishes as $1/\sqrt{N}$.

This is remarkable:

Monte Carlo error always diminishes as $1/\sqrt{N}$.

The trapezoidal rule error reduces as $N^{-2/d}$ and the Monte Carlo error reduces as $N^{-1/2}$, so the two methods are equally efficient for four-dimensional integrals, but beyond that, Monte Carlo wins.

1.3.3 Integrals in quantum mechanics

Quantum mechanical expectation values are multidimensional integrals, and the points $\mathbf{x}$ are possible particle coordinates. The energy of a system in quantum state $\Psi(\mathbf{x})$ is

$$E = \frac{\langle \Psi | \hat{\mathcal{H}} | \Psi \rangle}{\langle \Psi | \Psi \rangle} = \frac{\int d\mathbf{x}\,\Psi^*(\mathbf{x})\hat{\mathcal{H}}\Psi(\mathbf{x})}{\int d\mathbf{x}\,|\Psi(\mathbf{x})|^2} = \frac{\int d\mathbf{x}\,|\Psi(\mathbf{x})|^2 \left(\dfrac{\hat{\mathcal{H}}\Psi(\mathbf{x})}{\Psi(\mathbf{x})} \right)}{\int d\mathbf{x}\,|\Psi(\mathbf{x})|^2}. \qquad (1.62)$$

Division by $\Psi(\mathbf{x})$ does not cause any problems because points where one would divide by zero have zero measure in the integral. The normalization integral is a real number, so we can write the energy in the form

$$E = \int d\mathbf{x} \underbrace{\left(\frac{|\Psi(\mathbf{x})|^2}{\int d\mathbf{x}\,|\Psi(\mathbf{x})|^2} \right)}_{P(\mathbf{x})} \underbrace{\left(\frac{\hat{\mathcal{H}}\Psi(\mathbf{x})}{\Psi(\mathbf{x})} \right)}_{E_L(\mathbf{x})} = \int d\mathbf{x}\,P(\mathbf{x})E_L(\mathbf{x}) \approx \frac{1}{M}\sum_{i=1}^{M} E_L(\mathbf{x}_i). \qquad (1.63)$$

The function $P(\mathbf{x})$ is positive, so we can call it a probability distribution and sample points from it. It is also normalized:

$$\int d\mathbf{x}\,P(\mathbf{x}) = \frac{\int d\mathbf{x}\,|\Psi(\mathbf{x})|^2}{\int d\mathbf{x}\,|\Psi(\mathbf{x})|^2} = 1. \qquad (1.64)$$

The function $E_L(\mathbf{x})$ is the so-called *local energy*, a quantity (with units of energy) that varies with position. The QMC algorithm used to evaluate the energy is similar to the example we saw above:

> Sample N random points $\mathbf{x}_i$ from the distribution $P(\mathbf{x})$ and calculate the average of E_Ls at these points.

The algorithm works equally well for both boson and fermion wave functions because $P(\mathbf{x}) \geqslant 0$ in both cases.

The local energy is a sum of the local kinetic energy and the local potential energy:

$$E_L(\mathbf{x}) = -\frac{\hbar^2}{2m}\frac{\nabla_x^2\Psi(\mathbf{x})}{\Psi(\mathbf{x})} + \frac{V(\mathbf{x})\Psi(\mathbf{x})}{\Psi(\mathbf{x})} = \underbrace{-\frac{\hbar^2}{2m}\frac{\nabla_x^2\Psi(\mathbf{x})}{\Psi(\mathbf{x})}}_{T_L(\mathbf{x})} + V(\mathbf{x}). \tag{1.65}$$

The potential and kinetic energies are the expectation values

$$\langle V \rangle = \int d\mathbf{x} P(\mathbf{x}) V(\mathbf{x}), \qquad \langle T \rangle = \int d\mathbf{x} P(\mathbf{x}) T_L(\mathbf{x}). \tag{1.66}$$

Although the kinetic energy $\langle T \rangle$ must always be positive, the local kinetic energy T_L usually has negative values at some points. This can be seen by evaluating the ground-state energy using the exact wave function,

$$E_L(\mathbf{x}) = \frac{\hat{\mathcal{H}}\phi_0(\mathbf{x})}{\phi_0(\mathbf{x})} = \frac{E_0\phi_0(\mathbf{x})}{\phi_0(\mathbf{x})} = E_0 \qquad \text{exact ground state}, \tag{1.67}$$

so that the local energy is the ground-state energy at every point $\mathbf{x}$. But then

$$E_L(\mathbf{x}) = T_L(\mathbf{x}) + V(\mathbf{x}) = E_0 \qquad \text{exact ground state}, \tag{1.68}$$

so if $V(\mathbf{x}) > 0$ at some point, then at that point, $T_L(\mathbf{x}) < 0$.

1.3.4 Zero variance principle

If the wave function is exactly the ith eigenstate of $\hat{\mathcal{H}}$, then the local energy is

$$E_L(\mathbf{x}) = \frac{\hat{\mathcal{H}}\phi_i(\mathbf{x})}{\phi_i(\mathbf{x})} = E_i = \text{constant} \quad \text{exact eigenstate.} \tag{1.69}$$

These constant values occur only at exact eigenstates; in any other state, $E_L(\mathbf{x})$ is position dependent.[9] The energy variance describes how much the local energy fluctuates w.r.t. the average. Its definition is

$$\sigma_E^2 := \langle E_L - (\langle E_L \rangle)^2 \rangle = \langle E_L^2 \rangle - \langle E_L \rangle^2 = \langle E_L^2 \rangle - E^2 \quad \text{energy variance.} \tag{1.70}$$

---

[9] In principle, one could find energy eigenstates by scanning through states and locating variance minima. As far as I am aware, this has not been successfully implemented.

The variance is trivially zero for eigenstates and positive for any other state. This is known as the *zero-variance principle*, and it is an important guideline in searching for better approximate wave functions on our way to more efficient QMC methods. In chapter 3, we write parameterized forms of wave functions and optimize them either for smaller variance or lower energy.

1.3.5 Sampling from a distribution

Any normalizable positive function can be interpreted as a probability density $P(\mathbf{x})$. The reason that we are interested in probability densities is that they can be readily used in Monte Carlo. We have two viable candidates for $P(\mathbf{x})$, one that is always valid,

$$P(\mathbf{x}) = |\Psi(\mathbf{x})|^2, \tag{1.71}$$

and one that is valid for the boson ground state,

$$P(\mathbf{x}) = \phi_0(\mathbf{x}). \tag{1.72}$$

The fermion wave function changes sign and cannot be interpreted as a probability density $\mathbf{x}$. One could keep track of the sign of the wave function,

$$\Psi(\mathbf{x}) = \mathrm{sign}_{\Psi(\mathbf{x})} \underbrace{|\Psi(\mathbf{x})|}_{P(\mathbf{x})}, \tag{1.73}$$

and use sign and P separately. In QMC, this leads to the so-called *fermion sign problem*, which we discuss further in chapters 4 and 5.

References

[1] Born. M 1926 Quantenmechanik der Stoßvorgänge *Z. Phys.* **38** 803–27
[2] Fierz M 1939 Über die relativistische Theorie kräftefreier Teilchen mit beliebigem Spin *Helv. Phys. Acta* **12** 3–37
[3] Pauli W 1940 The connection between spin and statistics *Phys. Rev.* **58** 716–22
[4] Fierz M, Pauli W E and Dirac P A M 1939 On relativistic wave equations for particles of arbitrary spin in an electromagnetic field *Proc. R. Soc. Lond.* A **173** 211–32
[5] Slater J C 1929 The theory of complex spectra *Phys. Rev.* **34** 1293–322
[6] Slater J C and Phillips J C 1974 Quantum theory of molecules and solids, vol. 4: the self-consistent field for molecules and solids *Phys. Today* **27** 49–50
[7] Szabo A and Ostlund N S 1996 *Modern Quantum Chemistry: Introduction to Advanced Electronic Structure Theory* 1st edn (Minneola, MN: Dover)
[8] Hartree D R 1928 The wave mechanics of an atom with a non-Coulomb central field. Part I. Theory and methods *Math. Proc. Cambridge Phil. Soc.* **24** 89–110
[9] Fock V 1930 Näherungsmethode zur Lösung des quantenmechanischen Mehrkörperproblems *Z. Phys.* **61** 126–48
[10] Penrose O and Onsager L 1956 Bose-Einstein condensation and liquid helium *Phys. Rev.* **104** 576–84
[11] Matsumoto M and Nishimura T 1998 Mersenne twister: a 623-dimensionally equidistributed uniform pseudo-random number generator *ACM Trans. Model. Comput. Simul.* **8** 3–30

[12] O'Neill M E 2014 *PCG: a family of simple fast space-efficient statistically good algorithms for random number generation* HMC-CS-2014-0905 Harvey Mudd College, Claremont, CA https://www.cs.hmc.edu/tr/hmc-cs-2014-0905.pdf

[13] Blackman D and Vigna S 2021 Scrambled linear pseudorandom number generators *ACM Trans. Math. Softw.* **47** 36

[14] Harris C R *et al* 2020 Array programming with NumPy *Nature* **585** 357–62

[15] Virtanen P *et al* SciPy 1.0 Contributors 2020 SciPy 1.0: fundamental algorithms for scientific computing in Python *Nat. Meth.* **17** 261–72

IOP Publishing

A Practical Course on Quantum Monte Carlo

Vesa Apaja

Chapter 2

Variational Monte Carlo (VMC)

Variational Monte Carlo (VMC) is the computation of physical properties from a guessed, time-independent wave function,

$$\varphi_T(\mathbf{x}) \qquad \text{trial wave function.} \tag{2.1}$$

There is actually nothing variational in VMC, not in the sense that you could use variations $\delta\varphi_T$ to find an optimal φ_T. At this point, you decide the functional form of the wave function, with some freedom in terms of adjustable parameters. It is far from trivial to see what a many-body wave function *implies* for, say, particle density or energy. This task is left to VMC.

The wave function gives the probability density,

$$P(\mathbf{x}) = |\varphi_T(\mathbf{x})|^2, \tag{2.2}$$

so the spatial structure of the system is set. We can only go so far without specifying the Hamiltonian. We will use

$$\hat{\mathcal{H}} = -\frac{\hbar^2}{2m}\sum_{i=1}^{N}\nabla_i^2 + V(\mathbf{x}) \tag{2.3}$$

with a system-specific potential V and mass m. The trial state energy is the expectation value (equation (1.63) with explicit normalization):

$$E = \int d\mathbf{x}\left(\frac{P(\mathbf{x})}{\int d\mathbf{x}P(\mathbf{x})}\right)E_L(\mathbf{x}). \tag{2.4}$$

To compute the $\mathbf{x}$-integral, we need to sample points $\mathbf{x}$ from the distribution:

$$\frac{P(\mathbf{x})}{\int d\mathbf{x}P(\mathbf{x})}. \tag{2.5}$$

The calculation of the normalization integral $\int d\mathbf{x}P(\mathbf{x})$ would be an effort of the same magnitude as computing E itself. It is not practical to compute it separately

doi:10.1088/978-0-7503-6310-5ch2

using, for example, Monte Carlo, because that would introduce another source of error into the results. We are saved by the *Metropolis algorithm*, which lets us sample from an *unnormalized distribution*.

2.0.1 Metropolis algorithm

Metropolis *et al* [1, 2] published an algorithm that can be used to sample from an almost arbitrary multidimensional distribution. Here, we are applying it to sample $\mathbf{x}$ from the distribution P. The point is that P is *not* a probability distribution, since we have not normalized it.

The algorithm generates a sequence of positions (points for short),

$$\mathbf{x}_1 \rightarrow \mathbf{x}_2 \rightarrow \mathbf{x}_3 \rightarrow ... \rightarrow, \tag{2.6}$$

where the rule indicated by $\rightarrow$ depends on P. The sequence is a *Markov chain*, named after the mathematician A A Markov. The defining feature of a Markov chain is that the next point $\mathbf{x}_{k+1}$ depends only on the previous point $\mathbf{x}_k$.

The collection of points $\{\mathbf{x}_i\}$ can be thought of as a *representation* of the distribution P, although it is usually impossible to infer the mathematical form of P from a finite collection of points. In VMC, the order of evaluation is

$$\text{known } P(\mathbf{x}) \colon = |\varphi_T(\mathbf{x})|^2 \Rightarrow \text{ points } \{\mathbf{x}_i\} \Rightarrow \text{ compute QMC integrals.} \tag{2.7}$$

The QMC integrals are what we actually want; so, looking ahead, suppose we have absolutely no idea what the wave function or P is, but we have an algorithm that correctly produces points $\{\mathbf{x}_i\}$. The evaluation would go through the following stages:

$$\text{Miraculous algorithm} \Rightarrow \text{points}\{\mathbf{x}_i\} \Rightarrow \text{ compute QMC integrals,} \tag{2.8}$$

and we would be just as happy. One such algorithm is the diffusion Monte Carlo (DMC), the topic of chapter 4. One downside of DMC is that there is no explicit wave function you can write on paper and share with colleagues.

Since it is unnecessary to remember old points in a Markov chain, it is sufficient to keep the current point, mark it $\mathbf{x}$, and mark the next point $\mathbf{x}'$. We start with particles at $\mathbf{x}$. The only criteria are that the initial potential energy $V(\mathbf{x})$ must be finite, and the initial weight $P(\mathbf{x})$ must be a number that fits in your floating-point variable.[1] A relatively high $P(\mathbf{x})$ is preferable because then the walk will *thermalize*, i.e. lose memory of its origin, a bit faster.

> The coordinates $\mathbf{x}$ define a *walker*.

---

[1] If the wave function squared becomes too small or too big, it is best to use $\ln(|\varphi_T(\mathbf{x})|^2)$.

In VMC, coordinates are possible simultaneous positions of particles; however, in general, they can have a more abstract meaning.

Given $\mathbf{x}$, the next point $\mathbf{x}'$ is found using the Metropolis algorithm:

Metropolis algorithm

1. Randomly pick a point $\mathbf{x}'$ from the neighborhood of $\mathbf{x}$, $\mathbf{x}' = \mathbf{x} + \boldsymbol{\delta}$, where $\boldsymbol{\delta}$ is a random shift. For example, in 3D Cartesian coordinates,

$$\boldsymbol{\delta} = \left((r_1, r_2, r_3) - \frac{1}{2} \right) * \texttt{step}, \qquad (2.9)$$

where the random numbers $r_i \in U[0, 1]$ and $\texttt{step}$ is a free parameter. Start with something small, say, $\texttt{step} = 0.1$.

2. Calculate

$$ratio = \frac{P(\mathbf{x}')}{P(\mathbf{x})}. \qquad (2.10)$$

In VMC, $P(\mathbf{x}) = |\varphi_T(\mathbf{x})|^2$ and $P(\mathbf{x}') = |\varphi_T(\mathbf{x}')|^2$.

3. Ask the *Metropolis question*:
   if *ratio* > 1, accept the move; the walker moves to $\mathbf{x}'$.
   else pick another random number $z \in U[0, 1]$;
      if *ratio* $> z$, accept the move; the walker moves to $\mathbf{x}'$,
      else reject the move; the walker stays at $\mathbf{x}$.

4. Compute the *acceptance*,

$$\texttt{acceptance} = \frac{\text{number of accepted moves}}{\text{number of attempted moves}}. \qquad (2.11)$$

If $\texttt{acceptance}$ is less than 50%, reduce $\texttt{step}$; if larger than 60%, increase $\texttt{step}$.

5. Use the current walker in calculations, e.g. evaluate $E_L(\mathbf{x})$.

6. Go to step 1.

The Metropolis question in step 3 means that moves to higher P are always accepted, while moves to lower P are sometimes accepted. After a rejected move, the walker stays where it was, and at step 5, that *is* the current walker.

The free parameter $\texttt{step}$ dictates the acceptance ratio and the effectivity of sampling. The walker should move around in the available space (which is

dN-dimensional for N particles in d dimensions) as fast as possible in as few cycles as possible. An overlarge `step` leads to low acceptance, and the walker mostly refuses to move. An undersized `step` leads to very high acceptance, but the walker moves slowly. A rule of thumb is that for the most effective sampling, *the acceptance should be about 60 percent*.

The Metropolis decision is made solely based on a ratio of weights, where the normalization integrals cancel out:

$$ratio = \frac{\frac{P(\mathbf{x}')}{\int d\mathbf{x} P(\mathbf{x})}}{\frac{P(\mathbf{x})}{\int d\mathbf{x} P(\mathbf{x})}} = \frac{P(\mathbf{x}')}{P(\mathbf{x})}. \tag{2.12}$$

This is why we never need to compute $\int d\mathbf{x} P(\mathbf{x})$ in QMC, and we do not have to pay much attention to the normalization of wave functions. We have the Markov chain no-memory property to thank, but it comes with a price: the point-to-point motion of the walker is *highly correlated*; it takes many steps to travel far from the current point $\mathbf{x}$. This means that the samples we are collecting for, say, local energy, are correlated, and we have to take that into account in estimating the statistical error. Details of error estimation are given in appendix F, but for now, we can concentrate on getting some results.

2.0.2 Application to He atom ground state

The experimental ionization energies of helium atoms are [3, 4]

$$E^{(1)} = 24.587\ 387\ 936(25)\,\text{eV} = 0.903\ 569\ 5844(9)\,\text{a.u.} \qquad \text{1st electron} \tag{2.13}$$

$$E^{(2)} = 54.417\ 763\ 11(2)\,\text{eV} = 1.999\ 815\ 9170(7)\,\text{a.u.} \qquad \text{2nd electron.} \tag{2.14}$$

Therefore, the ground-state energy of the He atom is

$$E_0^{\text{exp.}} = -79.005\ 154\ 539(25)\ \text{eV} = -2.903\ 385\ 83(13)\ \text{a.u.} \tag{2.15}$$

The energy conversions are

$$1\ \text{a.u.} = 1\ \text{Ha} = 27.211\ 386\ 131\ 79\ \text{eV} \tag{2.16}$$

$$= 4.359\ 7482 \times 10^{-18}\ \text{J} = 6.579\ 689\ 744\ 79 \times 10^9\ \text{MHz}. \tag{2.17}$$

For theoretical calculations, we use the Born–Oppenheimer approximation; that is, we assume the nucleus is a point-like particle at rest (in this case, at the origin). We use the nonrelativistic electron Hamiltonian

$$\hat{\mathcal{H}} = -\frac{1}{2}\left(\boldsymbol{\nabla}_1^2 + \boldsymbol{\nabla}_2^2\right) - \frac{Z}{r_1} - \frac{Z}{r_2} + \frac{1}{r_{12}}. \tag{2.18}$$

Here, $Z = 2$, and the electrons are at $\mathbf{r}_1$ and $\mathbf{r}_2$. The electron–nucleus distances are r_1 and r_2, and the electron–electron distance is $r_{12} = |\mathbf{r}_{12}|: = |\mathbf{r}_1 - \mathbf{r}_2|$. The corrections due to finite nucleon mass and relativistic effects are smaller than the accuracy of our calculations.

In the ground state, electrons have opposite spins, and the spin wave function is antisymmetric,

$$\Xi(1, 2) = \varphi_\uparrow(1)\varphi_\downarrow(2) - \varphi_\downarrow(1)\varphi_\uparrow(2) \quad \text{spin singlet state.} \tag{2.19}$$

The spatial part is symmetric; therefore, once we fix the spins, the He atom ground-state problem turns into a boson ground-state calculation. The boson ground state is nonnegative, so we can use the form

$$\varphi_T(\mathbf{x}) = e^{J(\mathbf{x})}, \tag{2.20}$$

where J is a symmetric, real function. This exponential form is very frequently encountered in QMC, in a special form known as the *Jastrow factor*. We will soon look at Jastrow factors in detail; for now, let us concentrate on the He atom problem.

The local energy is

$$E_L(\mathbf{x}) = \frac{\hat{\mathcal{H}}\varphi_T(\mathbf{x})}{\varphi_T(\mathbf{x})} = -\frac{1}{2}\sum_{i=1,2}((\nabla_i J(\mathbf{x}))^2 + \nabla_i^2 \ J(\mathbf{x})) - \frac{Z}{r_1} - \frac{Z}{r_2} + \frac{1}{r_{12}}. \tag{2.21}$$

The local energy depends only on the gradients of $J(\mathbf{x})$.[2] A gradient needs a direction, and the directions in a He atom are the vectors $\mathbf{r}_1$, $\mathbf{r}_2$, and $\mathbf{r}_{12}$, given w.r.t. the α-particle nucleus in the origin. From the identities

$$(\nabla r)^2 = \hat{\mathbf{r}}^2 = 1, \quad \text{unit vector squared} \tag{2.22}$$

$$\nabla^2 r = \frac{d-1}{r} = \frac{2}{r}, \quad \text{in 3D, where } d = 3, \tag{2.23}$$

we see that a suitable symmetric J is

$$J(\mathbf{x}) = -\alpha(r_1 + r_2) + \alpha_{12} r_{12}, \tag{2.24}$$

with gradients and Laplacians

$$\nabla_1 J(\mathbf{x}) = -\alpha\hat{\mathbf{r}}_1 + \alpha_{12}\hat{\mathbf{r}}_{12}, \quad \nabla_2 J(\mathbf{x}) = -\alpha\hat{\mathbf{r}}_2 - \alpha_{12}\hat{\mathbf{r}}_{12} \tag{2.25}$$

$$\nabla_1^2 \ J(\mathbf{x}) = \frac{-2\alpha}{r_1} + \frac{2\alpha_{12}}{r_{12}}, \quad \nabla_1^2 \ J(\mathbf{x}) = \frac{-2\alpha}{r_2} + \frac{2\alpha_{12}}{r_{12}}. \tag{2.26}$$

The local energy is

$$E_L(\mathbf{x}) = -\alpha^2 - \alpha_{12}^2 + \alpha\alpha_{12}(\hat{\mathbf{r}}_1 - \hat{\mathbf{r}}_2)\cdot\hat{\mathbf{r}}_{12} + \underbrace{\frac{1 - 2\alpha_{12}}{r_{12}} - \frac{Z-\alpha}{r_1} - \frac{Z-\alpha}{r_2}}_{\equiv 0 \ \Rightarrow \textbf{cusp conditions}}. \tag{2.27}$$

---

[2] Adding any constant c to $J(\mathbf{x})$ multiplies the trial wave function by e^c, which only affects the wave function normalization.

The factors of the unit vectors leave the constant energy terms $-\alpha^2 - \alpha_{12}^2$, while $\nabla^2 J$ gives a term that was grouped with the Coulomb terms. The dot product term comes from $(\nabla_i J(\mathbf{x}))^2$, which 'mixes' directions.

In equation (2.27), we indicated that the Coulomb terms cancel for a well-chosen trial wave function. The motivation is that electrons can be on top of the nucleus, and the Coulomb repulsion is so weak that two antiparallel-spin electrons can occupy the same exact location. From this point of view, the only sensible wave function is one that leads to the **exact cancellation of the Coulomb singularities in E_L**. At singular points, the wave function is not zero but has a cusp (a kink), and the wave function satisfies the so-called Kato **cusp conditions** [5].

From equation (2.27), we now see that the cusp conditions require that

$$\alpha = Z = 2, \ \alpha_{12} = \frac{1}{2} \qquad \text{He atom cusp conditions (antiparallel spins)}, \quad (2.28)$$

which leads to the functions

$$\varphi_T(\mathbf{x}) = e^{J(\mathbf{x})} \tag{2.29}$$

$$J(\mathbf{x}) = e^{-2(r_1 + r_2) + \frac{1}{2} r_{12}} \tag{2.30}$$

$$\nabla_1 J(\mathbf{x}) = -2\hat{\mathbf{r}}_1 + \frac{1}{2}\hat{\mathbf{r}}_{12}, \ \nabla_2 J(\mathbf{x}) = -2\hat{\mathbf{r}}_2 - \frac{1}{2}\hat{\mathbf{r}}_{12} \tag{2.31}$$

$$E_L(\mathbf{x}) = -4 - \frac{1}{4} + (\hat{\mathbf{r}}_1 - \hat{\mathbf{r}}_2) \cdot \hat{\mathbf{r}}_{12}. \tag{2.32}$$

The local energy is not a constant, so the expression $\varphi_T(\mathbf{x})$ is certainly not the exact ground state. However, we can immediately see that the local energy lies between the limits[3]

$$-4.25 \leqslant E_L(\mathbf{x}) \leqslant -2.25 \quad \text{(a.u.)}. \tag{2.33}$$

The gradients of $J(\mathbf{x})$ in equation (2.31) show that the trial wave function pushes electrons closer to the nucleus and away from each other. There is an apparent flaw; namely, the latter does not die out with distance r_{12}. This does not cause any problems, and electrons do not fly away, but this flaw will nevertheless be corrected later in equation (2.37).

VMC results for He atom ground-state energy

The approximate ground-state energy can now be calculated using VMC and the Metropolis algorithm. The Julia code vmc.heatom.jl gives the result

$$E_0 = -2.8556(1) \quad \text{a.u.} \quad \sim 130 \text{ million VMC steps}. \tag{2.34}$$

---

[3] The dot product is a difference of two cosines; on the other hand, it equals $\frac{r_1 + r_2}{r_{12}}(1 - \hat{\mathbf{r}}_1 \cdot \hat{\mathbf{r}}_2) \geqslant 0$.

Run the code again with $\alpha = 27/16$ and $\beta = 0$,

$$E_0 = -2.8476(1) \qquad \text{a.u.} \qquad \sim 700 \text{ million VMC steps}. \qquad (2.35)$$

The analytical expectation value in the latter case is $E_0 = 2.847\ 656\ 25$ a.u., which is useful in debugging VMC code. Hitting a singularity dead on is extremely rare, even in long Monte Carlo calculations, so using non-cusp values practically never causes the simulation to crash due to division by zero.

Figure 2.1 shows how the local energy fluctuates when the cusp conditions are satisfied compared to the fluctuation when the cusp conditions are not satisfied. Now and then, Coulomb singularities cause huge peaks in the local energy, which dramatically increase the variance and QMC error.

Like all Monte Carlo methods, VMC also converges as $1/\sqrt{N}$, where N is the number of steps. Looking at the results, if we wished to improve our error from 1×10^{-4} a.u. to 1×10^{-5} a.u., we would need roughly 100 times more Monte Carlo steps. A short 20 second computation of 130 million steps would turn into a rather boring 30 minute computation of 130 billion steps. Getting a faster machine and optimizing the code for speed could help, but not nearly as much as careful trial wave function optimization would. Chapter 3 is devoted to optimization.

Other trial wave functions for Coulomb systems

As long as it satisfies the cusp conditions, the trial wave function can be freely modified. One popular choice (in Hartree atomic units) is

$$\varphi_T(\mathbf{x}) = e^{J(\mathbf{x})} \qquad (2.36)$$

$$J(\mathbf{x}) = -\alpha(r_1 + r_2) + \frac{r_{12}}{(2|4)(1 + \beta r_{12})}, \qquad (2.37)$$

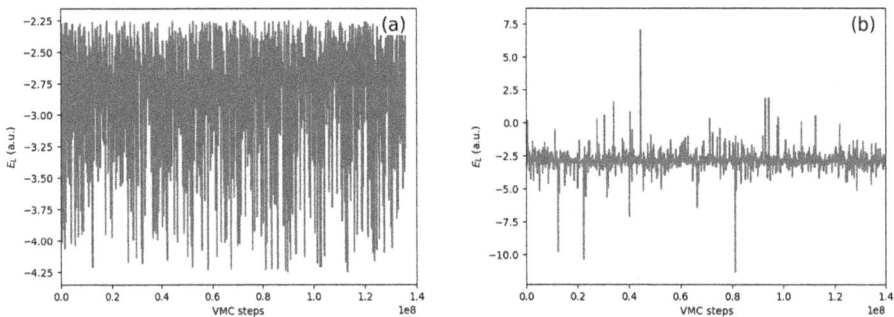

Figure 2.1. The local energy of a He atom trial wave function that satisfies the cusp conditions (a), and one that does not (b). In the latter case, the upward peaks are caused by electrons getting very close to each other, while the downward peaks are caused by an electron getting very close to the nucleus. If the cusp conditions are satisfied, these events cause no problems because they are accurately compensated by the kinetic part of the local energy, and the local energy remains between -4.25 and -2.25, in accordance with equation (2.33).

where the factors of two and four (short notation (2|4)) correspond to antiparallel and parallel spins, respectively [6]. The cusp condition value is $\alpha = Z$ with a nuclear charge of Z. The helium local energy is[4]

$$b := 1 + \beta r_{12}$$

$$E_L(\mathbf{x}) = -\alpha^2 + \frac{1}{2b^2}\left(\frac{\alpha(r_1 + r_2)}{r_{12}}(1 - \hat{\mathbf{r}}_1 \cdot \hat{\mathbf{r}}_2) - \frac{1}{2b^2} - \frac{2}{r_{12}} + \frac{2\beta}{b}\right) \qquad (2.38)$$

$$+ \frac{1}{r_{12}} - \frac{2}{r_1} - \frac{2}{r_2}.$$

We will come back to the optimization of this trial wave function in the next chapter.

Another popular choice is a method first described by David Ceperley and Berni Alder [7],

$$J(\mathbf{x}) = -\alpha(r_1 + r_2) + \frac{A}{r_{12}}(1 - e^{-r_{12}/F}) \qquad (2.39)$$

$$F = \sqrt{(1|2)A}, \qquad (2.40)$$

where the factors of one and two correspond to parallel and antiparallel spins, respectively. These forms improve upon our expression $J(\mathbf{x})$ in equation (2.31) in that the electron–electron correlations die out as r_{12} increases.

Historical significance

In the late 1920s, the fate of quantum mechanics hung in the balance. The hydrogen atom results were accurate, but the helium atom's first electron ionization energy calculated from perturbation theory was off by 4 eV compared to the best value of 24.46 eV obtained from spectroscopic measurements in 1929 [8]. The Norwegian mathematician and physicist Egil Hylleraas devised a more complete wave function and pushed the theoretical value to the convincing 24.35 eV [9]. In elliptic coordinates $s := r_1 + r_2$, $t := r_1 - r_2$, $u := r_{12}$, the Hylleraas He atom wave function is

$$\Psi(\mathbf{x}) = e^{-\alpha s}\sum_{i,j,k}c_{ijk}s^i t^j u^k, \qquad (2.41)$$

and now it is just a matter of how many terms one is able to keep in the sum. We will return to the Hylleraas wave functions for H_2 molecules in section 4.8.2. In 1959, C L Pekeris kept 1078 terms and obtained a value for the ground-state energy of -2.903 724 375 a.u. After correcting for finite nucleon mass and relativistic effects, this gives an ionization energy of 198 310.687 cm^{-1}, while the experimental value is 198 310.82 cm^{-1}[10]. Case closed.

Other methods

To date, the helium atom has been a popular benchmark. The so-called *scaled Schrödinger equation method* [11, 12] gives the nonrelativistic Born–Oppenheimer ground-state energy with an impressive accuracy of 40 decimal places:

---

[4] In programs, compute $\nabla_i J(\mathbf{x})$ and square it. There is no need to calculate the square analytically.

$$E_0 = -2.903\,724\,377\,034\,119\,598\,311\,159\,245\,194\,404\,446\,6969 \text{ a.u..} \quad (2.42)$$

The deviation from the experimental energy, 3.4×10^{-4} a.u. or 9.21×10^{-3} eV, is attributed to relativistic and nuclear recoil effects.

Relation to density functional theory (DFT)

The idea of cusps has profound consequences for computational physics. In the vicinity of the cusp points, the wave function is known exactly. The wave function squared gives the electron density, and from the electron density near the cusp points, one can read the nuclear charge. The conclusion is that from a given electron density, one can reverse engineer the Hamiltonian, and the Hamiltonian dictates the ground state. This is E Bright Wilson's famous explanation of DFT [13]. The *Hohenberg–Kohn theorem* [14] states that the ground-state energy E_0 is a unique functional of the electron density n, $E_0[n]$. We know a few good energy functionals, but they all contain an approximate exchange-correlation functional $E_{xc}[n]$, and we know of no way to systematically improve them. This reliance on approximation in DFT leaves a niche for QMC in computational chemistry.

2.1 Spin in expectation values

Particles have spatial and spin degrees of freedom, and the N-particle wave function is

$$\Psi(x) = \Psi(\mathbf{r}_1\sigma_1, \mathbf{r}_2\sigma_2, \ldots, \mathbf{r}_N\sigma_N). \quad (2.43)$$

Here, I discuss only multielectron systems, so the spins σ_i are either $\uparrow$ or $\downarrow$. The expectation value of a *spin-independent operator* $\hat{O}(\mathbf{x})$ in this state counts all possible quantum mechanical outcomes, so we integrate over spatial variables and sum over spins:

$$\langle \hat{O} \rangle = \frac{\sum_\sigma \int d\mathbf{x}\Psi^*(x)\hat{O}(\mathbf{x})\Psi(x)}{\sum_\sigma \int d\mathbf{x}\Psi^*(x)\Psi(x)}. \quad (2.44)$$

This needs some rearrangement to benefit from the fact that there is no spin in $\hat{O}(\mathbf{x})$.

The spin summation is restricted by the condition of fixed spin counts $N_\uparrow$ and $N_\downarrow$, so that $N_\uparrow + N_\downarrow = N$. Consider the case of three electrons, two with spins up and one with spin down. The denominator in the expectation value is

$$\sum_\sigma \int d\mathbf{x}\Psi^*(x)\hat{O}(\mathbf{x})\Psi(x) \quad (2.45)$$

$$= \int d\mathbf{r}_1 d\mathbf{r}_2 d\mathbf{r}_3 \Psi^*(\mathbf{x}_1\uparrow, \mathbf{x}_2\uparrow, \mathbf{x}_3\downarrow)\hat{O}(\mathbf{r}_1, \mathbf{r}_2, \mathbf{r}_3)\Psi(\mathbf{x}_1\uparrow, \mathbf{x}_2\uparrow, \mathbf{x}_3\downarrow)$$
$$+ \int d\mathbf{r}_1 d\mathbf{r}_2 d\mathbf{r}_3 \Psi^*(\mathbf{x}_1\uparrow, \mathbf{x}_2\downarrow, \mathbf{x}_3\uparrow)\hat{O}(\mathbf{r}_1, \mathbf{r}_2, \mathbf{r}_3)\Psi(\mathbf{x}_1\uparrow, \mathbf{x}_2\downarrow, \mathbf{x}_3\uparrow) \quad (2.46)$$
$$+ \int d\mathbf{r}_1 d\mathbf{r}_2 d\mathbf{r}_3 \Psi^*(\mathbf{x}_1\uparrow, \mathbf{x}_2\downarrow, \mathbf{x}_3\uparrow)\hat{O}(\mathbf{r}_1, \mathbf{r}_2, \mathbf{r}_3)\Psi(\mathbf{x}_1\downarrow, \mathbf{x}_2\uparrow, \mathbf{x}_3\uparrow).$$

Because of antisymmetry,

$$\Psi(\mathbf{x}_1 \uparrow, \mathbf{x}_2 \downarrow, \mathbf{x}_3 \uparrow) = -\Psi(\mathbf{x}_1 \uparrow, \mathbf{x}_3 \uparrow, \mathbf{x}_2 \downarrow) \tag{2.47}$$

$$\Psi^*(\mathbf{x}_1 \uparrow, \mathbf{x}_2 \downarrow, \mathbf{x}_3 \uparrow) = -\Psi^*(\mathbf{x}_1 \uparrow, \mathbf{x}_3 \uparrow, \mathbf{x}_2 \downarrow), \tag{2.48}$$

so their product remains the same. Particles can now be relabeled to follow the order 1, 2, 3, and we see that the three integrals in equation (2.46) are the same. Following a similar procedure for the numerator gives the result

$$\langle \hat{O} \rangle = \frac{\beta \int d\mathbf{r}_1 d\mathbf{r}_2 d\mathbf{r}_3 \Psi^*(\mathbf{x}_1 \uparrow, \mathbf{x}_2 \uparrow, \mathbf{x}_3 \downarrow)\hat{O}(\mathbf{r}_1, \mathbf{r}_2, \mathbf{r}_3)\Psi(\mathbf{x}_1 \uparrow, \mathbf{x}_2 \uparrow, \mathbf{x}_3 \downarrow)}{\beta \int d\mathbf{r}_1 d\mathbf{r}_2 d\mathbf{r}_3 \Psi^*(\mathbf{x}_1 \uparrow, \mathbf{x}_2 \uparrow, \mathbf{x}_3 \downarrow)\Psi(\mathbf{x}_1 \uparrow, \mathbf{x}_2 \uparrow, \mathbf{x}_3 \downarrow)}. \tag{2.49}$$

After *listing particles in the order spin-ups followed by spin-downs*, spatial integration also automatically completes the spin summation.

The expectation value of a spin-independent operator $\hat{O}$ in the N-electron state is [15]

$$\langle \hat{O} \rangle = \frac{\int d\mathbf{x} \Psi^*(\mathbf{x})\hat{O}(\mathbf{x})\Psi(\mathbf{x})}{\int d\mathbf{x} \Psi^*(\mathbf{x})\Psi(\mathbf{x})} \tag{2.50}$$

$$\Psi(\mathbf{x}) = \Psi(\underbrace{\mathbf{r}_1, \mathbf{r}_2, \ldots, \mathbf{r}_{N\uparrow}}_{\text{spin} \uparrow}, \underbrace{\mathbf{r}_{N\uparrow+1}, \ldots, \mathbf{r}_{N\uparrow+N_\downarrow}}_{\text{spin} \downarrow}) \tag{2.51}$$

$$N\uparrow + N_\downarrow = N. \tag{2.52}$$

This *QMC multielectron spin-free formulation* removes the need to explicitly keep count of electron spin. Indeed, some QMC articles never mention spin at all. The applicability of a spin-free formulation depends on the specific system and the physical properties of interest. If spin effects are crucial, a spin-free approach might not be appropriate. A spin-free formulation, if valid, can reduce computational cost and make calculations more tractable.

2.2 The multielectron wave function

As shown in the introduction, antisymmetry in a wave function $\Psi(x)$, with spatial and spin parts, can be expressed with the aid of Slater determinants of spin orbitals $\chi(x)$:

$$\Psi(\boldsymbol{x}) = \frac{1}{\sqrt{N!}} \begin{vmatrix} \chi_1(\boldsymbol{x}_1) & \chi_2(\boldsymbol{x}_1) & \cdots & \chi_N(\boldsymbol{x}_1) \\ \chi_1(\boldsymbol{x}_2) & \chi_2(\boldsymbol{x}_2) & \cdots & \chi_N(\boldsymbol{x}_2) \\ \vdots & & & \\ \chi_1(\boldsymbol{x}_N) & \chi_2(\boldsymbol{x}_N) & \cdots & \chi_N(\boldsymbol{x}_N) \end{vmatrix}. \tag{2.53}$$

Using the spin-free formulation in QMC, we would like to organize the wave function so that spin-up electrons are listed before spin-down electrons. For that purpose, we introduce two Slater determinants, $D^\uparrow(\mathbf{x}^\uparrow)$ for spin-up electrons and $D^\downarrow(\mathbf{x}^\downarrow)$ for spin-down electrons. These are determinants of spatial single-particle orbitals $\phi_j(\mathbf{r}_i)$, where particle i is in orbital j.

The notation used for Slater determinants varies. The Slater determinant of spin-up electrons could be denoted by

$$\frac{1}{\sqrt{N_\uparrow!}} \begin{vmatrix} \phi_1(\mathbf{r}_1) & \phi_2(\mathbf{r}_1) & \cdots & \phi_{N_\uparrow}(\mathbf{r}_1) \\ \phi_1(\mathbf{r}_2) & \phi_2(\mathbf{r}_2) & \cdots & \phi_{N_\uparrow}(\mathbf{r}_2) \\ \vdots & & & \\ \phi_1(\mathbf{r}_{N_\uparrow}) & \phi_2(\mathbf{r}_{N_\uparrow}) & \cdots & \phi_{N_\uparrow}(\mathbf{r}_{N_\uparrow}) \end{vmatrix}_\uparrow \tag{2.54}$$

$$= \mathrm{Det}\, S^\uparrow(\mathbf{r}_1, \ldots, \mathbf{r}_{N^\uparrow}) = \mathrm{Det}\, S^\uparrow(\mathbf{x}^\uparrow) \quad \text{Slater matrix } S^\uparrow \tag{2.55}$$

$$= D^\uparrow(\mathbf{r}_1, \ldots, \mathbf{r}_{N^\uparrow}) = D^\uparrow(\mathbf{x}^\uparrow) \tag{2.56}$$

$$= |\phi_1(\mathbf{r}_1), \phi_2(\mathbf{r}_2), \ldots, \phi_{N_\uparrow}(\mathbf{r}_{N_\uparrow})\rangle \quad \text{chemists' notation} \tag{2.57}$$

$$= |\phi_1, \phi_2, \ldots, \phi_{N_\uparrow}\rangle \quad \text{chemists' short notation.} \tag{2.58}$$

Chemists list only the diagonal elements in the determinant. The Slater matrix has orbitals as elements:

$$S_{i,j}^\uparrow = \phi_j(\mathbf{r}_i). \tag{2.59}$$

2.2.1 The Slater–Jastrow wave function

A simple Slater determinant product wave function,

$$D^\uparrow(\mathbf{x}^\uparrow) D^\downarrow(\mathbf{x}^\downarrow), \tag{2.60}$$

takes care of the Pauli exclusion principle, but it does not adequately represent the correlations between electrons. In addition to the Pauli exclusion principle, electron positions are correlated due to the interparticle potential $V(\mathbf{x})$. The missing correlations can be addressed by including a *Jastrow factor* $e^{J(\mathbf{x})}$,

$$\boxed{\Psi(\mathbf{x}) = \underbrace{e^{J(\mathbf{x})}}_{\text{Jastrow}}\ \underbrace{D^\uparrow(\mathbf{x}^\uparrow) D^\downarrow(\mathbf{x}^\downarrow)}_{\text{Slater}} \quad \text{Slater–Jastrow wave function}.} \tag{2.61}$$

The Jastrow factor is always positive, and all fermion nodes are in the Slater part. A well-chosen Jastrow factor can satisfy the electron–electron cusp conditions and reduce variance in QMC. This is a possible source of confusion because the electron–nucleon cusp conditions may already be satisfied by the orbitals in the Slater determinant; in this case, do not try to add them again to the Jastrow factor. The orbitals can be made to satisfy the cusp conditions using the so-called *cusp corrections*, see section 2.2.4.

The Slater–Jastrow wave function is surprisingly accurate, considering that it lacks direct spin coupling. It may not be adequate for strongly correlated electron systems or systems with quantum entanglement. In most cases, the Slater–Jastrow wave function includes only two-body correlations, meaning that it explicitly includes electron–electron correlations but only up to pairwise interactions. Three-body or higher-body correlations may be important in high-density systems. It is worth remembering that orbitals obtained from DFT have ions at fixed positions by default.

We take a closer look at the Jastrow factor in section 2.3. Before that, we can make one more generalization to the wave function.

2.2.2 The multideterminant and the configuration interaction wave function

The Slater–Jastrow wave function has the determinants $D^\uparrow$ and $D^\downarrow$, where electrons occupy $N^\uparrow$ and $N^\downarrow$ orbitals, respectively. However, these are not the only possibilities when choosing the occupied orbitals. Remember that the single-particle orbitals are basically eigenvectors of a simple eigenvalue problem, such as the Hartree–Fock equations, so there are many more orbitals to choose from. Each orbital corresponds to an eigenvalue, which is the single-particle energy of that orbital, and they can be used to organize the orbitals in order of increasing energy.

Let us enumerate the determinants using indices $k = 0, 1, \ldots$, so that each determinant D_k has a different set of occupied orbitals. For a ground-state calculation, it is reasonable to start with D_0 as the Slater determinant of the lowest-energy single-particle orbitals. For spin-up electrons (in chemists' notation),

$$D_0^\uparrow : = |\phi_1, \phi_2, \ldots, \phi_m, \phi_n, \ldots, \phi_{N_\uparrow}\rangle \qquad \text{lowest orbitals occupied.} \qquad (2.62)$$

In the case of Hartree–Fock orbitals, the energy of the state $\Psi(\mathbf{x}) = D_0^\uparrow D_0^\downarrow$ is the Hartree–Fock ground-state energy. Next, we choose each $D_{k+1}^\uparrow$ so that it differs from $D_0^\uparrow, \ldots, D_k^\uparrow$ in at least one orbital. Keeping $D_0^\uparrow$ as a base, $D_1^\uparrow$ can be viewed as the excitation of one electron from, say, orbital m to p. In the following notation, replaced orbitals are denoted by subscripts, and their replacements are denoted by superscripts:

$$(D^\uparrow)_m^p = |\phi_1, \phi_2, \ldots, \phi_p, \phi_n, \ldots, \phi_{N_\uparrow}\rangle \qquad \text{replaced } m \text{ with } p \qquad (2.63)$$

$$(D^\uparrow)_{mn}^{pq} = |\phi_1, \phi_2, \ldots, \phi_p, \phi_q, \ldots, \phi_{N_\uparrow}\rangle \qquad \text{replaced } m \text{ with } p \text{ and } n \text{ with } q. \quad (2.64)$$

Obviously, there are very many singly excited Slater determinants and even more doubly excited ones.

Take a linear combination of different Slater determinants, supplemented with a Jastrow factor:

$$\boxed{\Psi(\mathbf{x}) = e^{J(\mathbf{x})} \sum_k c_k D_k^\uparrow(\mathbf{x}^\uparrow) D_k^\downarrow(\mathbf{x}^\downarrow) \qquad \text{multideterminant wave function.}} \qquad (2.65)$$

The multideterminant wave function is a linear combination of all Slater determinants that can be constructed by distributing the available electrons over single-particle orbitals. Each determinant represents a different electronic configuration, and the name 'configuration interaction' (CI) reflects the fact that the superposition includes the interaction of many configurations. The wave function is not antisymmetric under the exchange of spin-up and spin-down electrons, but the expectation values (2.50) remain valid [15].

A related form is the CI wave function, proposed by Per-Olof Löwdin [16], 'father of quantum chemistry.' It is a linear combination of Slater determinants $|\Psi\rangle$ (chemists' notation; see, e.g. reference [17]),

$$\Psi = c_0|\Psi_0\rangle + \sum_{m,p} c_m^p|\Psi_m^p\rangle + \sum_{\substack{m<n \\ p<q}} c_{mn}^{pq}|\Psi_{mn}^{pq}\rangle + \cdots \quad \text{CI wave function.} \tag{2.66}$$

Here, the Slater matrix elements are spin orbitals. The all-electron, full configuration interaction (FCI) method is exact within the chosen basis set, but it is computationally very demanding, so results are available only for systems with a small number of electrons.

Be atom ground state

Sometimes, one gets lucky, and keeping just two terms gives a clear improvement over the Slater–Jastrow wave case. For example, the Be atom ground-state configuration is $1s^2 2s^2$, so a first approximation is to occupy those Hartree–Fock orbitals. It turns out, however, that just adding the Hartree–Fock configuration $1s^2 2p^2$ yields a much better ground state.

2.2.3 Configuration state functions (CSFs)

Apart from the all-important Pauli exclusion principle, there is another physical property that needs attention, namely that in atomic and molecular systems, spin angular momentum is conserved. The eigenstates of the total spin $\hat{S}^2$ and its z-component $\hat{S}_z$ are states with a well-defined magnitude and orientation of spin. In free space without any preferred direction, the Hamiltonian $\hat{\mathcal{H}}$ in these systems commutes with both operators, and they commute with each other:

$$[\hat{\mathcal{H}}, \hat{S}^2] = 0, \; [\hat{\mathcal{H}}, \hat{S}_z] = 0, \; [\hat{S}^2, \hat{S}_z] = 0, \tag{2.67}$$

so that all three operators have common eigenstates. Thus, the exact eigenstates of atoms and molecules should also be eigenstates of $\hat{S}^2$ and $\hat{S}_z$.

A single Slater determinant is not necessarily an eigenstate of $\hat{S}^2$, but a certain linear combination of them is.

> **CSFs are linear combinations of products of spin-up and spin-down Slater determinants that have a definite total spin and spin projection.**

A CSF can be written as an expansion:

$$|\Phi_i^{\text{CSF}}\rangle = \sum_k d_{ik}|D_k^{\uparrow}\rangle|D_k^{\downarrow}\rangle, \tag{2.68}$$

where the coefficients d_{ik} are fully determined by the spin and spatial symmetry of the target state. The final Jastrow–Slater trial state is a sum of CSFs, and the resulting sum is multiplied by the Jastrow factor:

$$|\varphi_T\rangle = e^J \sum_i c_i |\Phi_i^{\text{CSF}}\rangle. \tag{2.69}$$

Unlike the symmetry-fixed CSF coefficients d_{ik}, the coefficients c_i and the unspecified Jastrow parameters are free parameters. While a CSF is not an eigenstate of the Hamiltonian, it is an acceptable trial wave function candidate with the benefit that the number of free parameters is reduced by symmetry requirements. If you dig deeper, 'symmetry' brings in the principles of *group theory*, while the 'spin of a collection of electrons' hints that you will probably need the *coupling of angular momenta*. In some simple cases, you can manage without knowing much about group theory, less so without some idea of the coupling of angular momenta.

As an example, consider the ground state of a carbon atom. The ground-state electronic configuration of carbon is $1s^2 2s^2 2p^2$. The spherically symmetric 1s and 2s orbitals are full, so the two electrons in the 2p orbital dictate the spin and spatial symmetries. **Hund's rules** say that the lowest-energy configuration has the maximum total spin (this can be shown to minimize the exchange energy) and the maximum possible total angular momentum. The first rule takes precedence over the second one.[5]

- **Spin symmetry:** CSFs are eigenstates of angular momentum, so the way angular momenta are coupled is crucial. The two 2p electrons can form either a spin singlet or a spin triplet. The coupled states $|s, m\rangle$ with total spin s and projection m are

$$\text{singlet: } |0, 0\rangle = \frac{1}{\sqrt{2}}(|\uparrow\downarrow\rangle - |\downarrow\uparrow\rangle) \Big\} \quad S = 0 \tag{2.70}$$

$$\text{triplet:} \begin{array}{l} |1, 1\rangle = |\uparrow\uparrow\rangle \\ |1, 0\rangle = \frac{1}{\sqrt{2}}(|\uparrow\uparrow\rangle + |\downarrow\downarrow\rangle) \\ |1, -1\rangle = |\downarrow\downarrow\rangle \end{array} \Bigg\} \quad S = 1. \tag{2.71}$$

In general, the angular momentum state $|j, m\rangle$ is obtained using the coupling

---

[5] There is also a third Hund's rule ('fine structure splitting'): for less than half-filled subshells, the state with the lowest total angular momentum $J = |L - S|$ usually has the lowest energy. A subshell is a collection of orbitals with the same orbital quantum number l; a subshell may contain more than one orbital. The third Hund's rule is a trend rather than a strict rule.

$$|j, m\rangle = \sum_{m_1+m_2=m} C^{j_1 j_2 S}_{m_1 m_2 m} \underbrace{|j_1, m_1\rangle |j_2, m_2\rangle}_{|j_1, j_2; m_1, m_2\rangle}, \tag{2.72}$$

with the Clebsch–Gordan coefficients

$$C^{s_1 s_2 S}_{m_1 m_2 m} := \langle j_1, j_2; m_1, m_2 | j_1, j_2; j, m\rangle. \tag{2.73}$$

- **Spatial symmetry:** The 2p electrons have angular momenta $l_1 = 1$ and $l_2 = 1$, so their coupling gives the total angular momenta

$$L = |l_1 - l_2|, \ldots, |l_1 + l_2| = 0, 1, 2 = S, P, D. \tag{2.74}$$

The three suggested spatial symmetries are (in spectroscopic ^{2S+1}L notation, $2S + 1$ is the multiplicity, now singlet or triplet):
^1S ($S = 0$, $L = 0$), a spherically symmetric singlet state;
^3P ($S = 1$, $L = 1$), an elongated or cigar-shaped triplet state; and
^1D ($S = 0$, $L = 2$), a four-leaf clover, a four-lobed clover, or a dumbbell with a donut singlet state.

The $S = 1$ spin triplet has the maximum total spin; therefore, the carbon ground state is ^3P. Apart from $1s^2 2s^2 2p^2$, there are contributions from, for example, $2s^2 2p^4$ or $1s^2 2s^2 2p^1 3p^1$, which also have the same symmetry. In quantum chemistry, many such configurations are summed up in the CI, and the QMC trial wave function has a similar structure. In practice, the number of configurations in the sum (2.68) can be in the thousands.

If an atom has three p electrons, couple the first two of them and then couple the third. In this case, the outcome may be $S = 1/2$ or $S = 3/2$, and the spin and angular momentum combinations are $^2S,^2 P,^2 D$ or ^2F for the former spin symmetry, and $^4S,^4 P,^4 D$ or ^4F for the latter.

Where did the Slater determinants in the expansion come from? A practical way of constructing the Slater determinants is to start from the Hartree–Fock ground-state determinant, add determinants with a single electron excited to a higher-energy orbital (the so-called CIS, CI with singles), proceed to make two-electron excitations (CISD, CI with singles and doubles), and so on. The number of triple excitations is huge, which makes CISDT (singles, doubles, and triples) prohibitively expensive to compute; therefore, in many cases, only some of the triples are included. Here, 'excitation' means the promotion of an electron to an unoccupied ('virtual') Hartree–Fock orbital. This process is governed by the so-called Slater rules, a convenient notation for excitations w.r.t. the Hartree–Fock ground-state Slater determinant. Personally, I find it easier to think about an algorithm that kicks electrons to higher orbitals than a math formula. Many of the Slater determinants generated in this way have the wrong symmetry, so the CSF is a sum over a restricted set of determinants.

Molecular wave functions should also be size consistent to be physical.

Size consistency: The properties of two noninteracting molecules are the sum of those properties computed for the individual molecules.

When you think about it, size consistency is trivial. If you let the atoms in, say, an H_2 molecule separate so that they are far apart, you obviously get two H atoms and reach the so-called atomic limit. Problems arise when you expand a molecular quantity as a perturbative expansion and realize that only the full sum is size consistent. This happens, for example, in the GW approximation, a first-order perturbative expansion of the electronic self-energy in terms of a screened interaction. A major drawback of truncating the CSF expansion in equation (2.68) is that it breaks size consistency.

What about group theory? Some symmetry operations, such as rotations, reflections, and inversions, leave the overall structure of a molecule intact and form a point group. Each point group has a set of representations, mathematical operations that tell how functions, such as the molecular wave function, transform under a symmetry operation of the point group. Some representations are reducible and can be split into smaller symmetry operations. An irreducible representation is one that cannot be split; it is the smallest, fundamental, and indivisible symmetry operation. In short, if a wave function belongs to the wrong irreducible representation of the molecule's point group, it will have the wrong spatial symmetry, that is, not the symmetry of the target state. Such wave functions can be ignored from the start.

2.2.4 Cusp corrections to Gaussian orbitals

Slater orbitals have a cusp, which can be adjusted to fulfill the Kato cusp conditions,

$$\frac{\partial \langle \Psi \rangle}{\partial r_i}\bigg|_{r_i = 0} = Z \langle \Psi \rangle_{r_i = 0}, \tag{2.75}$$

where $\langle .. \rangle$ indicates the spherical average near $\mathbf{r}_i = 0$.[6] The popular Gaussian orbitals have no cusp, so near the nucleus, the uncorrected electronic local energy has rapid oscillations. In [18], the goal is to add *cusp corrections* to Gaussian orbitals. The Gaussian that is nonzero at the nucleus is replaced with a spherical wave function with a cusp, up to a chosen correction radius r_c. This has been shown to dramatically reduce the QMC variance for many tested small molecules. The improvement in local energy is similar to the He atom example in figure 2.1.

---

[6] A spherical average is an angular average on a sphere centered at the singularity. It removes all inconsequential angular dependencies.

2.2.5 Nodal surface and the backflow transformation

The multideterminant wave function (2.65) is an excellent trial wave function for electron problems, but it is computationally expensive. Hence, the preference leans toward the more straightforward Slater–Jastrow form (2.61). The price of using only one Slater determinant for each spin is that the nodes, or nodal surfaces, are not the exact ground-state nodes, and the variational principle says the energy will be greater than the ground-state energy.

There is no escape from the fact that once the Slater–Jastrow occupied orbitals are fixed, the nodes are fixed.[7] Nodes are many-electron effects, so one approach is to make the orbitals depend on the positions of the other electrons.[8] The Jastrow factor has no nodes, so this is the only way to alter the nodal surface of a single-determinant Slater–Jastrow wave function.

The task is then to find a parameterized function that depends on the total configuration of the system and optimize the parameters using QMC. One possibility is to add so-called *backflow transformations* to orbitals. The backflow transformations originate from improvements to the theory of liquid helium excitations introduced by Feynman and Cohen [20]. They argued that a moving ^{3}He impurity atom displaces ^{4}He atoms and creates a flow that fills the void left behind so that the effective mass of the impurity increases and there is momentum transfer between the moving atom and its surroundings. Backflow was first included in QMC in 1981 by Lee *et al* [21] and was later shown to be important in the Slater–Jastrow calculations of high-density (low r_s) electron gases by Kwon *et al* [22].

The single-particle orbitals depend on one coordinate, but nothing says they have to be particle positions. The nodal structure of orbitals can be altered by replacing the coordinates with new, generalized coordinates [19]:

$$\tilde{\mathbf{r}}_i = \mathbf{r}_i + \boldsymbol{\xi}_i(\mathbf{x}) \quad \text{backflow transformation}, \qquad (2.76)$$

so that the whole configuration $\mathbf{x} := (\mathbf{r}_1, \ldots, \mathbf{r}_N)$ is present in all orbitals through a vector field $\boldsymbol{\xi}$. With backflow transformation, the Slater–Jastrow and multidetermi-nant wave functions are no longer spatial angular momentum eigenstates, a problem termed *spatial angular momentum contamination*. One also has to take care that the many-body effects of backflow do not prevent the cancellation of singular potentials.

Applying a backflow transformation to the orbitals has been shown to significantly improve the nodal surfaces of atoms and solids [19, 23, 24]. It should be noted, however, that while the backflow transformation does allow the nodal surfaces to be moved, it cannot change the topology of the nodal cells (a nodal cell is a connected space of configurations $\mathbf{x}$ with the same sign of the wave function). For more than two spin-up and two spin-down electrons, the two Slater determinants have at least two nodal surfaces and four nodal domains. The number of nodal domains cannot be reduced below four, no matter which continuous backflow transformation is applied. On the other hand, the Be atom ground state is

---

[7] We will also encounter this later in fixed-node diffusion Monte Carlo.
[8] As pointed out in [19], this line of thinking dates back to 1935 and the work of Wigner and Seitz.

known to have only two nodal domains; therefore, no backflow transformation can yield its ground-state wave function [25].

2.3 The Jastrow factor

The correlation factor $e^{J(\mathbf{x})}$ is named after R Jastrow, who in 1955 proposed an ansatz for particle–particle correlation functions in presence of a strong potential [26]. He applied the method to boson and fermion hard-sphere fluids, aiming to find the pair distribution function $g(r)$ and the kinetic energy. The fermion wave function used in [26] would nowadays be called a Slater–Jastrow wave function.[9]

In a Coulomb system, a trial wave function satisfying the cusp conditions gives a local energy where the Coulomb singularities cancel exactly. For atoms, such an electronic wave function can easily be found by extending equation (2.37) to N electron atoms:

$$V(\mathbf{x}) = \sum_{i=1}^{N}\left(-\frac{Z}{r_i}\right) + \sum_{i<j}^{N}\frac{1}{r_{ij}} \tag{2.77}$$

$$\text{Jastrow factor } e^{J(\mathbf{x})},\ J(\mathbf{x}) = \sum_{i=1}^{N}(-\alpha_i r_i) + \sum_{i<j}^{N}\frac{r_{ij}}{(2|4)\left(1 + \beta r_{ij}\right)}. \tag{2.78}$$

In $J(\mathbf{x})$, the first term takes care of the electron–nucleon cusp conditions, while the second term is tuned to satisfy the electron–electron cusp conditions.

At some point, one has to decide which correlations are in the Slater determinant and what is left for the Jastrow factor. If, for example, the orbitals satisfy the electron–nucleon cusp conditions, then one possible Slater–Jastrow wave function would be

$$\Psi(\mathbf{x}) = e^{\sum_{i<j}^{N}\frac{r_{ij}}{(2|4)(1+\beta r_{ij})}} D^{\uparrow}(\mathbf{x}^{\uparrow})D^{\downarrow}(\mathbf{x}^{\downarrow}). \tag{2.79}$$

Evidently, there is some *spin contamination* because the Jastrow factor also contains spin information in the factors of two and four for antiparallel and parallel spins, respectively. The consequence is that the state is not an exact eigenstate of the total spin $\hat{S}^2$ [29].

2.3.1 Optimization of the Jastrow factor

Following Jastrow's original work, it is possible to generalize the Jastrow factor for a one-body interaction and a pair potential (these cover all cases encountered in this book):

$$\hat{\mathcal{H}} = -\frac{1}{2}\sum_{i=1}^{N}\nabla_i^2 + \underbrace{\sum_{i=1}^{N}V(r_i)}_{\text{one-body interaction}} + \underbrace{\sum_{i<j}^{N}V(r_{ij})}_{\text{pair potential}} \tag{2.80}$$

---

[9] A similar boson ground-state wave function was previously proposed by Arie Bijl [27] and Robert Dingle [28], and the wave function may be called the Bijl-Dingle-Jastrow wave function.

$$\text{Jastrow factor} = e^{J(\mathbf{x})}, \quad J(\mathbf{x}) = \underbrace{\sum_{i=1}^{N} u_1(\mathbf{r}_i)}_{\text{one-body correlations}} + \underbrace{\frac{1}{2}\sum_{i<j}^{N} u_2(\mathbf{r}_i, \mathbf{r}_j)}_{\text{two-body correlations}} . \tag{2.81}$$

The Jastrow factor ending at two-body correlations is an approximation. Following down the rabbit hole, the pair potential creates correlations between a pair and a third particle, between pairs of particles, and so on. A hierarchy of correlations can be expressed by adding higher-order correlation functions $u_3(\mathbf{r}_1, \mathbf{r}_2, \mathbf{r}_3)$, etc. Such a wave function goes under the name of the *Jastrow–Feenberg wave function*, and it has been elemental in the development of the diagrammatic theory of low-temperature quantum fluids. In QMC, one often stops at u_2 because either the higher-order correlation functions are intractable or the approximate forms are too slow to compute. Computationally cheap approximate three-body terms are usually the limit.

For the given Hamiltonian, the correlation functions $u_1(r_i)$ and $u_2(r_{ij})$ can be chosen so that *any* one-body potential and pair potential terms are canceled by the local kinetic energy $T_L(\mathbf{x})$ in the local energy $E_L(\mathbf{x}) = T_L(\mathbf{x}) + V(\mathbf{x})$. Compared to the cusp conditions, we now cancel the whole potential, not just the singularities. We can write the trial wave function for bosons in the form

$$\varphi_T(\mathbf{x}) = e^{J(\mathbf{x})}S(\mathbf{x}), \tag{2.82}$$

where $e^{J(\mathbf{x})}$ is symmetric and $S(\mathbf{x}) = 1$, while we can write an antisymmetric function (possibly a Slater determinant) for fermions. The local energy is

$$E_L(\mathbf{x}) = -\frac{1}{2}\frac{\sum_{i=1}^{N} \nabla_i^2 \left[e^{J(\mathbf{x})}S(\mathbf{x})\right]}{e^{J(\mathbf{x})}S(\mathbf{x})} + V(\mathbf{x})$$

$$= \underbrace{-\frac{1}{2}\frac{\sum_{i=1}^{N} \nabla_i^2 \, e^{J(\mathbf{x})}}{e^{J(\mathbf{x})}} + V(\mathbf{x})}_{:=E_L^J(\mathbf{x})} - \frac{1}{2}\frac{\sum_{i=1}^{N} \nabla_i^2 \, S(\mathbf{x})}{S(\mathbf{x})} + 2\sum_{i=1}^{N}\frac{\nabla_i S(\mathbf{x})}{S(\mathbf{x})} \cdot \frac{\nabla_i e^{J(\mathbf{x})}}{e^{J(\mathbf{x})}}. \tag{2.83}$$

The $E_L^J(\mathbf{x})$ part of the local energy is bosonic, and we can find a $J(\mathbf{x})$ such that the potentials cancel. With $J(\mathbf{x})$ as given in equation (2.81), the bosonic part is

$$E_L^J(\mathbf{x}) = \sum_{i=1}^{N}\left[-\frac{1}{2}\frac{\nabla_i^2 e^{u_1(\mathbf{r}_i)}}{e^{u_1(\mathbf{r}_i)}} + V(r_i)\right] \qquad \text{1-body term}$$

$$+ \frac{1}{2}\sum_{i\neq j}^{N}\left[-\frac{\nabla_i^2 e^{u_2(\mathbf{r}_i, \mathbf{r}_j)/2}}{e^{u_2(\mathbf{r}_i, \mathbf{r}_j)/2}} + 2\,\nabla_i u_1(\mathbf{r}_i) \cdot \nabla_j u_2(\mathbf{r}_i, \mathbf{r}_j) + V(r_{ij})\right] \text{2-body term} \tag{2.84}$$

$$+ \sum_{i\neq j\neq k}^{N}\left[-\frac{1}{4}\nabla_k u_2(\mathbf{r}_k, \mathbf{r}_i) \cdot \nabla_k u_2(\mathbf{r}_k, \mathbf{r}_j)\right] \qquad \text{3-body term.}$$

Since $E_L^J(\mathbf{x})$ is always integrated over all coordinates, we can choose $i = 1$ in one-body terms and multiply by N; we can choose $i, j = 1, 2$ in two-body terms and multiply by $N(N - 1)$, and so on. This expression shows how the one-body correlations spill into the two-body terms, and how a three-body term emerges that combines two two-body correlation functions with one common coordinate.

From the variance-optimization point of view, there is nothing here that can make the three-body term a constant, unless there are only two particles. On the other hand, there are no potential terms beyond the two-body term, so we could add a u_3 to cancel the three-body term. That would, however, leave a non-constant four-body term, so we are back in the rabbit hole. Nevertheless, the correlation functions $u_3, u_4, \ldots u_N$ would be functionals of just u_1 and u_2, but we do not know their form.

The one-body potential is easy to cancel. Set the quantity in the brackets equal to a constant E_1:

$$-\frac{\nabla^2 e^{u_1(\mathbf{r})}}{e^{u_1(\mathbf{r})}} + V(r) = E_1 \Leftrightarrow -\nabla^2 e^{u_1(\mathbf{r})} + V(r)e^{u_1(\mathbf{r})} = E_1 e^{u_1(\mathbf{r})}. \tag{2.85}$$

Atoms with $N = Z$ electrons have $V(r) = -Z/r$, and the lowest eigenvalue solution is, as expected, the electron–nucleus cusp condition solution $u_1(\mathbf{r}) = -Zr$. The resulting local energy has the constant term

$$NE_1 = -\frac{1}{2}NZ^2 = -\frac{1}{2}Z^3 \quad \text{1-body term for } N \text{ electrons}. \tag{2.86}$$

For the hydrogen atom, $Z = 1$, and $E_L(\mathbf{x}) = E_1 = -1/2$ is the ground-state energy.

Two or more particles

Suppose we are not willing to add three-body and higher-order correlation functions. We can still choose a u_2 that *minimizes* the remaining three-body fluctuations. It is convenient to mark the two-body piece with $W(\mathbf{r}_i, \mathbf{r}_j)$ and add and subtract it,

$$E_L^J(\mathbf{x}) = NE_1$$
$$+ \frac{1}{2}\sum_{i \neq j}^{N}\underbrace{\left[-\frac{\nabla_i^2 e^{u_2(\mathbf{r}_i, \mathbf{r}_j)/2}}{e^{u_2(\mathbf{r}_i, \mathbf{r}_j)/2}} + 2\nabla_i u_1(\mathbf{r}_i) \cdot \nabla_j u_2(\mathbf{r}_i, \mathbf{r}_j) + W(\mathbf{r}_i, \mathbf{r}_j) + V(r_{ij})\right]}_{= 0}$$
$$+ \frac{1}{2}\sum_{i \neq j}^{N}\underbrace{\left[\sum_{k(\neq i,j)}\left(-\frac{1}{2}\nabla_k u_2(\mathbf{r}_k, \mathbf{r}_i) \cdot \nabla_k u_2(\mathbf{r}_k, \mathbf{r}_j)\right) - W(\mathbf{r}_i, \mathbf{r}_j)\right]}_{\text{reduce fluctuations}}. \tag{2.87}$$

The reason why the quantity in the first bracket must be zero is that if it were nonzero, the local energy would have a term that scales with the particle number as $N(N - 1)$, which would violate energy extensivity.

The condition given for the last bracket is that W minimizes the local energy variance; this is a line of thought inspired by QMC. The functional form of W depends on where the third particle k can be. If it can be anywhere, we can average over all positions and set

$$W(\mathbf{r}_i, \mathbf{r}_j) = -\frac{1}{2}\frac{N}{V}\int d\mathbf{r}_k \, \nabla_k u_2(\mathbf{r}_k, \mathbf{r}_i) \cdot \nabla_k u_2(\mathbf{r}_k, \mathbf{r}_j) \quad \text{(approximation)}. \tag{2.88}$$

This is a reasonable starting point for homogeneous systems, where the one-body potential vanishes, and hence $V(\mathbf{r}_i) = 0$ and $u_1 = 0$. All functions depend on the distance $r_{12}\colon = |\mathbf{r}_1 - \mathbf{r}_2|$, and the two- and three-body equations read from equation (2.87) for homogeneous systems are

$$W(r_{12}) = -\frac{1}{2}\frac{N}{V}\int d\mathbf{r}_3\, \nabla_3 u_2(r_{31}) \cdot \nabla_3 u_2(r_{32}) \qquad \text{(approximation)} \qquad (2.89)$$

$$-\frac{(\nabla_1^2 + \nabla_2^2)e^{u_2(r_{12})/2}}{e^{u_2(r_{12})/2}} + W(r_{12}) + V(r_{12}) = 0. \qquad (2.90)$$

An interesting fact is that W has the density N/V as a factor, so if the functional form of W is the same at a very low density, then the approximations made in deriving the equation become quite accurate; in particular, the u_3 and higher contributions to the local energy are negligible (we already left them out). Because of cavitation, an interacting system does not remain homogeneous if the density is lowered below a certain threshold, usually below the local liquid vapor pressure. So a valid strategy would be to use W in the form of equation (2.89) at a low density, solve $u_2(r_{12})$ from equation (2.90), and use it in the trial wave function. The reward for all this work would be that the pair potential $V(r_{12})$—the whole potential—cancels from the local energy, leaving only a short-range function $W(r_{12})$. This approach is unpopular for an obvious reason: the equation for u_2 is a nonlinear differential equation.

Equation (2.90) has the form of an Euler–Lagrange equation, common in optimization problems. For example, in the hypernetted chain (HNC) diagrammatic theory, the radial distribution function $g(r)$ of a homogeneous isotropic system is found as a solution of the equation

$$-\frac{2\,\nabla^2\,\sqrt{g(r)}}{\sqrt{g(r)}} + W(r) + V(r) = 0. \qquad (2.91)$$

Here, $W(r)$ is the so-called induced potential, a functional of $g(r)$ which has a closed but only approximate form. Furthermore, $g(r) = e^{u_2(r)}\times$ corrections, so now you see the common origin of equations (2.90) and (2.91)—and why I chose to call the extra function W. Equation (2.89) was derived from a variance-optimization point of view, and it gives an approximate functional form for W, namely, $W[u_2](r)$.

I will not pursue the matter of optimizing the Jastrow factor and correlation functions u_1, u_2, and so on any further, because it is more in the spirit of QMC to keep one's hands off paper and pencil and let computers do the heavy work. Nevertheless, analytical expressions often show the physics more clearly and can speed up QMC quite a lot. A very potent Jastrow factor for atoms, molecules, and solids is given in [30]. For N electrons and N_{ions} ions, the correlation functions in the Jastrow factor are[10]

---

[10] A system can be homogeneous but not isotropic if it shows variations in different directions, and it can be isotropic but not homogeneous if the properties vary even though they do so uniformly in all directions.

$$J(\mathbf{x}) = \sum_{i<j}^{N} u(r_{ij}) \qquad \text{electron–electron (homogeneous \& isotropic)}$$

$$+ \sum_{I=1}^{N_{\text{ions}}} \sum_{i=1}^{N} \mathcal{X}_I(r_{iI}) \qquad \text{electron–ion (isotropic)}$$

$$+ \sum_{I=1}^{N_{\text{ions}}} \sum_{i<j}^{N} f_I(r_{iI}, r_{jI}, r_{ij}) \qquad \text{electron–electron–ion (isotropic)} \qquad (2.92)$$

$$+ \sum_{i<j}^{N} p(\mathbf{r}_{ij}) \qquad \text{periodic electron–electron}$$

$$+ \sum_{i=1}^{N} q(\mathbf{r}_i) \qquad \text{periodic electron position.}$$

References

[1] Metropolis N and Ulam. S 1949 The Monte Carlo method *J. Am. Stati. Assoc.* **44** 335–41
[2] Metropolis N, Rosenbluth A W, Rosenbluth M N, Teller A H and Teller E 2004 Equation of state calculations by fast computing machines *J. Chem. Phys.* **21** 1087–92
[3] Kramida A, Ralchenko Y and Reader JNIST ASD Team 2022 NIST atomic spectra database (ver. 5.10) [Online]. Available: https://physics.nist.gov/asd [2023 November 8]. National Institute of Standards and Technology, Gaithersburg, MD
[4] Kandula D Z, Gohle C, Pinkert T J, Ubachs W and Eikema K S E 2010 Extreme ultraviolet frequency comb metrology *Phys. Rev. Lett.* **105** 063001
[5] Kato T 1957 On the eigenfunctions of many-particle systems in quantum mechanics *Commun. Pure Appl. Math.* **10** 151–77
[6] Fahy S, Wang X W and Louie S G 1990 Variational quantum Monte Carlo nonlocal pseudopotential approach to solids: formulation and application to diamond, graphite, and silicon *Phys. Rev.* B **42** 3503–22
[7] Ceperley D M and Alder B J 1980 Ground state of the electron gas by a stochastic method *Phys. Rev. Lett.* **45** 566–9
[8] Hylleraas E A 1963 Reminiscences from early quantum mechanics of two-electron atoms *Rev. Mod. Phys.* **35** 421–30
[9] Hylleraas E A 1928 Über den Grundzustand des Heliumatoms *Z. Phys.* **48** 469–94
[10] Pekeris C L 1959 1^1S and 2^3S states of helium *Phys. Rev.* **115** 1216–21
[11] Nakatsuji H 2004 Scaled Schrödinger equation and the exact wave function *Phys. Rev. Lett.* **93** 030403
[12] Nakatsuji H, Nakashima H and Kurokawa Y I 2022 Accurate scaling functions of the scaled Schrödinger equation *J. Chem. Phys.* **156** 014113
[13] Wilson B E 1968 DFT explained *Structural Chemistry and Molecular Biology* ed A Rich and N Davidson (San Francisco, CA: W.H. Freeman) pp 753–60
[14] Hohenberg P and Kohn W 1964 Inhomogeneous electron gas *Phys. Rev.* **136** B864–71
[15] Foulkes W M C, Mitas L, Needs R J and Rajagopal G 2001 Quantum Monte Carlo simulations of solids *Rev. Mod. Phys.* **73** 33–83

[16] Löwdin P-O 1955 Quantum theory of many-particle systems. I. Physical interpretions by means of density matrices, natural spin-orbitals, and convergence problems in the method of configurational interaction *Phys. Rev.* **97** 1474–89

[17] Szabo A and Ostlund N S 1996 *Modern Quantum Chemistry: Introduction to Advanced Electronic Structure Theory* 1st edn (Minneola, MN: Dover)

[18] Ma A, Towler M D, Drummond N D and Needs R J 2005 Scheme for adding electron-nucleus cusps to Gaussian orbitals *J. Chem. Phys.* **122** 224322

[19] López Ríos P, Ma A, Drummond N D, Towler M D and Needs R J 2006 Inhomogeneous backflow transformations in quantum Monte Carlo calculations *Phys. Rev.* E **74** 066701

[20] Feynman R P and Cohen M 1956 Energy spectrum of the excitations in liquid helium *Phys. Rev.* **102** 1189–204

[21] Lee M A, Schmidt K E, Kalos M H and Chester G V 1981 Green's function Monte Carlo method for liquid ^{3}He *Phys. Rev. Lett.* **46** 728–31

[22] Kwon Y, Ceperley D M and Martin R M 1998 Effects of backflow correlation in the three-dimensional electron gas: quantum Monte Carlo study *Phys. Rev.* B **58** 6800–6

[23] Drummond N D, López Ríos P, Ma A, Trail J R, Spink G G, Towler M D and Needs R J 2006 Quantum Monte Carlo study of the Ne atom and the Ne+ ion *J. Chem. Phys.* **124** 224104

[24] Brown M F, Trail J R, López Ríos P and Needs R J 2007 Energies of the first row atoms from quantum Monte Carlo *J. Chem. Phys.* **126** 224110

[25] Bressanini D 2012 Implications of the two nodal domains conjecture for ground state fermionic wave functions *Phys. Rev.* B **86** 115120

[26] Jastrow R 1955 Many-body problem with strong forces *Phys. Rev.* **98** 1479–84

[27] Bijl. A 1940 The lowest wave function of the symmetrical many particles system *Physica* **7** 869–86

[28] Dingle R B 1949 LI. The zero-point energy of a system of particles *Lond. Edinb. Dubl. Philo. Mag. J. Sci.* **40** 573–8

[29] Huang C-J, Filippi C and Umrigar C J 1998 Spin contamination in quantum Monte Carlo wave functions *J. Chem. Phys.* **108** 8838–47

[30] Drummond N D, Towler M D and Needs R J 2004 Jastrow correlation factor for atoms, molecules, and solids *Phys. Rev.* B **70** 235119

IOP Publishing

A Practical Course on Quantum Monte Carlo

Vesa Apaja

Chapter 3

Principles of wave function optimization

A good trial wave function $\varphi_T(\mathbf{x})$ can improve quantum Monte Carlo (QMC) efficiency by several orders of magnitude. In one way or another, the quantity that is optimized is the energy or its fluctuation from sample to sample. The energy of a set of particle coordinates (walkers) $\{\mathbf{x}\}$ sampled from $|\varphi_T(\mathbf{x})|^2$ is the local energy,

$$E_L(\mathbf{x}) = -\frac{\hat{\mathcal{H}}\varphi_T(\mathbf{x})}{\varphi_T(\mathbf{x})} = -\frac{1}{2}\frac{\nabla_\mathbf{x}^2\varphi_T(\mathbf{x})}{\varphi_T(\mathbf{x})} + V(\mathbf{x}) = T_L(\mathbf{x}) + V(\mathbf{x}). \qquad (3.1)$$

In QMC, we try out a large number of positions $\{\mathbf{x}\}$; therefore, both $|\varphi_T(\mathbf{x})|^2$ and $E_L(\mathbf{x})$ must be functions that can be evaluated very fast. QMC favors functions that have simple algebraic gradients. The trial wave function is often a parameterized combination of such functions $\varphi_T(\mathbf{x}; \boldsymbol{\alpha})$ with parameters $\boldsymbol{\alpha} = (\alpha_1, \alpha_2, ..., \alpha_p)$.

The necessity for rapid function evaluation and accessible gradients imposes constraints on the function space, creating a perpetual challenge for QMC in achieving *completeness*. Does our function space overlook an important feature in the particle distribution? In other words, does it contain enough important physics? I am not saying that the optimization process is not an important factor, just that once the trial wave function has been written down, the ballpark is set and there are only so many games you can play within the premises. This is what makes choosing the trial wave function a task for dedicated experts.

The times of writing system-specific trial wave functions may be over one day, and we will all be glad when this happens. One way forward is machine learning and neural networks that can mimic functions [1, 2]. Recently, a machine learning algorithm for finding the ground states of both fermionic and bosonic quantum systems was proposed [3]. In the first proofs of the *universal approximation theorem* in 1989, A Hornik *et al* [4] demonstrated that neural networks with a single hidden layer containing a sufficient number of neurons could approximate any continuous function on a compact subset of Euclidean space to arbitrary precision. This flexibility comes from the nonlinearity of the so-called activation functions and the fact that even modest-sized neural networks

doi:10.1088/978-0-7503-6310-5ch3
3-1

have thousands of parameters. Neural networks are accompanied by programs that can do *automatic differentiation* (AD), that is, codes that can compute the derivatives of any combination of practically all elementary functions, such as $\sin(x)$ or $\exp(x)$, to machine precision accuracy (usually 10^{-16}). The AD component of machine learning enables the very efficient optimization of thousands of neural network parameters using *backward propagation*. Backward propagation, a simple yet clever application of the chain rule of differentiation, tells exactly which way to change each parameter for more optimal performance. Such all-at-once optimization is in stark contrast to common line-search optimization, which takes a single parameter at a time and evolves it in one direction until a local minimum is reached. However, one downside of a neural network is that it is an ultimate black box.

Once $\varphi_T(\mathbf{x}; \boldsymbol{\alpha})$ is chosen, one can start looking for optimal parameters. The properties that make or break an optimization algorithm in QMC are *stability*, *robustness*, and *reliability*:

- The method should converge to the same optimal parameters across different optimization runs and find the same *cost function* minimum or maximum. Examples of cost functions will be given shortly.
- The method should be impervious to numerical rounding errors and finite accuracy.
- After a testing period, the optimization method is inevitably (due to human laziness) applied as a black box, so it is assumed to be reliable, at least in scenarios not too different from the test cases.

QMC optimization is limited by the fact that it relies on a *finite number of sampled particle coordinates*, and the efficiency of an optimization method depends on how many stochastically independent walkers M are needed. This is especially likely to cause instabilities in energy minimization. Often, there is a set of parameters that gives a walker a ridiculously low local energy, and that single walker pulls down the whole approximate energy expectation value. Another factor in maximizing efficiency is the number of iterations needed for convergence. To summarize, the factor that sets QMC optimization apart from other optimization problems is its stochastic nature, which results from using a finite set of walkers $\{\mathbf{x}\}$, as well as physical constraints.

The trial wave function is system-specific, so let us first take a look at how optimal parameters can be found. To begin with, we define what we mean by 'optimal' and introduce a few cost functions.

3.1 Energy and variance optimization

The *objective or cost functions* in optimization can penalize high energy or large deviations from an average value [5–8]:

- Energy $E = \langle \hat{\mathcal{H}} \rangle$,

$$E(\boldsymbol{\alpha}) = \int d\mathbf{x} \left(\frac{P(\mathbf{x}; \boldsymbol{\alpha})}{\int d\mathbf{x} P(\mathbf{x}; \boldsymbol{\alpha})} \right) E_L(\mathbf{x}; \boldsymbol{\alpha}). \tag{3.2}$$

- Variance $\sigma^2 = \left\langle \hat{\mathcal{H}}^2 \right\rangle - \langle \hat{\mathcal{H}} \rangle^2$

$$\sigma^2(\boldsymbol{\alpha}) = \int d\mathbf{x} \left(\frac{P(\mathbf{x}; \boldsymbol{\alpha})}{\int d\mathbf{x} P(\mathbf{x}; \boldsymbol{\alpha})} \right) (E_L(\mathbf{x}; \boldsymbol{\alpha}) - E)^2. \tag{3.3}$$

Here, the energy expectation value E is unknown, and is in practice replaced by a reference energy based on the current best guess. Obviously, one can replace $(E_L(\mathbf{x}; \boldsymbol{\alpha}) - E)^2$ with any positive function $f(E_L(\mathbf{x}; \boldsymbol{\alpha}) - E)$ such that $f(0) = 0$.

Notice how I have kept the normalization integrals: changing the parameters $\boldsymbol{\alpha}$ changes the normalization as well, and therefore has to be kept in the optimization process.

The *variational principle of quantum mechanics* states that the ground-state energy is always less than or equal to the expectation value of the Hamiltonian in a trial state,

$$E[\varphi_T] \geqslant E_0. \tag{3.4}$$

Here, one implicitly assumes that the trial wave function has the correct symmetry for bosons or fermions. A trial wave function that has nodes is orthogonal to the boson ground-state wave function, so energy minimization gives the best low-energy excited state candidate in the chosen function space.

For the eigenstates of $\hat{\mathcal{H}}$, $E_L(\mathbf{x}) = E$; thus, one possible goal is to make $E_L(\mathbf{x})$ as constant as possible. This is the zero-variance principle mentioned in section 1.3.4. The variance of the local energy 'measures' the distance of the trial wave function from an energy eigenstate.

Variance minimization used to dominate in QMC, but recently emphasis has shifted to energy minimization due to the higher accuracy of its results. A feasible choice is to optimize a bit of both, say

$$\text{cost}(\boldsymbol{\alpha}) = 0.95 E(\boldsymbol{\alpha}) + 0.05 \sigma^2(\boldsymbol{\alpha}), \tag{3.5}$$

which emphasizes energy minimization while trying to keep the variance reasonably low.

3.1.1 H atom: analytical variance optimization

The hydrogen atom is a nice test case that can be used to see whether the reference energy in the variance in equation (3.3) has any effect on optimization. In most cases, we do not know the energy expectation value E, and so we calculate the variance using a guessed value E_{guess}. The H atom's normalized trial wave function, local energy, and variance are

$$\varphi_T(\mathbf{x}) = \frac{\alpha^{3/2}}{\sqrt{\pi}} e^{-\alpha r} \tag{3.6}$$

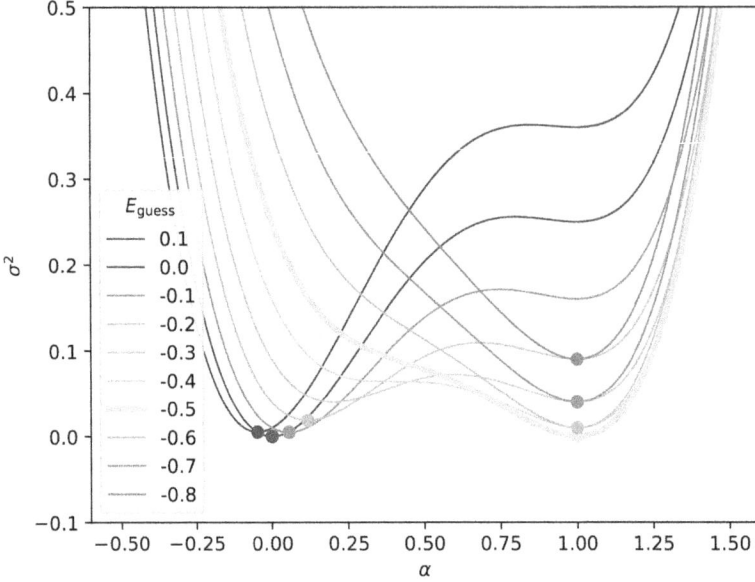

Figure 3.1. The exact variance σ^2 from equation (3.9) as a function of α for a hydrogen atom. The guessed energies E_{guess} are indicated in the figure. The bold line shows the ground-state case $E_{guess} = E_0 = -0.5$. Variance minima for $E_{guess} > -0.5$ are not at the ground-state value $\alpha = 1$, but near $\alpha = 0$.

$$E_L(\mathbf{x}) = -\frac{1}{2}\alpha^2 + \frac{\alpha - 1}{r} \tag{3.7}$$

$$\sigma^2 = \int d\mathbf{x}\,|\varphi_T(\mathbf{x})|^2\big(E_L(\mathbf{x}) - E_{guess}\big)^2. \tag{3.8}$$

The exact ground state has the cusp condition value $\alpha = 1$. The variance can be calculated analytically:

$$\sigma^2 = \frac{5}{4}\alpha^4 - 3\alpha^3 + \alpha^2\big(2 - E_{guess}\big) + 2\alpha E_{guess} + E_{guess}^2. \tag{3.9}$$

This would be the QMC variance with an infinite number of walkers. An obvious choice for E_{guess} would be the expectation value of the energy for the trial wave function with initial parameters, followed by updates as more optimized parameters become available [5]. Figure 3.1 shows σ^2 as a function of α for a few values of E_{guess}. Surprisingly, if one chooses $E_{guess} > E_0$, there is a minimum at a wrong α. This indicates that the variance optimization process can be unstable even with an infinite number of walkers! This is just a possibility to keep in mind if the variance optimization process fails.

3.2 Direct and correlated sampling

Direct sampling The direct sampling energy minimization proceeds along these steps:
    1 Choose wave function parameters α.
    2 Sample M walkers (points $\{\mathbf{x}_i\}$) from $|\varphi_T(\mathbf{x}; \alpha)|^2$.

3 Evaluate $E(\alpha) = \frac{1}{M} \sum_{i=1}^{M} E_L(\mathbf{x}_i; \alpha)$.

4 Repeat steps 1-3 and locate parameters α that give the lowest $E(\alpha)$.

While the parameter space is predominantly multidimensional, the minimization of $E(\alpha)$ would pose more of a technical challenge *if the energy values were exact*. However, there is a considerable amount of statistical uncertainty in each energy value. Consider the process of deciding whether parameters α_1 are better than parameters α_2. The QMC energies are, respectively, $E(\alpha_1) \pm \Delta E(\alpha_1)$ and $E(\alpha_2) \pm \Delta E(\alpha_2)$. Decreasing the statistical errors $\Delta E(\alpha_1)$ and $\Delta E(\alpha_2)$ is just a matter of increasing M and sampling more walkers. In optimization, however, the decision between parameters α_1 and α_2 is based solely on the approximate expectation values $E(\alpha_1)$ and $E(\alpha_2)$. The expectation values have stochastic fluctuations, and the decision as to whether $E(\alpha_1) < E(\alpha_2)$ is unclear if the energies differ by less than the accuracy. Increasing M helps but also makes direct sampling more time-consuming, because every proposed new α requires the sampling of M walkers.

Correlated sampling The energies $E(\alpha)$ evaluated using M walkers fluctuate, but the fluctuations may be somewhat 'systematic'. If this is the case, then it should be possible to reduce the fluctuations. In direct sampling, the energy $E(\alpha_1)$ is not correlated with the other energy $E(\alpha_2)$, because the former is evaluated using walkers sampled from $|\varphi_T(\mathbf{x}; \alpha_1)|^2$, while the latter uses walkers sampled from $|\varphi_T(\mathbf{x}; \alpha_2)|^2$. They are the results of two completely separate calculations done with random numbers. If there were any systematic pattern in the fluctuations of energies, this would inevitably blend it away.

If employing walkers sampled from a trial wave function with varying parameters α proves to be detrimental to optimization, then use *the same walkers* for every α. That is, sample the walkers from $|\varphi_T(\mathbf{x}; \alpha_o)|^2$ and use them for all parameters α. In this case, the energies $E(\alpha)$ with different α are correlated, warranting the name *correlated sampling*. The price we pay for using correlated samples is that the variances are larger than for uncorrelated ones. However, the fluctuations may 'cancel in the same direction', and the differences $E(\alpha_1) - E(\alpha_2)$ may appear more accurately than the values $E(\alpha_1)$ or $E(\alpha_2)$.

> *The differences between the correlated expectation values may be much more accurate than the expectation values themselves.*

This is exactly what we are counting on:

> *In optimization, we want to know whether we are going up or down, not how high we are.*

Sometimes, minute energy differences matter, as when dealing with ^{3}He impurities in liquid ^{4}He, or in pinpointing the Wigner crystallization or paramagnetic-to-ferromagnetic transitions in an electron gas. Separate computations of electron gas and Wigner crystal energies as a function of density easily lead to large uncertainties in the Wigner crystallization density. The principle of more accurate energy differences than energies themselves is widely used in density functional theory (DFT), in, for example, structural optimization. The DFT energy accuracy is about 0.1 eV, which is insufficient in deciding between two suggested structures, but the energy difference between the structures may be accurate enough.

To elucidate correlated sampling, consider two ways of writing the energy expectation value with the wave function parameters $\boldsymbol{\alpha}$:

$$E(\boldsymbol{\alpha}): = \underbrace{\frac{\int d\mathbf{x} |\varphi_T(\mathbf{x}, \boldsymbol{\alpha})|^2 E_L(\mathbf{x}; \boldsymbol{\alpha})}{\int d\mathbf{x} |\varphi_T(\mathbf{x}, \boldsymbol{\alpha})|^2}}_{\text{Direct Sampling}} = \underbrace{\frac{\int d\mathbf{x} |\varphi_T(\mathbf{x}, \boldsymbol{\alpha}_0)|^2 \left[\frac{|\varphi_T(\mathbf{x}, \boldsymbol{\alpha})|^2}{|\varphi_T(\mathbf{x}, \boldsymbol{\alpha}_0)|^2} E_L(\mathbf{x}; \boldsymbol{\alpha})\right]}{\int d\mathbf{x} |\varphi_T(\mathbf{x}, \boldsymbol{\alpha}_0)|^2 \frac{|\varphi_T(\mathbf{x}, \boldsymbol{\alpha})|^2}{|\varphi_T(\mathbf{x}, \boldsymbol{\alpha}_0)|^2}}}_{\text{Correlated Sampling}}. \quad (3.10)$$

While these expressions are mathematically equivalent (unless dividing by zero), their QMC values are not. In QMC, integrals are approximated by sums, so the direct sampling energy is

$$E^{\text{Direct sampling}}(\boldsymbol{\alpha}) \approx \frac{1}{M}\sum_{i=1}^{M} E_L(\mathbf{x}_i; \boldsymbol{\alpha}) = \frac{1}{M}\sum_{i=1}^{M}[T_L(\mathbf{x}_i; \boldsymbol{\alpha}) + V(\mathbf{x}_i)], \quad (3.11)$$

where walkers $\{\mathbf{x}_i\}$ are sampled from $|\varphi_T(\mathbf{x}; \boldsymbol{\alpha})|^2$. I wrote the local energy as the sum of kinetic local energy and potential energy to draw attention to the fact that the potential energy has no parameters; it changes only in resampling the walkers. Attempts to optimize without resampling optimize only the T_L part.

In correlated sampling, the first step is to sample M walkers $\{\mathbf{x}_i\}$ from the distribution

$$P(\mathbf{x}, \boldsymbol{\alpha}_0): = \frac{|\varphi_T(\mathbf{x}; \boldsymbol{\alpha}_0)|^2}{\int d\mathbf{x} |\varphi_T(\mathbf{x}; \boldsymbol{\alpha}_0)|^2}. \quad (3.12)$$

This is done using the Metropolis variational Monte Carlo (VMC) algorithm, so the normalization integral can be ignored. Next, we evaluate the energies as weighted sums

$$E(\boldsymbol{\alpha}) \approx \frac{\sum_{i=1}^{M} w(\mathbf{x}_i; \boldsymbol{\alpha}, \boldsymbol{\alpha}_0) E_L(\mathbf{x}_i; \boldsymbol{\alpha})}{\sum_{i=1}^{M} w(\mathbf{x}_i; \boldsymbol{\alpha}, \boldsymbol{\alpha}_0)}, \quad (3.13)$$

with weights

$$w(\mathbf{x}_i; \boldsymbol{\alpha}, \boldsymbol{\alpha}_0): = \frac{|\varphi_T(\mathbf{x}_i, \boldsymbol{\alpha})|^2}{|\varphi_T(\mathbf{x}_i, \boldsymbol{\alpha}_0)|^2}. \quad (3.14)$$

The sums in the numerator and in the denominator are evaluated separately. Compared to the direct sampling in equation (3.11), the potential part now depends on the parameters through the weight factor, even with a fixed set of walkers. Similarly, the variance in correlated sampling is calculated using the expression

$$\sigma^2(\alpha) \approx \frac{\sum_{i=1}^{M} w(\mathbf{x}_i; \alpha, \alpha_0)(E_L(\mathbf{x}_i; \alpha) - E)^2}{\sum_{i=1}^{M} w(\mathbf{x}_i; \alpha, \alpha_0)}. \tag{3.15}$$

Umrigar *et al* [5] optimized this expression for small atoms and molecules.

Reweighted and unreweighted variance The correlated sampling variance in equation (3.15) is a *reweighted variance*, with weights $w(\mathbf{x}_i; \alpha, \alpha_0)$. Instead of optimizing the reweighted variance, it is often better to set $w(\mathbf{x}_i; \alpha) = 1$ and use the *unreweighted variance*,

$$\sigma^2(\alpha) \approx \frac{1}{M}\sum_{i=1}^{M}(E_L(\mathbf{x}_i; \alpha) - E)^2 \qquad \text{unreweighted variance.} \tag{3.16}$$

The walkers are still sampled from $\varphi_T(\mathbf{r}; \alpha_0)$. The reason why the weight can be problematic is that it may fluctuate a lot from walker to walker, meaning that some walkers are overemphasized while others are useless. In [9], the conclusion is that in the model systems studied, the unreweighted variance is superior, showing better stability and allowing accelerated optimization.

Setting the weights to unity means that one finds parameters that give as constant as possible E_L for a fixed set of walkers. In other words, the optimization of unreweighted variance fits E_L to a constant. To test this for the He atom, set wratio = 1.0 in vmc.heatom_optimization_correlated_sampling. jl - but only for the variance.

Energy optimization is a completely different story, in which using a fixed set of walkers and *setting the weight to unity amounts to fixing the potential energy*. After that, energy minimization reduces to minimization of the local kinetic energy T_L, which is a notoriously unstable process. In a reasonably good local energy, the kinetic and potential parts approximately cancel, so positive potential energies for some walkers in our fixed list of walkers correspond to negative values of T_L, and the minimization of T_L exploits these T_L dips. This instability can be seen in the He atom optimization code: setting weights to unity finds a bogus energy minimum below -100 a.u., even with as many as 10^7 walkers.

3.2.1 He atom: comparing direct and correlated sampling

Direct and correlated sampling energy minimizations are compared in figure 3.2. The corresponding Julia code is vmc.heatom_optimization_correlated_-sampling.jl. The goal is to find the optimal parameter β in the He atom trial wave function in equation (2.37). The other parameter is kept fixed at $\alpha = 2$ to satisfy the cusp conditions. As shown in the figure, the direct sampling energies make up an irregular curve, and the location of the energy minimum is difficult to pinpoint. In comparison, the correlated sampling data shows much larger error bars, but the energies make up a smooth curve with a clear minimum. Another benefit of correlated sampling is speed, since the M walkers need to be generated only once.

In figure 3.2, the optimization using an algorithm is a separate calculation. The algorithm does not use gradients, which will be added in section 3.3.2. The algorithm

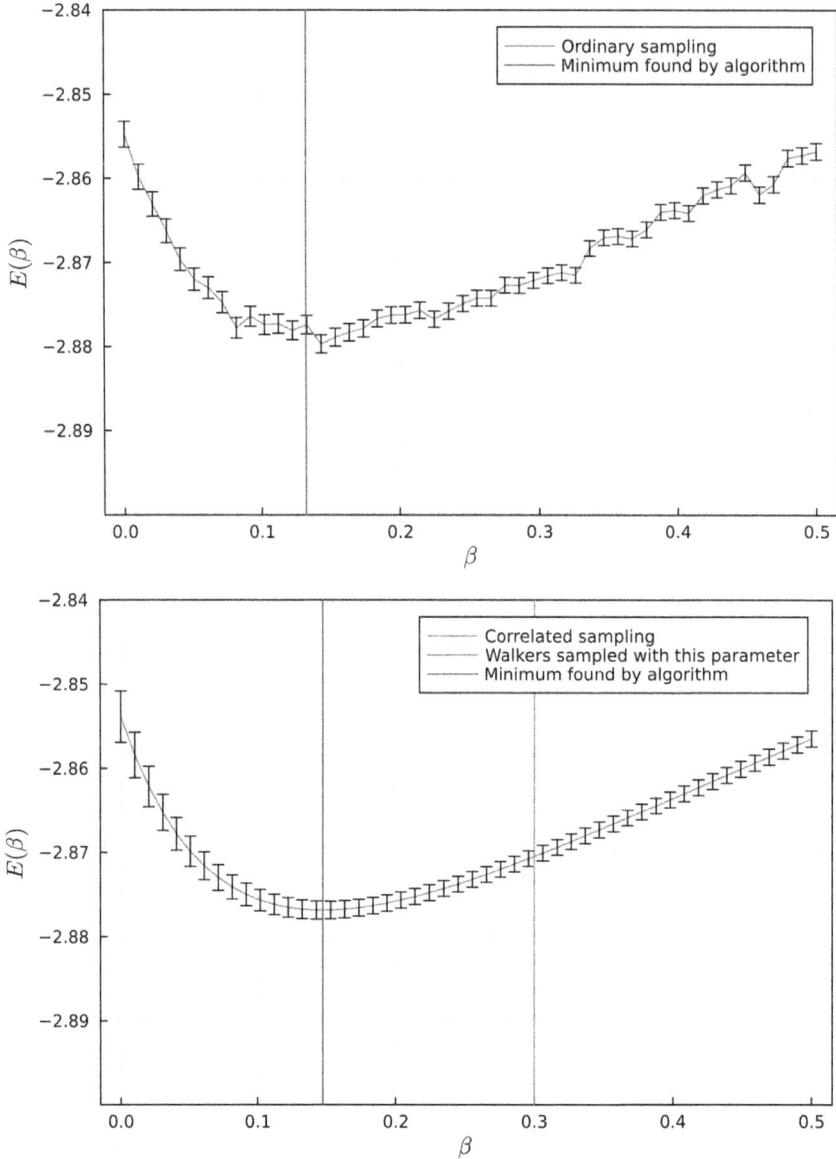

Figure 3.2. Upper figure: direct optimization of the parameter β. Every energy value is calculated using 100 000 walkers sampled from $|\varphi_T(\mathbf{x}; \beta)|^2$ with the β value indicated on the x-axis. Lower figure: correlated sampling optimization of the parameter β. First, 100 000 walkers are sampled from $|\varphi_T(\mathbf{x}; \beta_0)|^2$ with $\beta_0 = 0.3$, indicated by the vertical line. Next, the correlated energy values are computed using equation (3.13) for each β. In both figures, the vertical line labeled 'Minimum found by algorithm' indicates the optimal β found by a minimization algorithm.

starts from an initial guess for β, then moves around to locate the energy minimum, either a local or an absolute minimum.

In the direct sampling case, the algorithm happened to reach a minimum at the value of β indicated by the vertical line. This does not seem to coincide with the minimum of the points we obtained earlier, and a repeated run would probably converge to a distinctively different value. This lack of reproducibility is caused by stochastic fluctuations in the values of $E(\beta)$.

In correlated sampling, the minimum seems to be quite accurate. Repeating the correlated sampling calculation ten times gives a rather good idea of the optimal value, $\beta = 0.1437 \pm 0.0012$. Using the Julia code vmc.heatom.jl, VMC gives $E = -2.878\ 216 \pm 0.000\ 030$ a.u. As a rule of thumb, one should not trust optimization results without double-checking them against a longer VMC calculation.

3.3 He atom: energy and variance optimization

Optimizations for low energy and low variance are both solid concepts on paper, but, as we shall see, as soon as one moves away from the exact case, they disagree on the definition of a better trial wave function. In this section, *correlated sampling will be used in all optimizations.*

The differences between the energy and variance optimizations of the same trial wave function are illustrated in figure 3.3 for the He atom. The energy-optimized trial wave function has $\beta = 0.144$ and $E = -2.878\ 187(20)$ with a variance of $\sigma^2 = 0.114$. Variance optimization finds the optimal parameters $\beta = 0.327$ and $E = -2.878\ 187(20)$ with a variance of $\sigma^2 = 0.0843$. Clearly, in optimization, the two objective functions are not equivalent, and their ideas of the 'best' parameters are not the same.

One point worth remembering is that if your trial wave function has exactly the correct form with n parameters, then you need just n independent walkers to find out their values. Just solve the linear set of equations.

3.3.1 About cusp conditions and ground-state optimization

Energy and variance optimization converge toward the ground state from different directions. Table 3.1 shows a collection of optimizations of the trial state in equation (2.37) calculated using the code vmc.heatom_optimization.jl. The optimization was done with 10^7 walkers and correlated sampling, followed by energy and variance computation using vmc.heatom.jl. Unsurprisingly, variance optimization finds parameters that more closely satisfy the cusp conditions $\alpha = 2$ and $\alpha_{12} = 0.5$. However, relaxing the cusp conditions allows the calculation to reach a lower energy.

In a certain number of VMC steps, you can either compute a low energy less accurately or a somewhat higher energy more accurately. Such a trade-off is a consequence of Monte Carlo integration, not an inherent property of the trial wave function.

Consider the He atom case. With some algebra, one can solve the He atom energy as a function of parameters, $E(\alpha, \alpha_{12}, \beta)$, and find the optimal parameters by minimization. The result will be close to the numerical values given in the lowest row of table 3.1. *The energetically optimal trial wave function does not care about the cusp conditions, nor does it need to.*

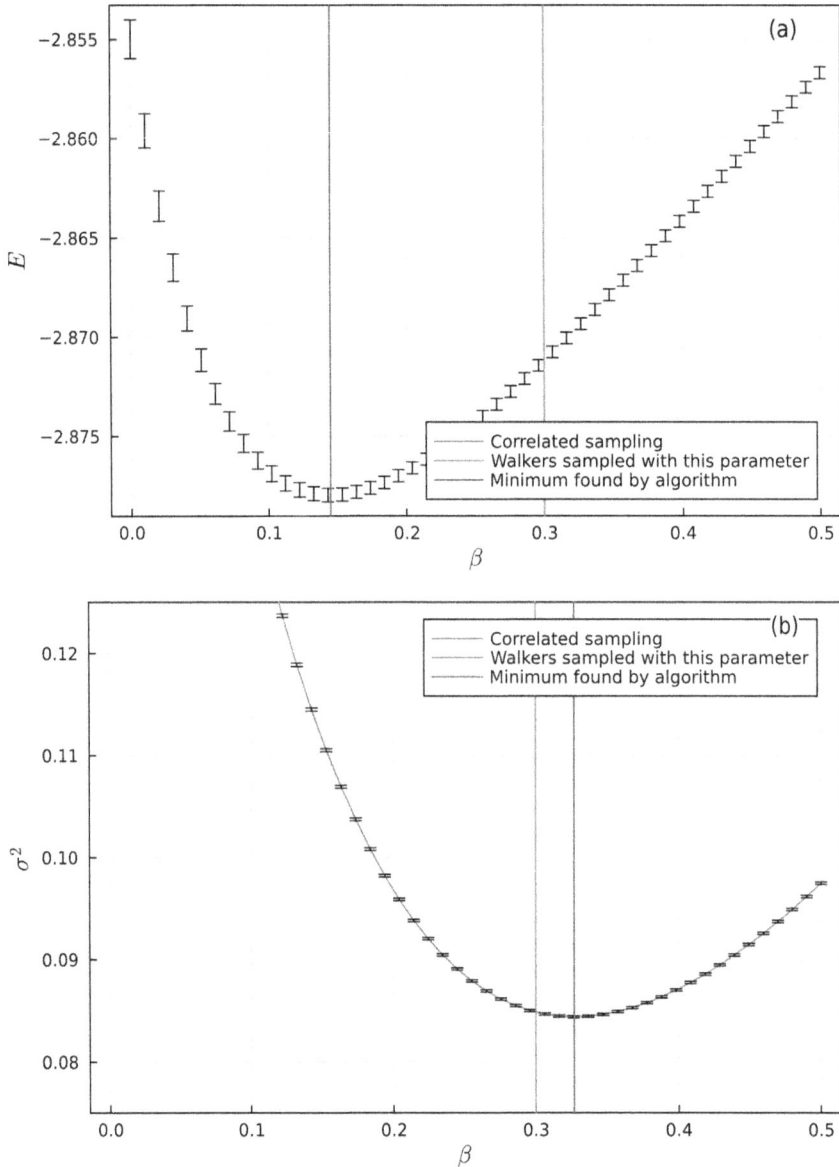

Figure 3.3. He atom energy optimization (a) and reweighted variance optimization (b) give notably different optimal values for the parameter β.

The cusp conditions are a physical property of the exact ground state, and they do stabilize QMC calculations. But are they that useful in trial wave functions? If we examine a He atom with a trial wave function that is the product of hydrogen 1s orbitals,

$$\varphi_T(\mathbf{x}) = e^{-\alpha(r_1 + r_2)} \tag{3.17}$$

Table 3.1. He atom energy and variance optimization results for the trial wave function in equation (2.37). Parameters in parentheses were kept fixed at their cusp condition values during optimization. In the last two rows, the cusp conditions were ignored, and all three parameters were freely optimized.

α	α_{12}	β	E	σ^2	Comments
(2.0000)	(0.5000)	0.3268	$-2.869\ 34(2)$	0.0843	Optimizes σ^2
(2.0000)	(0.5000)	0.1438	$-2.878\ 22(2)$	0.1142	Optimizes E
1.9495	(0.5000)	0.3885	$-2.877\ 90(2)$	0.0697	Optimizes σ^2
1.8422	(0.5000)	0.3485	$-2.890\ 28(2)$	0.1380	Optimizes E
1.9522	0.4581	0.3081	$-2.879\ 09(2)$	0.0678	Optimizes σ^2
1.8486	0.3676	0.1681	$-2.891\ 11(2)$	0.1347	Optimizes E

$$E_L(\mathbf{x}) = -\alpha^2 + (\alpha - 2)\left(\frac{1}{r_1} + \frac{1}{r_2}\right) + \frac{1}{r_{12}} \tag{3.18}$$

$$E(\alpha) = \alpha^2 - 4\alpha + \frac{5}{8}\alpha, \tag{3.19}$$

the minimum energy is at $\alpha_{\text{opt}} = \frac{27}{16}$, and

$$E(\alpha_{\text{opt}}) = -\frac{729}{256}. \tag{3.20}$$

I emphasize that these are exact values for the chosen trial wave function. However, α_{opt} does not satisfy the cusp conditions, so the local energy in a QMC calculation may even become infinite and yield no result at all. This situation arises because *in QMC, integrals are evaluated as finite sums*. Mathematically, the singularities in the energy integrand are integrable, but without enforcing cusp conditions, our numerical integration method is not quite up to the task.

As a final point, energy and variance minimizations are meaningful only in stochastic methods such as QMC; in algebraic energy minimization, the variance is always flat zero.

3.3.2 Gradients in variance and energy optimization

The He atom results with just a few parameters were computed using a gradient-free numerical package, but more efficient methods, e.g. steepest descent and the Newton–Raphson method, utilize gradients. The energy gradient has components [8]

$$E_i := \frac{\partial E}{\partial \alpha_i} = \frac{\partial}{\partial \alpha_i} \frac{\int d\mathbf{x} \varphi^* \hat{\mathcal{H}} \varphi}{\int d\mathbf{x} |\varphi|^2} \tag{3.21}$$

$$= \frac{\int d\mathbf{x} \left(\varphi_i^* \hat{\mathcal{H}} \varphi + \varphi^* \hat{\mathcal{H}} \varphi_i\right)}{\int d\mathbf{x} |\varphi|^2} - \frac{\int d\mathbf{x} \varphi^* \hat{\mathcal{H}} \varphi}{\left(\int d\mathbf{x} |\varphi|^2\right)^2} \int d\mathbf{x} \left(\varphi^* \varphi_i + \varphi_i^* \varphi\right), \tag{3.22}$$

where

$$\varphi: =\varphi(\mathbf{x}; \alpha) \text{ trial wave function} \qquad (3.23)$$

$$\varphi_i: =\frac{\partial}{\partial \alpha_i}\varphi(\mathbf{x}; \alpha). \qquad (3.24)$$

Based on Hermiticity,

$$\int d\mathbf{x}\varphi^*\hat{\mathcal{H}}\varphi_i = \int d\mathbf{x}(\hat{\mathcal{H}}\varphi^*)\varphi_i, \qquad (3.25)$$

and on the fact that E is real (take complex conjugates of integrands at will), we get

$$E_i = 2\frac{\int d\mathbf{x}\varphi_i^*\hat{\mathcal{H}}\varphi}{\int d\mathbf{x}|\varphi|^2} - 2\frac{\int d\mathbf{x}\varphi^*\hat{\mathcal{H}}\varphi}{\left(\int d\mathbf{x}|\varphi|^2\right)^2}\int d\mathbf{x}\varphi^*\varphi_i \qquad (3.26)$$

$$=2\frac{\int d\mathbf{x}|\varphi|^2\frac{\varphi_i^*}{\varphi^*}E_L}{\int d\mathbf{x}|\varphi|^2} - 2\frac{\int d\mathbf{x}|\varphi|^2 E_L}{\left(\int d\mathbf{x}|\varphi|^2\right)^2}\int d\mathbf{x}|\varphi|^2\frac{\varphi_i}{\varphi} \qquad (3.27)$$

$$=2\left\langle\frac{\varphi_i}{\varphi}E_L\right\rangle - 2\underbrace{\langle E_L\rangle}_{E}\left\langle\frac{\varphi_i}{\varphi}\right\rangle. \qquad (3.28)$$

The result is [8]

$$\boxed{E_i = 2\left\langle\frac{\varphi_i}{\varphi}(E_L - E)\right\rangle \qquad \text{energy gradient.}} \qquad (3.29)$$

Using Hermiticity helped to simplify a term with $\hat{\mathcal{H}}\varphi_i$ and saved us from evaluating the gradients of the local energy. The gradients E_i vanish if φ is an exact eigenstate ($E_L = E$). Looking at equation (3.28), one sees that the fluctuations in gradients are reduced because gradients are evaluated using an expression in the form

$$\langle AB\rangle - \langle A\rangle\langle B\rangle \qquad \text{covariance}, \qquad (3.30)$$

which has smaller fluctuations than either $\langle AB\rangle$ or $\langle A\rangle\langle B\rangle$. Typically, the subtracted term $\langle A\rangle\langle B\rangle$ would be zero in an infinite sample but not in a finite one. The covariance vanishes if A and B are statistically independent.

The gradient of variance is [8]

$$\sigma_i^2: =\frac{\partial(\sigma^2)}{\partial \alpha_i} = \frac{\partial}{\partial \alpha_i}\frac{\int d\mathbf{x}|\varphi|^2(E_L - E)^2}{\int d\mathbf{x}|\varphi|^2} \qquad (3.31)$$

$$= \frac{\int d\mathbf{x}\left[\varphi_i^*\varphi(E_L - E)^2 + \varphi^*\varphi_i(E_L - E)^2 + 2\varphi^*\varphi(E_L - E)E_{L,i}\right]}{\int d\mathbf{x}|\varphi|^2}$$

$$- \frac{\int d\mathbf{x}|\varphi|^2(E_L - E)^2}{\left(\int d\mathbf{x}|\varphi|^2\right)^2}\int d\mathbf{x}\left[\varphi_i^*\varphi + \varphi^*\varphi_i\right]. \tag{3.32}$$

Since σ^2 is real, we find

$$\sigma_i^2 = 2\frac{\int d\mathbf{x}|\varphi|^2\left[\frac{\varphi_i}{\varphi}(E_L - E)^2 + (E_L - E)E_{L,i}\right]}{\int d\mathbf{x}|\varphi|^2}$$

$$- 2\frac{\int d\mathbf{x}|\varphi|^2(E_L - E)^2}{\left(\int d\mathbf{x}|\varphi|^2\right)^2}\int d\mathbf{x}|\varphi|^2\frac{\varphi_i}{\varphi} \tag{3.33}$$

$$= 2\left\langle\frac{\varphi_i}{\varphi}(E_L - E)^2\right\rangle + 2\langle(E_L - E)E_{L,i}\rangle - 2\langle(E_L - E)^2\rangle\left\langle\frac{\varphi_i}{\varphi}\right\rangle. \tag{3.34}$$

In [8], the gradients φ_i are left out, based on the fact that variance optimization is actually fitting E_L to a constant value.[1] This leaves [8]

$$\boxed{\sigma_i^2 = 2<(E_L - E)E_{L,i}>} \qquad \text{variance gradient.} \tag{3.35}$$

Energy and variance gradients in correlated sampling In correlated sampling, the energy and the variance are

$$E = \frac{\int d\mathbf{x}|\varphi_0|^2\frac{|\varphi|^2}{|\varphi_0|^2}E_L}{\int d\mathbf{x}|\varphi_0|^2\frac{|\varphi|^2}{|\varphi_0|^2}}, \sigma^2 = \frac{\int d\mathbf{x}|\varphi_0|^2\frac{|\varphi|^2}{|\varphi_0|^2}(E_L - E)^2}{\int d\mathbf{x}|\varphi_0|^2\frac{|\varphi|^2}{|\varphi_0|^2}}, \tag{3.36}$$

where only φ and E_L contain adjustable parameters. Similar to equation (3.29), the gradients are found to be

$$E_i = \frac{\int d\mathbf{x}|\varphi_0|^2\frac{|\varphi|^2}{|\varphi_0|^2}2\frac{\varphi_i}{\varphi}(E_L - E)}{\int d\mathbf{x}|\varphi_0|^2\frac{|\varphi|^2}{|\varphi_0|^2}}, \sigma_i^2 = \frac{\int d\mathbf{x}|\varphi_0|^2\frac{|\varphi|^2}{|\varphi_0|^2}2\frac{\varphi_i}{\varphi}(E_L - E)^2}{\int d\mathbf{x}|\varphi_0|^2\frac{|\varphi|^2}{|\varphi_0|^2}}. \tag{3.37}$$

---

[1] See the earlier discussion about reweighted and unreweighted variances.

The results change slightly if the gradients are calculated for the QMC correlated sampling sums.

3.3.3 Multielectron optimizations with paper and pen

The hydrogen 1s wave function in equation (3.8) satisfies the electron–nucleus cusp condition. Let us apply a trial wave function based on a product of the 1s wave function to multielectron systems, namely He, Li, and Be atoms. The product wave function does not satisfy the electron–electron cusp condition. Moreover, it is bosonic, so for Li and Be, the Pauli exclusion principle is not satisfied. Comparison with known ground-state energies gives us insight into the significance of the Pauli exclusion principle.

The optimization can be done analytically. For He, the expectation values in a.u. are

$$\left\langle \varphi_T(\alpha) | -\frac{1}{2} \nabla_i^2 | \varphi_T(\alpha) \right\rangle = \frac{1}{2}\alpha^2 \tag{3.38}$$

$$\left\langle \varphi_T(\alpha) | -\frac{2}{r_i} | \varphi_T(\alpha) \right\rangle = -2\alpha \tag{3.39}$$

$$\left\langle \varphi_T(\alpha) | -\frac{1}{r_{ij}} | \varphi_T(\alpha) \right\rangle = \frac{5}{8}\alpha, \tag{3.40}$$

so the energy is[2]

$$E(\alpha) = \alpha^2 - 4\alpha + \frac{5}{8}\alpha \Rightarrow \text{minimum at } \alpha = \frac{27}{16}, E = -\frac{729}{256}. \tag{3.41}$$

The calculations for Li and Be atoms are similar (see the SymPy code sympy_energies.py). Table 3.3.3 shows the results for few-electron atoms. Violations of the Pauli exclusion principle are marked in red.

Element	α	$E(\alpha)$	Exact
H	1	$-1/2 = -0.5$	-0.5
He	$27/16 = 1.6875$	$-729/256 = -2.8476$	$-2.903\ 72$
Li	$19/8 = 2.375$	$-1083/128 = -8.4609$	$-7.478\ 06$
Be	$49/16 = 3.0625$	$-2401/128 = -18.757\ 81$	$-14.667\ 36$

Ignoring the electron–electron cusp condition *increases energy*, but the Li and Be results are still *below* the true ground-state energies. This shows that it is far more important that their wave function satisfies the Pauli exclusion principle.

---

[2] The optimal α is one option in the code Model_Heatom.jl.

3.3.4 Stochastic gradient descent

Harju *et al* [10][3] introduced the *stochastic gradient descent* method, where parameters are updated using the formula

$$\alpha_i^{\text{new}} = \alpha_i - \gamma E_i. \tag{3.42}$$

Evaluated directly, the energy gradients E_i are

$$E_i := \frac{\partial E}{\partial \alpha_i} = \frac{\partial}{\partial \alpha_i}\left[\frac{1}{M}\sum_{i=1}^{M}E_L(\mathbf{x}_i; \boldsymbol{\alpha})\right], \tag{3.43}$$

for energy optimization using walkers $\{\mathbf{x}_i\}$ sampled from $|\varphi_T(\mathbf{x}; \boldsymbol{\alpha})|$. Less fluctuating gradients can be computed using the covariance form in equation (3.29).

The gradient is stochastic, since it is an estimate based on M walkers, and the process converges only if one slowly decreases the damping factor γ during the iteration.[4] As few as $M = 5$ walkers are resampled after every parameter update, thus avoiding the thousands of walkers used in deterministic optimization methods.

The idea is to use the randomness in the gradient for something beneficial, namely to avoid getting stuck at local minima. This is closely related to simulated annealing, with stochastic noise acting as temperature and, unfortunately, with the same downsides. Tuning the cooling schedule, now damping, is a nontrivial task; convergence is slow, and the result depends on the initial choice of parameters. For a small sample, the energy gradients can become very large, and the parameters may reach unreasonable values. A more stable optimization can be done using normalized gradients or using only the direction of the gradient:

$$\alpha_i^{\text{new}} = \alpha_i - \gamma \ \text{sign}(E_i). \tag{3.44}$$

Although deterministic optimization methods have emerged as winners in subsequent comparisons [11], the stochastic gradient descent method showed that energy optimization can be done in QMC with a limited set of walkers.

3.3.5 Linearized optimization method

The Taylor expansion of a function $f = f(x)$ around point x is

$$f(x + \Delta x) \approx f(x) + \Delta x f'(x) + \frac{1}{2}(\Delta x)^2 f''(x). \tag{3.45}$$

If the new point $x + \Delta x$ is a local minimum or maximum, it can be found by solving the update Δx using

$$\frac{df(x + \Delta x)}{d(\Delta x)} = f'(x) + \Delta x f''(x) = 0 \qquad \text{Newton–Raphson.} \tag{3.46}$$

---

[3] Again, the He atom was used to benchmark the method before applying it to positrons in a homogeneous electron gas.
[4] In machine learning, γ is known as the learning rate.

In many-variable scenarios, one gets a linear set of equations for updates Δx_i.

Diagonalizing the Hamiltonian in a finite basis Let us apply the Newton–Raphson method to a trial wave function, assuming it is a real function that can be approximated by a linear function of the parameters. If parameters change by only a small amount, for example, from α_0 to α, one can Taylor expand the wave function around the initial parameters,

$$\Psi(\mathbf{x}, \alpha) \approx \Psi(\mathbf{x}, \alpha_0) + \sum_{i=1}^{p} (\alpha_i - (\alpha_0)_i) \frac{\partial \Psi(\mathbf{x}, \alpha)}{\partial \alpha_i}\bigg|_{\alpha=\alpha_0}, \qquad (3.47)$$

where the sum is over the parameters. In a shorter notation, the linear updated state is

$$|\Psi\rangle^{\text{lin}} = |\Psi_0\rangle + \sum_{i=1}^{p} \Delta \alpha_i |\Psi_i\rangle. \qquad (3.48)$$

Before continuing, it is important to realize that this is not the whole story. As noticed by Umrigar et al [12, 13], the normalization of the linear-update state $|\Psi\rangle^{\text{lin}}$ has some freedom, and a better 'renormalized' linear-update state might actually be one with the derivative orthogonal to the linear combination of $|\Psi_0\rangle$ and the $|\Psi\rangle^{\text{lin}}$ given in equation (3.48). This turns out to be the case for nonlinear parameters, and we will return to them later, after we have solved the updates $\Delta \alpha_i$.

It is convenient to include Ψ_0 in the list of derivatives, $\{\Psi_0, \Psi_1, \ldots, \Psi_p\}$:

$$|\Psi\rangle^{\text{lin}} = \sum_{i=0}^{p} \Delta \alpha_i |\Psi_i\rangle, \qquad (3.49)$$

where $\Delta \alpha_0 := 1$. After the update, the energy is

$$E(\alpha) = \frac{\langle \Psi^{\text{lin}} | \hat{\mathcal{H}} | \Psi^{\text{lin}} \rangle}{\langle \Psi^{\text{lin}} | \Psi^{\text{lin}} \rangle} \Leftrightarrow \langle \Psi^{\text{lin}} | \hat{\mathcal{H}} | \Psi^{\text{lin}} \rangle = E(\alpha) \langle \Psi^{\text{lin}} | \Psi^{\text{lin}} \rangle, \qquad (3.50)$$

where

$$\begin{aligned} \langle \Psi^{\text{lin}} | \hat{\mathcal{H}} | \Psi^{\text{lin}} \rangle &= \sum_{i,j=0}^{p} \Delta \alpha_i \Delta \alpha_j \langle \Psi_i | \hat{\mathcal{H}} | \Psi_j \rangle := \sum_{i,j=0}^{p} \Delta \alpha_i \Delta \alpha_j H_{ij} \\ \langle \Psi^{\text{lin}} | \Psi^{\text{lin}} \rangle &= \sum_{i,j=0}^{p} \Delta \alpha_i \Delta \alpha_j \langle \Psi_i | \Psi_j \rangle := \sum_{i,j=0}^{p} \Delta \alpha_i \Delta \alpha_j S_{ij}, \end{aligned} \qquad (3.51)$$

which defines the elements of two $(p + 1) \times (p + 1)$ matrices H and S. The optimal parameters satisfy

$$\frac{\partial E(\alpha)}{\partial \alpha_k} = 0 \; \forall \, k, \qquad (3.52)$$

and equation (3.50) can be written as an optimization problem with normalization as a condition and energy as a Lagrange multiplier E^{lin} [13],

$$\frac{\partial}{\partial \alpha_i} \left[\underbrace{\langle \Psi^{\text{lin}} | \hat{\mathcal{H}} | \Psi^{\text{lin}} \rangle}_{\text{optimize}} - \underbrace{E^{\text{lin}}}_{\text{Lagrange multiplier}} \underbrace{\langle \Psi^{\text{lin}} | \Psi^{\text{lin}} \rangle}_{\text{normalization condition}} \right] \qquad (3.53)$$

$$= \frac{\partial}{\partial \alpha_i} \langle \Psi^{\mathrm{lin}} | \hat{\mathcal{H}} | \Psi^{\mathrm{lin}} \rangle - E^{\mathrm{lin}} \frac{\partial}{\partial \alpha_i} \langle \Psi^{\mathrm{lin}} | \Psi^{\mathrm{lin}} \rangle \tag{3.54}$$

$$= \sum_{j=0}^{p} H_{ij} \Delta \alpha_j - E^{\mathrm{lin}} \sum_{j=0}^{p} S_{ij} \Delta \alpha_j = 0, \tag{3.55}$$

where the last form comes from equations (3.51). Hence, the parameter updates $\Delta \alpha_j$ are the solution of the generalized eigenvalue equation

$$\boxed{H \Delta \boldsymbol{\alpha} = E^{\mathrm{lin}} S \Delta \boldsymbol{\alpha}.} \tag{3.56}$$

First, one has to approximate the matrices using a finite set of walkers [14]:

$$H_{ij} = \langle \Psi_i | \hat{\mathcal{H}} | \Psi_j \rangle = \int d\mathbf{x} \Psi_i(\mathbf{x}) \hat{\mathcal{H}} \Psi_j(\mathbf{x}) = \int d\mathbf{x} |\Psi_0(\mathbf{x})|^2 \frac{\Psi_i(\mathbf{x})}{\Psi_0(\mathbf{x})} \frac{\hat{\mathcal{H}} \Psi_j(\mathbf{x})}{\Psi_0(\mathbf{x})} \tag{3.57}$$

$$\approx \frac{1}{M} \sum_{k=1}^{M} \frac{\Psi_i(\mathbf{x}_k)}{\Psi_0(\mathbf{x}_k)} \frac{\hat{\mathcal{H}} \Psi_j(\mathbf{x}_k)}{\Psi_0(\mathbf{x}_k)}, \tag{3.58}$$

where walkers $\{\mathbf{x}_k\}$ are sampled from $|\Psi_0(\mathbf{x})|^2$. Similarly,

$$S_{ij} \approx \frac{1}{M} \sum_{k=1}^{M} \frac{\Psi_i(\mathbf{x}_k)}{\Psi_0(\mathbf{x}_k)} \frac{\Psi_j(\mathbf{x}_k)}{\Psi_0(\mathbf{x}_k)}. \tag{3.59}$$

In [12], the matrix elements are written in a compact notation,

$$H_{ij} = \left\langle \frac{\Psi_i}{\Psi_0} \frac{\hat{\mathcal{H}} \Psi_j}{\Psi_0} \right\rangle, \quad S_{ij} = \left\langle \frac{\Psi_i}{\Psi_0} \frac{\Psi_j}{\Psi_0} \right\rangle, \tag{3.60}$$

where averages are computed over M walkers sampled from $|\Psi_0|^2$. The quantum mechanical H is symmetric, whereas the estimated H is nonsymmetric. Moreover, the estimated H is the only nonsymmetric form that satisfies the zero-variance principle: $\Delta \boldsymbol{\alpha} = 0$ if either $|\Psi_0\rangle$ or the linear update $\sum_{i=0}^{p} \Delta \alpha_i |\Psi_i\rangle$ is an eigenstate [14]. Symmetrizing the estimated H would break the zero-variance principle. Parameter update using $\Delta \boldsymbol{\alpha}$ has many possibilities; one is

$$\boldsymbol{\alpha}^{\mathrm{new}} = \boldsymbol{\alpha} + \Delta \boldsymbol{\alpha}. \tag{3.61}$$

Reducing variance in estimated matrices H and S The matrix elements of H and S given in equations (3.58) and (3.59) have large fluctuations for a finite set of samples M, which results in an inaccurate parameter update $\Delta \boldsymbol{\alpha}$. Fluctuations can be suppressed using the covariance $\langle AB \rangle - \langle A \rangle \langle B \rangle$ introduced in equation (3.30). The recipe is now the following:

$$\boxed{\begin{array}{c} \text{Covariances: to reduce variance, replace} \\ \frac{\Psi_i}{\Psi_0} \rightarrow \frac{\Psi_i}{\Psi_0} - \left\langle \frac{\Psi_i}{\Psi_0} \right\rangle, \quad \text{for} \quad i > 0 \\ \text{in all averages taken over a finite sample.} \end{array}} \tag{3.62}$$

The reduced-variance estimator of the overlap matrix is (we indicate use of covariances with a bar)

$$\bar{S}_{ij} := \left\langle \left(\frac{\Psi_i}{\Psi_0} - \left\langle \frac{\Psi_i}{\Psi_0} \right\rangle \right) \left(\frac{\Psi_j}{\Psi_0} - \left\langle \frac{\Psi_j}{\Psi_0} \right\rangle \right) \right\rangle \quad (3.63)$$

$$= \left\langle \frac{\Psi_i}{\Psi_0} \frac{\Psi_j}{\Psi_0} \right\rangle - \left\langle \frac{\Psi_i}{\Psi_0} \right\rangle \left\langle \frac{\Psi_j}{\Psi_0} \right\rangle, \text{ for } i, j > 0, \quad (3.64)$$

and the rest of the elements are

$$\bar{S}_{00} = 1, \ \bar{S}_{0i} = \bar{S}_{i0} := \left\langle \left(\frac{\Psi_i}{\Psi_0} - \left\langle \frac{\Psi_i}{\Psi_0} \right\rangle \right) \right\rangle = 0, \text{ for } i > 0. \quad (3.65)$$

Similarly, the reduced-variance estimator of the Hamiltonian matrix elements for $i, j > 0$ is

$$\bar{H}_{ij} := \left\langle \left(\frac{\Psi_i}{\Psi_0} - \left\langle \frac{\Psi_i}{\Psi_0} \right\rangle \right) \frac{\hat{\mathcal{H}}\left[\Psi_0 \left(\frac{\Psi_j}{\Psi_0} - \left\langle \frac{\Psi_j}{\Psi_0} \right\rangle \right) \right]}{\Psi_0} \right\rangle \quad (3.66)$$

$$= \left\langle \left(\frac{\Psi_i}{\Psi_0} - \left\langle \frac{\Psi_i}{\Psi_0} \right\rangle \right) \left(\frac{\hat{\mathcal{H}}\Psi_j}{\Psi_0} - \left\langle \frac{\Psi_j}{\Psi_0} \right\rangle E_L \right) \right\rangle \quad (3.67)$$

$$= \left\langle \frac{\Psi_i}{\Psi_0} \frac{\hat{\mathcal{H}}\Psi_j}{\Psi_0} \right\rangle - \left\langle \frac{\Psi_i}{\Psi_0} \right\rangle \left\langle \frac{\hat{\mathcal{H}}\Psi_j}{\Psi_0} \right\rangle - \left\langle \frac{\Psi_i}{\Psi_0} E_L \right\rangle \left\langle \frac{\Psi_j}{\Psi_0} \right\rangle + \left\langle \frac{\Psi_i}{\Psi_0} \right\rangle \left\langle \frac{\Psi_j}{\Psi_0} \right\rangle \langle E_L \rangle. \quad (3.68)$$

Inserting

$$\left\langle \frac{\Psi_i}{\Psi_0} \frac{\hat{\mathcal{H}}\Psi_j}{\Psi_0} \right\rangle \equiv \left\langle \frac{\Psi_i}{\Psi_0} \frac{\hat{\mathcal{H}}\left(\frac{\partial}{\partial \alpha_j} \Psi_0 \right)}{\Psi_0} \right\rangle = \left\langle \frac{\Psi_i}{\Psi_0} \frac{\frac{\partial}{\partial \alpha_j}(\hat{\mathcal{H}}\Psi_0)}{\Psi_0} \right\rangle \quad (3.69)$$

$$= \left\langle \frac{\Psi_i}{\Psi_0} \frac{\frac{\partial}{\partial \alpha_j}(\Psi_0 E_L)}{\Psi_0} \right\rangle = \left\langle \frac{\Psi_i}{\Psi_0} \frac{\Psi_j}{\Psi_0} E_L \right\rangle + \left\langle \frac{\Psi_i}{\Psi_0} E_{L,j} \right\rangle \quad (3.70)$$

$$\left\langle \frac{\hat{\mathcal{H}}\Psi_j}{\Psi_0} \right\rangle = \left\langle \frac{\frac{\partial}{\partial \alpha_j}(\Psi_0 E_L)}{\Psi_0} \right\rangle = \left\langle \frac{\Psi_j}{\Psi_0} E_L \right\rangle + \langle E_{L,j} \rangle, \tag{3.71}$$

gives the matrix elements

$$\bar{H}_{ij} := \left\langle \frac{\Psi_i}{\Psi_0} \frac{\Psi_j}{\Psi_0} E_L \right\rangle - \left\langle \frac{\Psi_i}{\Psi_0} \right\rangle \left\langle \frac{\Psi_j}{\Psi_0} E_L \right\rangle - \left\langle \frac{\Psi_i}{\Psi_0} E_L \right\rangle \left\langle \frac{\Psi_j}{\Psi_0} \right\rangle + \left\langle \frac{\Psi_i}{\Psi_0} \right\rangle \left\langle \frac{\Psi_j}{\Psi_0} \right\rangle \langle E_L \rangle$$
$$+ \left\langle \frac{\Psi_i}{\Psi_0} E_{L,j} \right\rangle - \left\langle \frac{\Psi_i}{\Psi_0} \right\rangle \langle E_{L,j} \rangle, \text{ for } i, j > 0. \tag{3.72}$$

The rest of the elements are

$$\bar{H}_{00} := \langle E_L \rangle \tag{3.73}$$

$$\bar{H}_{0j} := \left\langle \frac{\hat{\mathcal{H}} \left[\Psi_0 \left(\frac{\Psi_j}{\Psi_0} - \left\langle \frac{\Psi_j}{\Psi_0} \right\rangle \right) \right]}{\Psi_0} \right\rangle = \left\langle \frac{\hat{\mathcal{H}}\Psi_j}{\Psi_0} \right\rangle - \left\langle \frac{\Psi_j}{\Psi_0} \right\rangle \langle E_L \rangle \tag{3.74}$$

$$= \left\langle \frac{\Psi_j}{\Psi_0} E_L \right\rangle + \langle E_{L,j} \rangle - \left\langle \frac{\Psi_j}{\Psi_0} \right\rangle \langle E_L \rangle, \text{ for } j > 0 \tag{3.75}$$

$$\bar{H}_{i0} := \left\langle \left(\frac{\Psi_i}{\Psi_0} - \left\langle \frac{\Psi_i}{\Psi_0} \right\rangle \right) \frac{\hat{\mathcal{H}}\Psi_0}{\Psi_0} \right\rangle = \left\langle \frac{\Psi_i}{\Psi_0} E_L \right\rangle - \left\langle \frac{\Psi_i}{\Psi_0} \right\rangle \langle E_L \rangle, \text{ for } i > 0. \tag{3.76}$$

These reduced-variance estimators for the overlap and Hamiltonian matrices were introduced in [13, 15]. Again, $\bar{H}$ is nonsymmetric.

Looking back, using covariances in matrices $\bar{S}$ and $\bar{H}$ changes the parameter updates and, consequently, also the states themselves. The covariance replacement

$$\frac{\Psi_i}{\Psi_0} \rightarrow \frac{\Psi_i}{\Psi_0} - \left\langle \frac{\Psi_i}{\Psi_0} \right\rangle \tag{3.77}$$

implies that

$$|\Psi_i\rangle \rightarrow |\Psi_i\rangle - |\Psi_0\rangle \left\langle \frac{\Psi_i}{\Psi_0} \right\rangle = |\Psi_i\rangle - |\Psi_0\rangle \underbrace{\left\langle \frac{\Psi_i}{\Psi_0} \frac{\Psi_0}{\Psi_0} \right\rangle}_{S_{i0}}, \tag{3.78}$$

so using covariances means that derivatives are modified to the values

$$\boxed{|\bar{\Psi}_i\rangle = |\Psi_i\rangle - S_{i0}|\Psi_0\rangle.} \tag{3.79}$$

Consequently, the parameter updates $\Delta\bar{\alpha}_i$ solved using $\bar{S}$ and $\bar{H}$ give the linear updated state

$$|\Psi\rangle^{\text{lin}} = |\Psi_0\rangle + \sum_{i=1}^{p}\Delta\alpha_i|\Psi_i\rangle \rightarrow |\Psi_0\rangle + \sum_{i=1}^{p}\Delta\bar{\alpha}_i(|\Psi_i\rangle - S_{i0}|\Psi_0\rangle), \quad (3.80)$$

so the linear update in the covariance scheme is actually

$$\boxed{|\bar{\Psi}\rangle^{\text{lin}} = \left(1 - \sum_{i=1}^{p}S_{i0}\Delta\bar{\alpha}_i\right)|\Psi_0\rangle + \sum_{i=1}^{p}\Delta\bar{\alpha}_i|\Psi_i\rangle.} \quad (3.81)$$

A natural line of thought is that since using covariances reduces variance in the expectation values and gives a better $|\bar{\Psi}\rangle^{\text{lin}}$, then why not push further and generalize the idea (and add another bar on top)? That is, replace S_{i0} with some $\mathcal{N}_i$ and, correspondingly, parameter updates $\Delta\bar{\alpha}_i$ with $\Delta\bar{\bar{\alpha}}_i$? This is the essence of the next paragraph.

Optimization of nonlinear parameters using parameter-dependent normalization
The update (3.61) is adequate as long as the linear approximation in equation (3.48) is valid. For nonlinear parameters, however, the length of $\Delta\alpha_j$ may be poor, and even its sign may be wrong. Umrigar *et al* [12] (and with a slightly different notation in [13]) found a way to improve the optimization of nonlinear parameters. They added the normalization $\mathcal{N}(\alpha)$ of the trial wave function, chosen so that the normalization before the update is $\mathcal{N}(\alpha_0) = 1$. With this new freedom, the trial wave function and its derivatives are

$$|\bar{\bar{\Psi}}_0\rangle := |\bar{\bar{\Psi}}(\alpha_0)\rangle = \mathcal{N}(\alpha_0)|\Psi(\alpha_0)\rangle = |\Psi_0\rangle \quad (3.82)$$

$$|\bar{\bar{\Psi}}_i\rangle := |\bar{\bar{\Psi}}(\alpha_0)_i\rangle = \mathcal{N}(\alpha_0)|\Psi(\alpha_0)_i\rangle + \mathcal{N}(\alpha_0)_i|\Psi(\alpha_0)\rangle := |\bar{\Psi}_i\rangle + \mathcal{N}_i|\Psi_0\rangle. \quad (3.83)$$

For linear parameters, the derivatives $|\Psi_i\rangle$ were perfect, and from equation (3.79) we see that

$$|\Psi_i\rangle = |\bar{\Psi}_i\rangle + S_{i0}|\Psi_0\rangle, \quad (3.84)$$

so it is reasonable to choose [13]

$$\boxed{|\bar{\bar{\Psi}}_i\rangle = |\Psi_i\rangle} \quad (3.85)$$

$$\boxed{\mathcal{N}_i = S_{i0} \qquad \text{linear parameters.}} \quad (3.86)$$

What about nonlinear parameters? The new normalization gives the linearized solution

$$|\bar{\bar{\Psi}}\rangle^{\text{lin}} = |\Psi_0\rangle + \sum_{i=1}^{p}\Delta\bar{\bar{\alpha}}_i|\bar{\bar{\Psi}}_i\rangle = |\Psi_0\rangle + \sum_{i=1}^{p}\Delta\bar{\bar{\alpha}}_i(|\bar{\Psi}_i\rangle + \mathcal{N}_i|\Psi_0\rangle) \quad (3.87)$$

$$= \left(1 + \sum_{i=1}^{p}\mathcal{N}_i\Delta\bar{\bar{\alpha}}_i\right)|\Psi_0\rangle + \sum_{i=1}^{p}\mathcal{N}_i\Delta\bar{\bar{\alpha}}_i|\bar{\Psi}_i\rangle. \quad (3.88)$$

Since only the normalization was modified, the optimal solution is still found within the same variational space, i.e. the functional form of the trial wave function is unchanged. Since both updates converge to the same state, one must have scaling relations,

$$|\bar{\bar{\Psi}}\rangle^{\text{lin}} \equiv C|\bar{\Psi}\rangle^{\text{lin}}, \text{ and } \Delta\bar{\bar{\alpha}}_i \equiv C\Delta\bar{\alpha}_i \; \forall \; i, \tag{3.89}$$

with the same scaling factor C for every parameter. Inserting the states, one finds that

$$\left(1 + \sum_{i=1}^{p} C\Delta\bar{\alpha}_i \mathcal{N}_i\right)|\Psi_0\rangle + \sum_{i=1}^{p} C\Delta\bar{\alpha}_i|\Psi_i\rangle \equiv C\left(|\Psi_0\rangle + \sum_{i=1}^{p}\Delta\bar{\alpha}_i|\bar{\Psi}_i\rangle\right), \tag{3.90}$$

so the solution is

$$C = \left(1 + \sum_{i=1}^{p} C\mathcal{N}_i\Delta\bar{\alpha}_i\right) \Leftrightarrow C = \frac{1}{1 - \sum_{i=1}^{p}\mathcal{N}_i\Delta\bar{\alpha}_i}, \tag{3.91}$$

and the relation of the updates of the jth parameter is

$$\boxed{\Delta\bar{\bar{\alpha}}_j = \frac{\Delta\bar{\alpha}_j}{1 - \sum_{i=1}^{p}\mathcal{N}_i\Delta\bar{\alpha}_i}.} \tag{3.92}$$

The choice of derivatives $\mathcal{N}_i$ may affect not only the size of the parameter update, but, remarkably, also its sign.

One way to choose $\mathcal{N}_i$ for nonlinear parameters i is to think that if the 'old' linear update $|\bar{\Psi}\rangle^{\text{lin}}$ is not working properly, a logical choice is to pick the derivative $|\bar{\bar{\Psi}}_i\rangle$ in a direction orthogonal to some direction in the space spanned by $|\Psi_0\rangle$ and $|\bar{\Psi}\rangle^{\text{lin}}$ [12] (choose $\langle\Psi_0|\Psi_0\rangle = 1$),

$$\left\langle \bar{\bar{\Psi}}_i \middle| \xi|\Psi_0\rangle + (1-\xi)\frac{|\bar{\Psi}\rangle^{\text{lin}}}{||\bar{\Psi}^{\text{lin}}||} \right\rangle = 0, \tag{3.93}$$

where $\xi \in [0, 1]$ controls the direction ($\xi = 1/2$ is a good choice). Toulouse and Umrigar [13] arrive at the formula

$$\boxed{\mathcal{N}_i = -\frac{(1-\xi)\sum_j^{\text{nonlin}} \bar{S}_{ij}\Delta\bar{\alpha}_j}{1 - \xi + \xi\sqrt{1 + \sum_{j,k}^{\text{nonlin}} \bar{S}_{jk}\Delta\bar{\alpha}_j\Delta\bar{\alpha}_k}}} \quad \text{nonlinear parameters.} \tag{3.94}$$

In the actual code, the modification is fairly simple because we only need the scaling sum $1 - \sum_{i=1}^{p}\mathcal{N}_i\Delta\bar{\alpha}_i$ (see equation (3.92)):

$$1 - \sum_{i=1}^{p}\mathcal{N}_i\Delta\bar{\alpha}_i = 1 + \sum_i^{\text{lin}} S_{0i}\Delta\bar{\alpha}_i + \frac{(1-\xi)\sum_{i,j}^{\text{nonlin}} \bar{S}_{ij}\Delta\bar{\alpha}_i\Delta\bar{\alpha}_j}{1 - \xi + \xi\sqrt{1 + \sum_{j,k}^{\text{nonlin}} \bar{S}_{jk}\Delta\bar{\alpha}_j\Delta\bar{\alpha}_k}}, \tag{3.95}$$

so only two sums are needed.

Finally, it is important to stabilize the optimization by letting the parameters first change to the 'safe' steepest descent direction before using the 'unsafe' Newton–Raphson update. The parameter updates can be turned to the steepest descent directions by adding a positive constant a_{opt} to the diagonals of H while keeping H_{00} fixed, similar to the scheme used in [8]. A few test values for a_{opt} reveal the energetically best choice at each iteration [12].

3.3.5.1 Practical considerations

The linearized optimization method uses the Hamiltonian matrix H and the overlap matrix S in the basis $\{\Psi_0, \Psi_1, \Psi_2, ..., \Psi_p\}$; that is, the trial wave function itself and the derivatives w.r.t. each parameter. The matrices are estimated using M walkers, and the Hamiltonian matrix elements and the overlap matrix elements are computed using the reduced-variance estimators $\bar{H}$ and $\bar{S}$ given in equations (3.72), (3.76), (3.64), and (3.65).

The parameter updates are obtained by solving the generalized eigenvalue problem in equation (3.56), using the reduced-variance estimators

$$\bar{H}\Delta\alpha_k = E_k^{\text{lin}}\bar{S}\Delta\alpha_k. \tag{3.96}$$

The one eigenvector we need is

$$\Delta\alpha = \left(1, \Delta\alpha_1, \Delta\alpha_2, ..., \Delta\alpha_p\right), \tag{3.97}$$

where the first update was fixed by definition, $\Delta\alpha_0 = 1$ (see equation (3.49)). This fixes the normalization of the eigenvector. Due to numerical issues, the lowest eigenvalue(s) may be way below $\langle E_L \rangle$, and one should instead pick the lowest physically reasonable eigenvalue and the corresponding eigenvector [13].

When we start with parameters that are far from optimal, the first parameter updates $\Delta\alpha$ may not improve the wave function. This is a problem of the underlying Newton or Newton–Raphson method, which assumes that the off-diagonal elements of the matrix $\bar{H}$ capture the curvature in the parameter space. In such cases, it is safer to turn the update further toward the steepest descent direction. The steepest descent direction is given by the diagonal elements of the Hamiltonian matrix, so one can turn $\Delta\alpha$ more toward that safe direction by adding a constant $a_{opt} \geqslant 0$ to the diagonal elements [8],

$$\bar{H}_{ii} \rightarrow \bar{H}_{ii} + a_{opt}, \text{ for } i > 0. \tag{3.98}$$

This modification alone also scales the parameter updates by a factor of $1/a_{opt}$. If this is not desirable, add a_{opt} also to the corresponding elements of the overlap matrix $\bar{S}$. When the parameters are close to optimal, the fastest descent to an energy minimum is guaranteed by $a_{opt} = 0$. Some QMC codes find the best a_{opt} for each optimization step, which is time well spent in cases involving thousands of parameters.

He atom ground state Let us consider a trial wave function made of a *Slater geminal* (notation Ψ instead of φ_T),

$$\Psi_0: =\Psi_0(\mathbf{x}) = \alpha_1 e^{-\alpha_2 r_1 - \alpha_3 r_1 - \alpha_4 r_{12}} + \alpha_1 e^{-\alpha_2 r_2 - \alpha_3 r_1 - \alpha_4 r_{12}}. \tag{3.99}$$

As required, this is symmetric under particle exchange. The parameter α_1 is a linear parameter because the derivative Ψ_1 does not depend on α_1, and

$$\Psi_0 = \alpha_1 \Psi_1. \tag{3.100}$$

The linear approximation in equation (3.48) is exact,

$$\Psi^{(\alpha_1 + \Delta \alpha_1)} = \Psi_0 + (\alpha_1 - \alpha_1^0)\Psi_1 = \Psi_0 + \Delta \alpha_1 \Psi_1 \qquad \text{parameter } \alpha_1, \tag{3.101}$$

and there exists a $\Delta \alpha_1$ that takes the parameter directly to its optimal value.

The parameters in the exponent are nonlinear parameters because the derivatives depend on the parameters themselves. For example, Ψ_2 depends on α_2:

$$\Psi_2 = -r_1 \alpha_1 e^{-\alpha_2 r_1 - \alpha_3 r_1 - \alpha_4 r_{12}} - r_2 \alpha_1 e^{-\alpha_2 r_2 - \alpha_3 r_1 - \alpha_4 r_{12}}. \tag{3.102}$$

Changing $\alpha_2 \to \alpha_2 + \Delta \alpha_2$ changes the wave function to

$$\Psi^{(\alpha_2 + \Delta \alpha_2)} = \underbrace{\Psi_0 + \Delta \alpha_2 \Psi_2}_{\text{linear approximation}} + \frac{1}{2}(\Delta \alpha_2)^2 \underbrace{\frac{\partial^2 \Psi}{\partial (\alpha_2)^2}}_{\neq 0} + \cdots. \tag{3.103}$$

The linear approximation does not work well if the higher-order corrections are large.

To add more freedom to the trial wave function, one can combine multiple Slater geminals,

$$\Psi_0(\mathbf{x}) = \sum_{k=1}^{K} c_k e^{-\alpha_k r_1 - \beta_k r_2 - \gamma_k r_{12}} + (1 \leftrightarrow 2), \tag{3.104}$$

where $(1 \leftrightarrow 2)$ exchanges the two electrons.

Automatic differentiation Computing the derivatives w.r.t. all parameters is a simple but rather boring task. The derivatives of the trial wave function w.r.t. coordinates or parameters can also be calculated using AD. Unlike finite differences, AD gives the derivatives with machine precision accuracy (usually $\sim 10^{-16}$). Compared to AD, code that calculates the values of an algebraic derivative may be faster, but it is a numerical evaluation all the same, so its accuracy will not be any higher than the accuracy of AD. Algebraic differentiation is prone to errors, and so is writing code that computes the derivatives. AD gets the derivatives right as soon as the function itself is programmed correctly, so if you do not want to use it in production code, you might want to check the results against AD.

AD is based on the notion that if one can evaluate a function, one can also evaluate its derivatives. Any coded function is a computational graph made of elementary functions (sin(), exp(), log(),...) and basic operations (+, /, *,...). AD packs together each known elementary function *and* its derivative. After traversing the computational graph forward, from arguments to function evaluation, one can follow the graph in reverse order to find the partial derivatives w.r.t. all arguments - this is called *forward AD*.

Many programming languages, such as Python and Julia, have easy-to-use AD libraries. AD is heavily used in searches for optimal neural network parameters in machine learning because the derivatives needed to train the network would be very complicated expressions.

References

[1] Carleo G, Cirac I, Cranmer K, Daudet L, Schuld M, Tishby N, Vogt-Maranto L and Zdeborová L 2019 Machine learning and the physical sciences *Rev. Mod. Phys.* **91** 045002

[2] Carleo G and Troyer M 2017 Solving the quantum many-body problem with artificial neural networks *Science* **355** 602–6

[3] Atanasova H, Bernheimer L and Cohen G 2023 Stochastic representation of many-body quantum states *Nat. Commun.* **14** 3601

[4] Hornik K, Stinchcombe M and White H 1989 Multilayer feedforward networks are universal approximators *Neural Netw.* **2** 359–66

[5] Umrigar C J, Wilson K G and Wilkins J W 1988 Optimized trial wave functions for quantum Monte Carlo calculations *Phys. Rev. Lett.* **60** 1719–22

[6] Kent P R C, Needs R J and Rajagopal G 1999 Monte Carlo energy and variance-minimization techniques for optimizing many-body wave functions *Phys. Rev.* B **59** 12344–51

[7] Assaraf R and Caffarel M 1999 Zero-variance principle for Monte Carlo algorithms *Phys. Rev. Lett.* **83** 4682–5

[8] Umrigar C J and Filippi C 2005 Energy and variance optimization of many-body wave functions *Phys. Rev. Lett.* **94** 150201

[9] Drummond N D and Needs R J 2005 Variance-minimization scheme for optimizing Jastrow factors *Phys. Rev.* B **72** 085124

[10] Harju A, Barbiellini B, Siljamäki S, Nieminen R M and Ortiz G 1997 Stochastic gradient approximation: an efficient method to optimize many-body wave functions *Phys. Rev. Lett.* **79** 1173–7

[11] Foulkes W M C, Mitas L, Needs R J and Rajagopal G 2001 Quantum Monte Carlo simulations of solids *Rev. Mod. Phys.* **73** 33–83

[12] Umrigar C J, Toulouse J, Filippi C, Sorella S and Hennig R G 2007 Alleviation of the Fermion-sign problem by optimization of many-body wave functions *Phys. Rev. Lett.* **98** 110201

[13] Toulouse J and Umrigar C J 2007 Optimization of quantum Monte Carlo wave functions by energy minimization *J. Chem. Phys.* **126** 084102

[14] Nightingale M P and Melik-Alaverdian V 2001 Optimization of ground- and excited-state wave functions and van der Waals clusters *Phys. Rev. Lett.* **87** 043401

[15] Toulouse J and Umrigar C J 2008 Full optimization of Jastrow-Slater wave functions with application to the first-row atoms and homonuclear diatomic molecules *J. Chem. Phys.* **128** 174101

IOP Publishing

A Practical Course on Quantum Monte Carlo

Vesa Apaja

Chapter 4

Diffusion Monte Carlo

In quantum Monte Carlo (QMC), optimization aims to adjust the parameters of a *chosen* trial wave function to minimize a cost function, such as variance or energy. The cost function is evaluated among walkers sampled using variational Monte Carlo (VMC), but neither optimization nor VMC can improve the trial wave function beyond the boundaries of the chosen functions. One can make educated guesses about the required functions, but attempts to faithfully describe the physics of a many-body system may still fail. The fact that that even thousands of parameters do not yield the known ground-state energy sends a clear signal that the trial wave function is missing something.

Luckily, a wave function is not needed in QMC for the computation of a quantum mechanical expectation value, because integrals are always approximated by sums, such as

$$E \approx \frac{1}{M}\sum_{i=1}^{M}E_L(\mathbf{x}_i). \tag{4.1}$$

To evaluate the ground-state energy, all we need are walkers $\{\mathbf{x}_i\}$ that are *known* to sample the exact ground state. The task in this chapter is to derive and implement one algorithm that samples walkers from the ground state: the diffusion Monte Carlo (DMC).

Why diffusion?

What does diffusion have to do with a quantum mechanical ground state? One way to make the connection plausible is to think about the ground state as the state where we have the least information about the system, under the assumptions in the Hamiltonian and a boson/fermion symmetry. A ground-state wave function provides the least information about particle positions under the same premises. Particle positions, on the other hand, are like a cloud, and we know the least about positions if we spread the cloud as thinly as possible—as in diffusion. This is pretty vague, but in the next section, the diffusion equations are derived, starting from the Schrödinger equation.

4.1 Projection on the ground state

The time-dependent eigenstates of the Hamiltonian are $\Phi_n(\mathbf{x}, t)$, and even though we cannot solve them, we can use them. The Hamiltonians used in this book are time independent. We have

$$\hat{\mathcal{U}}(t, 0): = e^{-i\hat{\mathcal{H}}t/\hbar} \qquad \text{time evolution operator,} \qquad (4.2)$$

and the eigenstates have the time-dependent phase,

$$\Phi_n(\mathbf{x}, t) = e^{-i\hat{\mathcal{H}}t/\hbar}\Phi_n(\mathbf{x}, 0) = e^{-iE_n t/\hbar}\Phi_n(\mathbf{x}, 0). \qquad (4.3)$$

The complex phase factors oscillate with a frequency that depends on E_n, the energy of the eigenstate. The wave function of the system is a superposition,

$$\Psi(\mathbf{x}, t) = \sum_n c_n \Phi_n(\mathbf{x}, t). \qquad (4.4)$$

The time evolution is unitary, meaning that the norm of the wave function does not change over time, so the system does not evolve to the ground state by itself, no matter how long we wait. Usually, when we want to bring a system to its ground state, we cool it to a temperature close to absolute zero. Cooling removes energy from the system, so in addition to the system Hamiltonian $\hat{\mathcal{H}}$, we need another Hamiltonian for the environment and a coupling between them to describe energy transfer.

Here, we want to keep things simple and reduce temperature, that is, remove the excited states from the superposition $\Psi(\mathbf{x}, t)$ by manipulating only the system without any reference to the environment. The options are very limited. We cannot change $\hat{\mathcal{H}}$—that would change the ground state we are looking for, so we cannot change any of the energies E_n. Looking at equation (4.3), the last quantity left is time t, and we can make it imaginary,

$$\boxed{-it/\hbar \rightarrow -\tau \Leftrightarrow t \rightarrow -i\tau/\hbar.} \qquad (4.5)$$

where the *imaginary time* $\tau \in \mathfrak{R}$ has the unit of inverse energy. This transformation has the desired effect because oscillating phases in the eigenstates turn into decay or amplification,

$$\Phi_n(\mathbf{x}, \tau) = \Phi_n(\mathbf{x}, 0)e^{-\tau E_n}. \qquad (4.6)$$

As τ increases, all amplitudes decay, so we need to *stabilize the ground state* by making an energy shift,

$$\Phi_n(\mathbf{x}, 0)e^{-\tau(E_n - E_0)} \xrightarrow{\tau \to \infty} \Phi_0(\mathbf{x}, 0), \qquad (4.7)$$

because excitations have exponentially fast decaying amplitudes. In practice, E_0 is unknown, and we have to replace it with the so-called *trial energy* $E_T(\tau)$, which is a parameter that has to be adjusted during DMC calculations so that

$$\boxed{E_T(\tau \to \infty) \approx E_0.} \tag{4.8}$$

Let us write the time evolution in equations (4.2) and (4.3) as imaginary-time evolution,

$$\boxed{\begin{aligned} \Phi_n(\mathbf{x}, \tau) &= e^{-\tau(\hat{H} - E_T)}\Phi_n(\mathbf{x}, 0) \\ &= e^{-\tau(E_n - E_T)}\Phi_n(\mathbf{x}, 0) \\ &\xrightarrow{\tau \to \infty} \Phi_0(\mathbf{x}, 0) \qquad \text{imaginary time evolution.} \end{aligned}} \tag{4.9}$$

The operator that evolves the system in imaginary time projects out the ground state,

$$\boxed{e^{-\tau(\hat{H} - E_0)} \qquad \text{projection operator.}} \tag{4.10}$$

The projection operator on any state $|\Phi\rangle$ is commonly expressed in Dirac's bra-ket notation as $|\Phi\rangle\langle\Phi|$; so, marking the ground state with $|0\rangle$, we obtain

$$e^{-\tau(\hat{H} - E_0)} \xrightarrow{\tau \to \infty} |0\rangle\langle 0|. \tag{4.11}$$

4.2 Imaginary-time Schrödinger equation

The imaginary-time Schrödinger equation describes evolution that takes the system to its ground state, and we anticipate that this evolution is related to diffusion. Diffusion results from a particle current that moves toward lower-density regions, given by the phenomenological Fick's law,

$$\mathbf{j}(\mathbf{r}, t) = -D \, \boldsymbol{\nabla} n(\mathbf{r}, t), \tag{4.12}$$

with particle density $n(\mathbf{r}, t)$ and diffusion constant D. Density also obeys the continuity equation

$$\frac{\partial}{\partial t} n(\mathbf{r}, t) + \boldsymbol{\nabla} \cdot \mathbf{j}(\mathbf{r}, t) = \text{source term}, \tag{4.13}$$

where the source term takes into account the possibility that particles may be created or destroyed in a volume element. Inserting Fick's law gives the diffusion equation

$$-\frac{\partial}{\partial t} n(\mathbf{r}, t) = -D \, \nabla^2 n(\mathbf{r}, t) + \text{source term}. \tag{4.14}$$

On the other hand, the imaginary-time evolution in equation (4.9) is the solution of

$$-\frac{\partial}{\partial \tau} \Psi(\mathbf{x}, \tau) = (\hat{H} - E_T)\Psi(\mathbf{x}, \tau), \tag{4.15}$$

which for the N-particle Hamiltonian is ($D: = \hbar^2/(2m)$)

$$-\frac{\partial}{\partial \tau}\Psi(\mathbf{x}, \tau) = -D\sum_{i=1}^{N}\nabla_i^2\ \Psi(\mathbf{x}, \tau) + (V(\mathbf{x}) - E_T)\Psi(\mathbf{x}, \tau). \qquad (4.16)$$

Comparing this with equation (4.14), we conclude that it is a *multidimensional diffusion equation*, provided the wave function $\Psi(\mathbf{x}, \tau)$ can be interpreted as a distribution $P(\mathbf{x}, \tau)$. If that is possible, we can identify terms corresponding to diffusion:

$$\boxed{-\frac{\partial}{\partial \tau}P(\mathbf{x}, \tau) = (\hat{\mathcal{H}} - E_T)P(\mathbf{x}, \tau)} \qquad (4.17)$$

$$\boxed{\Leftrightarrow\ -\frac{\partial}{\partial \tau}\underbrace{P(\mathbf{x}, \tau)}_{\text{distribution}} = \underbrace{-D\sum_{i=1}^{N}\nabla_i^2\ P(\mathbf{x}, \tau)}_{\text{diffusion to low-density regions}} + \underbrace{(V(\mathbf{x}) - E_T)P(\mathbf{x}, \tau)}_{\text{source term}},} \qquad (4.18)$$

and the formal solution is

$$\boxed{P(\mathbf{x}, \tau) = e^{-\tau(\hat{\mathcal{H}} - E_T)}P(\mathbf{x}, 0).} \qquad (4.19)$$

Before getting carried away, we should find out when and how a wave function $\Psi(\mathbf{x}, \tau)$ can be written as a distribution $P(\mathbf{x}, \tau)$.

4.3 Stochastic representation of a wave function

The imaginary-time Schrödinger equation for the wave function can be interpreted as a diffusion equation for the distribution as follows (omitting the arguments $(\mathbf{x}, \tau)$):

1. If $\Psi \geqslant 0$, we can immediately call it a distribution, $P: = \Psi$. The distribution evolves according to the diffusion equation (4.18). Although straightforward, in QMC, this approach turns out to be very inefficient and has been superseded by case 3.
2. If $\Psi \in \mathfrak{R}$ but changes sign somewhere, then we can always write

$$\Psi = \Psi_+ - \Psi_-, \qquad (4.20)$$

where $\Psi_+ \geqslant 0$ and $\Psi_- \geqslant 0$. The imaginary-time Schrödinger equation is a linear partial differential equation, so

$$\begin{cases} -\dfrac{\partial}{\partial \tau}\Psi_+ = (\hat{\mathcal{H}} - E_T)\Psi_+ \\ -\dfrac{\partial}{\partial \tau}\Psi_- = (\hat{\mathcal{H}} - E_T)\Psi_- \end{cases} \Rightarrow -\frac{\partial}{\partial \tau}\Psi = (\hat{\mathcal{H}} - E_T)\Psi. \qquad (4.21)$$

We can now identify two distributions $P_1: = \Psi_+$ and $P_2: = \Psi_-$, which both evolve according to the diffusion equation (4.9). From their solutions, we can collect a solution Ψ that changes sign. This approach has been superseded by the approximation made in case 3.

3. If $\Psi = \varphi_T f$, where φ_T is a known trial wave function and $f \geqslant 0$, then insert Ψ into the imaginary-time Schrödinger equation and find the equation for f, which can be taken as the distribution, $P(\mathbf{x}, \tau): = f(\mathbf{x}, \tau)$. This *importance sampling* is the topic of section 4.4. The trial wave function takes care of possible sign changes in Ψ.

4. If $\Psi(\mathbf{x}, \tau)$ has complex values, then we can first separate the real and imaginary parts,

$$\Psi = \Re\Psi + i\Im\Psi, \qquad (4.22)$$

where both $\Re\Psi(\mathbf{x}, \tau)$ and $\Im\Psi(\mathbf{x}, \tau)$ satisfy the imaginary-time Schrödinger equation, because for real $\hat{\mathcal{H}}$,

$$-\frac{\partial}{\partial\tau}(\Re\Psi + i\Im\Psi) = (\hat{\mathcal{H}} - E_T)(\Re\Psi + i\Im\Psi)$$

$$\Leftrightarrow \begin{cases} -\dfrac{\partial}{\partial\tau}\Re\Psi = (\hat{\mathcal{H}} - E_T)\Re\Psi \\[2mm] -\dfrac{\partial}{\partial\tau}\Im\Psi = (\hat{\mathcal{H}} - E_T)\Im\Psi. \end{cases} \qquad (4.23)$$

Note that this is not valid in real time, because there is an imaginary unit in the time-dependent Schrödinger equation, and the real and imaginary parts of the wave function become interdependent. Next, separate the positive and negative parts,

$$\Psi = (\Re\Psi_+ - \Re\Psi_-) + i(\Im\Psi_+ - \Im\Psi_-). \qquad (4.24)$$

The four functions on the right-hand side of the equation are all suitable for identification as distributions, and if they all satisfy the imaginary-time Schrödinger equation separately, then so does Ψ. As you can see, this case involves quite a bit of bookkeeping and is given here only for completeness.

The relation

$$\text{wave function} \leftrightharpoons \text{distribution}, \qquad (4.25)$$

assures us we are computing something that has physical relevance. The curse of dimensionality is still there because, just like the wave function, a distribution is a function of dN positional variables and τ. What makes all the difference is that the distribution $P(\mathbf{x}, \tau)$ satisfies a diffusion equation, and instead of solving for $P(\mathbf{x}, \tau)$, we can say that the objects that are diffusing are walkers. One more relationship needs to be established, namely

$$\text{distribution} \rightleftharpoons \text{walkers}. \tag{4.26}$$

The coordinates $\mathbf{x}$ can be sampled from the distribution $P(\mathbf{x}, \tau)$ at a certain τ, so obviously, sampling enough coordinates tells us pretty accurately what the distribution is. In DMC, we really do not know what $P(\mathbf{x}, \tau)$ is, so we let a large number of walkers represent it.

4.3.1 Stochastic representation of a distribution

Let us write the distribution in terms of an integral using a Dirac delta function and evaluate the integral using Monte Carlo:

$$P(\mathbf{x}, \tau) = \int d\mathbf{y} P(\mathbf{y}, \tau)\delta(\mathbf{y} - \mathbf{x}) \approx \frac{1}{M}\sum_{j=1}^{M}\delta(\mathbf{y}_j(\tau) - \mathbf{x}), \tag{4.27}$$

where the points $\mathbf{y}_j(\tau)$ are sampled from $P(\mathbf{y}, \tau)$. At first glance, it seems that we have achieved nothing; however, on the left-hand side of the equation, we have the distribution, and on the right-hand side, a finite set of configurations $\{\mathbf{y}_j\}$. In other words, we have represented a distribution as a set of walkers. Instead of $\mathbf{y}_j$, we can just as well call the walkers $\mathbf{x}_j$:

Representation of a distribution as a finite set of walkers:

$$P(\mathbf{x}, \tau) \approx \frac{1}{M}\sum_{j=1}^{M}\delta(\mathbf{x}_j(\tau) - \mathbf{x}). \tag{4.28}$$

As usual, the Dirac delta function notation makes sense only inside an integral. Although rather technical, this expression makes it straightforward to derive the equations for walkers from the equation for a distribution. In other words, we write stochastic equations of motion for walkers so that when looking at the motion of very many walkers from afar, we see them distributed according to $P(\mathbf{x}, \tau)$. This idea is due to Paul Langevin, who, in 1908, described Brownian motion in terms of a stochastic differential equation (see the translation in [1]).

The walkers representing P are a collection of M possible configurations of the N particles in the system. If you now think that it must take lots of Dirac delta function peaks to get a smooth function, you are absolutely correct: M has to be large. Just how large remains to be seen; however, note that we are not trying to get an accurate distribution but rather accurate quantum expectation values.

4.3.2 Diffusion

Without the source term in equation (4.16), the evolution of P is given by the second-order differential equation

$$-\frac{\partial}{\partial \tau} P(\mathbf{x}, \tau) = -D \sum_{i=1}^{N} \nabla_i^2 \; P(\mathbf{x}, \tau). \tag{4.29}$$

Separating variables with the product ansatz

$$P(\mathbf{x}, \tau) = \prod_{i=1}^{dN} P_i((\mathbf{x})_i, \tau) \tag{4.30}$$

gives dN independent 1D diffusion equations. For a solution, we need to specify two boundary conditions; the first is:

$$P(\mathbf{x}, \tau = 0) = \delta(\mathbf{x}) \qquad \text{1st boundary condition.} \tag{4.31}$$

This just states that particles start their diffusion from the point where they are at $\tau = 0$; in other words, we are not squeezing them into a zero-dimensional point.

The second boundary condition depends on the domain where diffusion takes place. The particles can be, for example, in free space, in a box, or in a box with periodic boundary conditions. For particles in free space, we can set $P(\mathbf{x}, \tau) \to 0$ if $\mathbf{x}$ is very far from the starting points, and with the aid of Fourier transforms, we find that the solution is a dN-dimensional Gaussian with standard deviation $\sigma = \sqrt{2D\tau}$,

$$P(\mathbf{x}, \tau) = \frac{1}{(4\pi D\tau)^{dN/2}} e^{-\frac{\mathbf{x}^2}{4D\tau}}. \tag{4.32}$$

This is not the solution for particles in a box, because freely diffusing particles can leave the box. Hard boundaries such as walls, where $P(\mathbf{x}, \tau)$ goes to zero, can be either introduced as infinite potentials (this turns out to kill walkers that cross the boundary), or we can go back and solve the diffusion equation with the walls in place. Walls in diffusion can be easily added using a mirror image of the system on the other side of the wall, but I am not going to give the details here. Periodic boundary conditions reduce the effects of boundaries in simulations of infinite systems with a finite computational box. These will be important in path integral Monte Carlo (PIMC) in chapter 5.

4.3.2.1 Diffusion Green's function G_D
Defining the kinetic energy operator,

$$\hat{T} := -D \sum_{i=1}^{N} \nabla_i^2 , \tag{4.33}$$

we can write the imaginary-time diffusion equation (4.29) in a shorter form,

$$-\frac{\partial}{\partial \tau} P(\mathbf{x}, \tau) = \hat{T} P(\mathbf{x}, \tau). \tag{4.34}$$

The solution can be expressed as an integral equation,

$$\boxed{P(\mathbf{x}', \tau) = \int d\mathbf{x} \langle \mathbf{x}' | e^{-\tau \hat{T}} | \mathbf{x} \rangle P(\mathbf{x}, 0) \qquad \text{only diffusion},} \qquad (4.35)$$

as one can check by taking the derivative w.r.t. τ,

$$\frac{\partial}{\partial \tau} P(\mathbf{x}', \tau) = \int d\mathbf{x} \langle \mathbf{x}' | -\hat{T} e^{-\tau \hat{T}} | \mathbf{x} \rangle P(\mathbf{x}, 0) = -\hat{T} \int d\mathbf{x} \langle \mathbf{x}' | e^{-\tau \hat{T}} | \mathbf{x} \rangle P(\mathbf{x}, 0) \quad (4.36)$$

$$= -\hat{T} P(\mathbf{x}', \tau). \qquad (4.37)$$

The integral equation defines the diffusion Green's function,

$$P(\mathbf{x}', \tau) = \int d\mathbf{x} G_D(\mathbf{x}' \leftarrow \mathbf{x}; \tau) P(\mathbf{x}, 0), \qquad (4.38)$$

where the propagator or Green's function moves walkers from $\mathbf{x}$ to $\mathbf{x}'$ in imaginary time τ. The free-space diffusion solution was already given in equation (4.32), where walkers moved from the origin to $\mathbf{x}$ in imaginary time τ, so obviously

$$\boxed{G_D(\mathbf{x}', \mathbf{x}; \tau) := \langle \mathbf{x}' | e^{-\tau \hat{T}} | \mathbf{x} \rangle = \frac{1}{(4\pi D\tau)^{dN/2}} e^{-\frac{(\mathbf{x}'-\mathbf{x})^2}{4D\tau}} \qquad \text{Free-space diffusion.}} \qquad (4.39)$$

A more detailed derivation is given in Appendix B. We insert this into equation (4.35), obtaining

$$\boxed{P(\mathbf{x}', \tau) = \int d\mathbf{x} \frac{1}{(4\pi D\tau)^{dN/2}} e^{-\frac{(\mathbf{x}'-\mathbf{x})^2}{4D\tau}} P(\mathbf{x}, 0).} \qquad (4.40)$$

In QMC, the integral is evaluated stochastically using the following algorithm:

---

1. Sample M walkers $\{\mathbf{x}_i\}$ from distribution $P(\mathbf{x}, 0)$.
2. For N particles in d dimensions, sample $d \times N$ normally distributed random numbers, collecting them in the vector $\boldsymbol{\eta}$.
3. For each walker i, move coordinates according to

$$\mathbf{x}'_i = \mathbf{x}_i + \sqrt{2D\tau} \, \boldsymbol{\eta}_i. \qquad (4.41)$$

The walkers $\{\mathbf{x}'\}$ sample $P(\mathbf{x}', \tau)$ in equation (4.40).

The proof can be most conveniently given with the aid of a Dirac delta function. Let us write the integral (4.40) in the form

$$P(\mathbf{x}', \tau) = \frac{1}{(4\pi D\tau)^{dN/2}} \int d\mathbf{x} \int d\mathbf{y} e^{-\frac{\mathbf{y}^2}{4D\tau}} \delta(\mathbf{y} - (\mathbf{x}' - \mathbf{x})) P(\mathbf{x}, 0). \quad (4.42)$$

Change variables and define $\mathbf{y} = \sqrt{2D\tau}\,\boldsymbol{\eta}$; then,

$$P(\mathbf{x}', \tau) = \int d\mathbf{x} \int d\boldsymbol{\eta} \frac{1}{(4\pi)^{dN/2}} e^{-\frac{\eta^2}{2}} \delta(\sqrt{2D\tau}\,\boldsymbol{\eta} - (\mathbf{x}' - \mathbf{x})) P(\mathbf{x}, 0) \quad (4.43)$$

$$= \underbrace{\int d\mathbf{x} P(\mathbf{x}, 0)}_{\text{1: sample } \mathbf{x} \text{ from } (\mathbf{x},0)} \underbrace{\int d\boldsymbol{\eta} \frac{1}{(2\pi)^{dN/2}} e^{-\frac{\eta^2}{2}}}_{\text{2: sample } \eta s \text{ from } \mathcal{N}(0,1)} \underbrace{\delta(\mathbf{x}' - (\mathbf{x} + \sqrt{2D\tau}\,\boldsymbol{\eta}))}_{\text{3: set argument to zero, solve } \mathbf{x}'}. \quad (4.44)$$

Here, $\mathcal{N}(0, 1)$ is the standard normal distribution with zero mean and unit variance.[1] The stages of the given algorithm are indicated in equation (4.44). If one wishes, the integrals can also be written explicitly as sums over walkers (f and g are test functions):

$$\int d\mathbf{x} P(\mathbf{x}, 0) f(\mathbf{x}) \approx \frac{1}{M} \sum_{i=1}^{M} f(\mathbf{x}_i), \quad \{\mathbf{x}_i\} \text{ sampled from } P(\mathbf{x}, 0) \quad (4.45)$$

$$\int d\boldsymbol{\eta} \frac{1}{(2\pi)^{dN/2}} e^{-\frac{\eta^2}{2}} g(\boldsymbol{\eta}) \approx \frac{1}{M} \sum_{i=1}^{M} g(\boldsymbol{\eta}_i), \quad \{\boldsymbol{\eta}_i\} \text{ sampled from } \mathcal{N}(0, 1). \quad (4.46)$$

The diffusion integral is evaluated as the sum

$$P(\mathbf{x}', \tau) \approx \frac{1}{M} \sum_{i=1}^{M} \delta(\mathbf{x}' - (\mathbf{x}_i + \sqrt{2D\tau}\,\boldsymbol{\eta}_i)) \approx \frac{1}{M} \sum_{i=1}^{M} \delta(\mathbf{x}' - \mathbf{x}'_i), \quad (4.47)$$

where the last equality is the representation in equation (4.28) ($\mathbf{x}_i := \mathbf{x}_i(\tau)$). Multiply both sides by $\mathbf{x}'$ and integrate over $\mathbf{x}'$:

$$\frac{1}{M} \sum_{i=1}^{M} \mathbf{x}'_i = \frac{1}{M} \sum_{i=1}^{M} (\mathbf{x}_i + \sqrt{2D\tau}\,\boldsymbol{\eta}_i), \quad (4.48)$$

so *one* solution is

$$\mathbf{x}'_i = \mathbf{x}_i + \sqrt{2D\tau}\,\boldsymbol{\eta}_i, \quad (4.49)$$

in agreement with the last step of the algorithm. The reason why we need to pick one solution is that the distribution $P(\mathbf{x}, \tau)$ does not change if we relabel the walker coordinates in the walker representation; only the sum over walkers matters.

---

[1] Random numbers from the standard normal distribution $\mathcal{N}(0, 1)$ are often a built-in feature in a program suite. They can be generated from random numbers in the unit interval [0, 1] using, for example, the Ziggurat algorithm or the Box–Muller algorithm.

4.3.3 The source term

The source-term imaginary-time evolution is

$$-\frac{\partial}{\partial \tau} P(\mathbf{x}, \tau) = (V(\mathbf{x}) - E_T)P(\mathbf{x}, \tau) \qquad \text{only source,} \qquad (4.50)$$

and the solution can be written as an integral equation,

$$P(\mathbf{x}', \tau) = \int d\mathbf{x}\, G_V(\mathbf{x}', \mathbf{x}, \tau)P(\mathbf{x}, 0), \qquad (4.51)$$

where the Green's function is

$$G_V(\mathbf{x}', \mathbf{x}, \tau) := \langle \mathbf{x}' | e^{-\tau(V(\mathbf{x}) - E_T)} | \mathbf{x} \rangle = e^{-\tau(V(\mathbf{x}) - E_T)}\delta(\mathbf{x}' - \mathbf{x}). \qquad (4.52)$$

This gives the evolution (change from $\mathbf{x}'$ to $\mathbf{x}$ for clarity)

$$P(\mathbf{x}, \tau) = e^{-\tau(V(\mathbf{x}) - E_T)}P(\mathbf{x}, 0) \qquad \text{source term evolution,} \qquad (4.53)$$

so the distribution is multiplied by a positive weight given by the exponent. The evolution described by the source term can be performed by two algorithms:

1. *Reweighting algorithm* or *pure DMC algorithm* Solve equation (4.53) by adding a weight $e^{-\tau(V(\mathbf{x}_i) - E_T)}$ to each walker i. The weights accumulate as products during the evolution, and they finally either magnify or diminish a walker's impact on DMC expectation values [2]. In effect, walkers remember their history.
   1. Pros: fixed number of walkers; unbiased source term evolution.
   2. Cons: instability issues [3]; variance increases rapidly for long projection times (for a detailed analysis, see [4]).
2. *Branching algorithm* Solve equation (4.53) by letting walkers spawn new walkers or die. If the weight $e^{-\tau(V(\mathbf{x}_i) - E_T)}$ is not an integer, the remainder is treated as a probability; this procedure guarantees that, on average, a walker's impact on the DMC expectation values is correct. In practice, the number of offspring or copies of walker i can be set to [5]

$$M_i^{\text{copies}} = \text{int}(e^{-\tau(V(\mathbf{x}_i) - E_T)} + \xi), \qquad (4.54)$$

where ξ is a uniform random number between zero and one, $\xi = U[0, 1]$, and int() gives the integer part of the argument (floor() in many programming languages). If $M_i^{\text{copies}} \geqslant 2$, then copies of the walker will be created. If $M_i^{\text{copies}} < 1$, then the walker is removed.

Weight control in the reweighting algorithm

Walker weights have to be controlled because a walker with a huge weight would dominate expectation values, and QMC integrals would be approximated by a sum with just one term. A walker with a tiny weight, on the other hand, is practically useless and a waste of computational effort. One way to control weights is to adjust the trial energy E_T (increasing E_T shifts all weights downward), but this does not solve the problem of disproportional weights. To get more evenly distributed weights, one can use the wheel of fortune. Start with an empty list of new walkers. Compute the sum of all weights. Pick a random number r between zero and the sum, and add the walker where r points to the list of new walkers; repeat the process M times. Some walkers may be added to the list several times. The selection process systematically favors large-weight walkers, but we have to live with that; there is no way to control weights without bias.

Population control in the branching algorithm

The reweighting algorithm leads to larger variance in results than the branching algorithm. The overwhelming majority of QMC codes use branching with a stabilizing step that maintains the average number of walkers. Walker population has to be controlled because a very large number of walkers is computationally very time-consuming, and, on the other hand, a small number of walkers cannot represent the distribution $P(\mathbf{x}, \tau)$. In a limiting case, a single walker would give a very poor idea of the distribution we have, and quantum integrals would be approximated by a sum with just one term. Reynolds *et al* [5] introduced what is now the de facto algorithm for adjusting E_T,

$$E_T = E_0^{\text{guess}} + \kappa \ln\left(\frac{M_{\text{target}}}{M_{\text{current}}}\right), \tag{4.55}$$

where E_0^{guess} is the current best guess of the ground-state energy, M_{target} is the walker population we would like to maintain, and M_{current} is the current number of walkers. The positive parameter κ needs to be adjusted so that the population neither varies too much nor too little. The former leads to increased variance in the results, while the latter biases the results. The trial energy increases if there are too few walkers and decreases if there are too many. Branching is so widely used that the source term is often called the **branching term**.

4.3.4 A crude DMC algorithm

Having solved the imaginary-time evolution separately for diffusion and source terms, we are ready to combine them. However, quantum mechanics throws a spanner in the works: the kinetic operator $\hat{\mathcal{T}}$ and the potential energy $V(\mathbf{x})$, or the potential operator $\hat{\mathcal{V}}$, do not commute:

$$[\hat{\mathcal{T}}, \hat{\mathcal{V}}] \neq 0. \tag{4.56}$$

The non-commutation manifests itself in the fact that it makes a difference whether walkers move in diffusion and then branch, or the other way around.[2] In the

---

[2] Anecdotally, our everyday life is full of non-commutative acts. It matters whether you open the garage door before driving the car in, or you drive the car in and....

exponent operator, the non-commutation shows up in a pre-quantum-era theorem, invented multiple times between 1890 and 1906,

$$e^{\hat{A}}e^{\hat{B}} = e^{\hat{A}+\hat{B}+\frac{1}{2}[\hat{A},\hat{B}]+\frac{1}{12}[\hat{A}-\hat{B},[\hat{A},\hat{B}]]+\cdots} \qquad \text{Baker–Campbell–Hausdorff.} \quad (4.57)$$

As a consequence, the imaginary-time evolution operator cannot be neatly split into kinetic and potential parts:

$$e^{-\tau\hat{T}}e^{-\tau(\hat{V}-E_T)} = e^{-\tau(\hat{H}-E_T)+\frac{\tau^2}{2}[\hat{T},\hat{V}]-\frac{\tau^3}{12}[\hat{T}-\hat{V},[\hat{T},\hat{V}]]+\cdots}. \qquad (4.58)$$

However, for a short τ, the kinetic and potential parts do approximately separate:

$$\boxed{\text{Primitive approximation } e^{-\tau\hat{H}} = e^{-\tau(\hat{T}+\hat{V})} \approx e^{-\tau\hat{T}}e^{-\tau\hat{V}} \quad (\text{error } \mathcal{O}(\tau^2)).} \qquad (4.59)$$

The primitive approximation is an example of *operator splitting*, or, when applied to quantum problems, *trotterization*. We shall look at the technique in more detail in section 4.4.7. Any τ can be divided into sufficiently short periods, and stacking primitive approximations *ad infinitum* gives the formula [6]

$$\boxed{\text{Trotter } e^{-\tau\hat{H}} = e^{-\tau(\hat{T}+\hat{V})} = \lim_{M\to\infty}\left[e^{-\frac{\tau}{M}\hat{T}}e^{-\frac{\tau}{M}\hat{V}}\right]^M.} \qquad (4.60)$$

Suzuki [7] showed that this formula is valid only for potentials V bounded from below, so it cannot be applied to systems with attractive Coulomb potential.

Armed with the primitive approximation, we are now ready to write the first DMC algorithm, where imaginary-time evolution is separated into diffusion and branching:

A Crude DMC Algorithm
    Before running, set M_{target}, τ, κ, and E_T.
1. Generate M_{target} (a large number, ~ 1000 or more) walkers. You can pick a suitable trial wave function, and during a VMC run for a single walker, pick coordinates now and then and store them in the walker list.
2. Diffusion: move walkers according to (see equation (4.41))

$$\mathbf{x}' = \mathbf{x} + \sqrt{2D\tau}\,\boldsymbol{\eta}. \qquad (4.61)$$

3. Branching: create copies (see equation (4.54))

$$M_i^{\text{copies}} = \text{int}(e^{-\tau(V(\mathbf{x}_i)-E_T)} + \xi). \qquad (4.62)$$

---

4. Now and then, adjust the trial energy according to (see equation (4.54))

$$E_T = E_0^{\text{guess}} + \kappa \ln\left(\frac{M_{\text{target}}}{M}\right), \qquad (4.63)$$

5. (Thermalization: Repeat steps 2–4 until excitations have died out.) Measure quantities of interest, such as the potential energy, and collect averages.
6. Repeat 2-5 until the statistical errors are small enough.

The crude DMC algorithm has a few peculiarities:
- The primitive approximation has an error that behaves like $\mathcal{O}(\tau^2)$, so τ needs to be very small.
- The algorithm tends to be **unstable** because of wild branching if the potential energy has a large positive or negative value.
- The ground-state energy is the expectation value of the potential energy,

$$E_0 \approx \langle V \rangle_{\text{walkers}} = \frac{1}{M}\sum_{i=1}^{M} V(\mathbf{x}_i) \qquad \text{in crude DMC algorithm.} \qquad (4.64)$$

On average, walkers visit positions where the *potential energy* is E_0. The algorithm has a hard time keeping walkers near the potential energy E_0 with just branching, considering they diffuse freely in step 2. The expectation value, if one is able to execute the algorithm at all, has a large variance due to wild branching. As we shall see,

E_0 is the average walker potential energy *only for self–bound systems*.

Notably, in the particle-in-a-box problem without importance sampling in section 4.6, the ground-state energy is not directly related to the potential energy. Another consequence of the relation $E_0 = \langle V \rangle$ is that **the walker coordinates have nothing to do with the ground-state positions of particles**. Consider, for example, a H atom and the ground-state expectation value $\langle 1/r \rangle$ of the inverse electron–proton distance. We know it to be the inverse Bohr radius in Hartree atomic units $\langle 1/r \rangle = 1.0$, but the crude DMC algorithm *should* give the result

$$E_0 = -0.5 = \langle V \rangle_{\text{walkers}} = \langle -1/r \rangle_{\text{walkers}} \Rightarrow \langle 1/r \rangle_{\text{walkers}} = 0.5, \qquad (4.65)$$

so walkers tend to be further away from the origin than the electron is from the proton. In general, the fact that walker coordinates are not particle coordinates causes a measurement problem that affects all observables $\hat{O}$ that do not commute with the Hamiltonian. We shall find ways to evaluate the expectation values of such observables in section 4.4.6.

The crude DMC energy $E_0 = \langle V \rangle_{\text{walkers}}$ can be made plausible by considering the Schrödinger equation for the ground state,

$$-D\sum_{i=1}^{N} \nabla_i^2 \; \Phi_0(\mathbf{x}) + V(\mathbf{x})\Phi_0(\mathbf{x}) = E_0\Phi_0(\mathbf{x}). \qquad (4.66)$$

Integrate both sides over $\mathbf{x}$,

$$-D\sum_{i=1}^{N} \int d\mathbf{x} \; \nabla_i^2 \; \Phi_0(\mathbf{x}) + \int d\mathbf{x} V(\mathbf{x})\Phi_0(\mathbf{x}) = E_0 \int d\mathbf{x}\Phi_0(\mathbf{x}). \qquad (4.67)$$

If the integrals exist, and if the potential is such that the system is self-bound (implying $\Phi_0(\mathbf{x}) = 0$ far away), then the first integral vanishes, and we get

$$E_0 = \frac{\int d\mathbf{x} V(\mathbf{x})\Phi_0(\mathbf{x})}{\int d\mathbf{x}\Phi_0(\mathbf{x})} = \int d\mathbf{x}\left(\frac{\Phi_0(\mathbf{x})}{\int d\mathbf{x}\Phi_0(\mathbf{x})}\right) V(\mathbf{x}) \approx \frac{1}{M}\sum_{i=1}^{M} V(\mathbf{x}_i), \qquad (4.68)$$

because in the crude DMC algorithm, the M walkers sample the ground-state distribution $\Phi_0(\mathbf{x})$, not $|\Phi_0(\mathbf{x})|^2$.

4.4 Importance sampling DMC

The weakest point of the crude DMC algorithm is the branching (source) term,

$$P(\mathbf{x}, \tau) = e^{-\tau(V(\mathbf{x})-E_T)}P(\mathbf{x}, 0). \qquad (4.69)$$

The branching term forces the crude DMC walkers to sample near a constant-potential energy surface while diffusing without aim. Since diffusion and branching go hand in hand, any modifications to branching must modify diffusion, too. The limitation is that we have to find a practical way to simulate modified, guided diffusion. I will outline the approach known as **importance sampling DMC**, which has become the workhorse of the majority of DMC simulations. It mitigates the potential energy branching problem by replacing the source term with

$$P(\mathbf{x}, \tau) = e^{-\tau(E_L(\mathbf{x})-E_T)}P(\mathbf{x}, 0), \qquad (4.70)$$

where the local energy fluctuates much less than the potential. If the trial wave function is an exact eigenstate of the Hamiltonian, then the source term is no longer needed at all.

The general idea behind importance sampling is that Monte Carlo integration is most efficient with a near-constant integrand. A one-dimensional integral can be written as

$$\int_0^1 dx f(x) = \int_0^1 dx g(x) \frac{f(x)}{g(x)}, \text{ where } g(x) > 0 \ \forall \ x \in [0, 1].$$ (4.71)

The expressions are mathematically equivalent, but in Monte Carlo, the integrals can be approximated as finite sums over two different sets of points. The integral of $f(x)$ is sampled at uniformly distributed random points between zero and one, which the last integral can sample from $g(x)$:

$$\int_0^1 dx f(x) \approx \frac{1}{M} \sum_{j=1}^{M} f(x_j), \ x_j = U[0, 1]$$ (4.72)

$$\int_0^1 dx g(x) \frac{f(x)}{g(x)} \approx \frac{1}{M} \sum_{j=1}^{M} \frac{f(x_j)}{g(x_j)}, \ x_j \text{ sampled from } g(x).$$ (4.73)

Therefore, if g can be chosen so that $f(x_j)/g(x_j)$ fluctuates less than $f(x_j)$, the latter sum converges to the result faster as M increases. From a geometrical point of view, the importance sampling function $g(x)$ is used as a point density in space, so that there are more points in regions where $f(x)$ is largest.

Figure 4.1 shows the classic Monte Carlo evaluation of π, which is inefficient beyond repair but serves as an example of importance sampling. If points are randomly picked from an area A, which envelops a circle with area πr^2, then the probability of hitting the circle is

$$P(\text{hit}) = \frac{\pi r^2}{A} \Leftrightarrow \pi = \frac{A P(\text{hit})}{r^2}.$$ (4.74)

The ideal case would be to have $A = \pi r^2$, but we pretend we do not know π, so we instead choose points in an $a \times a$ square. Let a point inside a circle have $f(x, y) = 1$ and zero elsewhere. Mathematically, the area of the circle is given by

$$\pi r^2 = \int_{-\infty}^{\infty} dx \int_{-\infty}^{\infty} dy f(x, y) = \int_{-\infty}^{\infty} dx \int_{-\infty}^{\infty} dy g_a(x, y) \frac{f(x, y)}{g_a(x, y)}$$ (4.75)

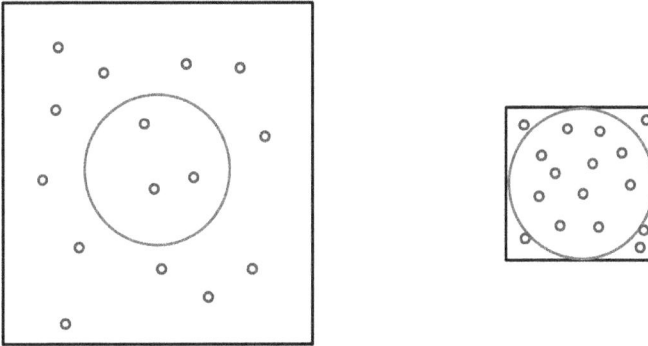

Figure 4.1. The aim is to hit the circle with random points picked from the square. Both squares illustrate importance sampling, but the right-hand one is the better of the two.

$$= \int_{-a}^{a} dx \int_{-a}^{a} dy f(x, y), \tag{4.76}$$

where $g_a(x, y)$ is the importance sampling function, which is one inside the square and zero elsewhere. Sampling M random points (x_i, y_i) from the square computes the integral as a sum over step functions,

$$\pi r^2 = \int_{-a}^{a} dx \int_{-a}^{a} dy f(x, y) \approx \sum_{i=1}^{M} \theta(r^2 - (x_i^2 + y_i^2)). \tag{4.77}$$

The evolution equation can be described either as an integral equation using a Green's function formalism or as a differential equation derived from the imaginary-time Schrödinger equation. I will show how importance sampling is introduced in these two formalisms.

4.4.1 Importance sampling in the evolution integral equation

The evolution equation for the distribution $P(\mathbf{x}, \tau)$, equation (4.9), can be expressed as an integral equation,

$$P(\mathbf{x}'; \tau) = \int d\mathbf{x} \langle \mathbf{x}' | e^{-\tau(\hat{\mathcal{H}} - E_T)} | \mathbf{x} \rangle P(\mathbf{x}; 0): = \int d\mathbf{x} G(\mathbf{x}' \leftarrow \mathbf{x}; \tau) P(\mathbf{x}; 0), \tag{4.78}$$

which we can integrate using importance sampling. The trial wave function $\varphi_T(\mathbf{x})$ is positive everywhere and can be used as an importance sampling function. Instead of evolving the distribution $P(\mathbf{x}, \tau)$ in imaginary time, we can evolve the product distribution $f(\mathbf{x}, \tau): = \varphi_T(\mathbf{x}) P(\mathbf{x}, \tau)$. We solve for $P(\mathbf{x}, \tau)$ and insert it into the evolution integral,

$$\frac{f(\mathbf{x}', \tau)}{\varphi_T(\mathbf{x}')} = \int d\mathbf{x} G(\mathbf{x}' \leftarrow \mathbf{x}; \tau) \frac{f(\mathbf{x}, 0)}{\varphi_T(\mathbf{x})} \tag{4.79}$$

$$\Leftrightarrow f(\mathbf{x}', \tau) = \int d\mathbf{x} \underbrace{\frac{\varphi_T(\mathbf{x}')}{\varphi_T(\mathbf{x})} G(\mathbf{x}' \leftarrow \mathbf{x}; \tau)}_{G_F(\mathbf{x}' \leftarrow \mathbf{x}; \tau)} f(\mathbf{x}, 0). \tag{4.80}$$

We define a new Green's function and write the evolution equation for $f(\mathbf{x}, \tau)$:

$$
\boxed{
\begin{array}{l}
\text{Importance sampling evolution} \\[4pt]
f(\mathbf{x}', \tau) = \int d\mathbf{x} G_F(\mathbf{x}' \leftarrow \mathbf{x}; \tau) f(\mathbf{x}, 0) \\[4pt]
f(\mathbf{x}, \tau): = \varphi_T(\mathbf{x}) P(\mathbf{x}, \tau) \\[4pt]
G_F(\mathbf{x}' \leftarrow \mathbf{x}; \tau): = \frac{\varphi_T(\mathbf{x}')}{\varphi_T(\mathbf{x})} G(\mathbf{x}' \leftarrow \mathbf{x}; \tau): = \frac{\varphi_T(\mathbf{x}')}{\varphi_T(\mathbf{x})} \langle \mathbf{x}' | e^{-\tau(\hat{\mathcal{H}} - E_T)} | \mathbf{x} \rangle.
\end{array}
}
\tag{4.81}
$$

While the original Green's function is symmetric, $G(\mathbf{x}' \leftarrow \mathbf{x}'; \tau) = G(\mathbf{x} \leftarrow \mathbf{x}; \tau)$, the importance sampling Green's function $G_F(\mathbf{x}' \leftarrow \mathbf{x}; \tau)$ is not. This is expected because G_F propagates walker coordinates and should guide diffusion so that more important walker coordinates are visited more frequently. A symmetrical Green's function describes reversible motion and cannot favor any configuration. Retrospectively, this is the reason why the symmetrical G had to leave all the hard work to the source term.

4.4.2 Importance sampling in the evolution differential equation

Traditionally, the importance sampling evolution equation is derived by inserting $P(\mathbf{x}, \tau) = \varphi_T(\mathbf{x})^{-1} f(\mathbf{x}, \tau)$ into the original differential equation (4.9),

$$-\frac{\partial}{\partial \tau} P(\mathbf{x}, \tau) = -D \sum_{i=1}^{N} \nabla_i^2 \ P(\mathbf{x}, \tau) + (V(\mathbf{x}) - E_T) P(\mathbf{x}, \tau) \tag{4.82}$$

$$\Leftrightarrow -\frac{\partial}{\partial \tau} f(\mathbf{x}, \tau) = -D \sum_{i=1}^{N} \varphi_T(\mathbf{x}) \nabla_i^2 \ [\varphi_T(\mathbf{x})^{-1} f(\mathbf{x}, \tau)] + (V(\mathbf{x}) - E_T) f(\mathbf{x}, \tau), \tag{4.83}$$

so at this stage, there is no apparent improvement to the potential term. However, the kinetic term is

$$-D \varphi_T(\mathbf{x}) \sum_{i=1}^{N} \nabla_i^2 \ [\varphi_T(\mathbf{x})^{-1} f(\mathbf{x}, \tau)] \tag{4.84}$$

$$= -D \varphi_T(\mathbf{x}) \sum_{i=1}^{N} \nabla_i \cdot [\varphi_T(\mathbf{x})^{-1} \nabla_i f(\mathbf{x}, \tau) - f(\mathbf{x}, \tau) \varphi_T(\mathbf{x})^{-2} \nabla_i \varphi_T(\mathbf{x})] \tag{4.85}$$

$$= -D \sum_{i=1}^{N} \Big[\nabla_i^2 f(\mathbf{x}, \tau) - 2 \varphi_T(\mathbf{x})^{-1} \nabla_i \varphi_T(\mathbf{x}) \cdot \nabla_i f(\mathbf{x}, \tau)$$
$$+ 2 \varphi_T(\mathbf{x})^{-2} (\nabla_i \varphi_T(\mathbf{x}))^2 f(\mathbf{x}, \tau) - 2 f(\mathbf{x}, \tau) \varphi_T(\mathbf{x})^{-1} \nabla_i^2 \ \varphi_T(\mathbf{x}) + f(\mathbf{x}, \tau) \varphi_T(\mathbf{x})^{-1} \nabla_i^2 \ \varphi_T(\mathbf{x}) \Big] \tag{4.86}$$

$$= -D \sum_{i=1}^{N} \left[\nabla_i^2 f(\mathbf{x}, \tau) - \nabla_i \cdot \left(\underbrace{2 \frac{\nabla_i \varphi_T(\mathbf{x})}{\varphi_T(\mathbf{x})}}_{\mathbf{F}_i(\mathbf{x})} f(\mathbf{x}, \tau) \right) \right] + T_L(\mathbf{x}) f(\mathbf{x}, \tau), \tag{4.87}$$

so the last term proportional to $f(\mathbf{x}, \tau)$ can be attached to the potential term. The resulting importance sampling evolution is

$$\boxed{\begin{aligned} -\frac{\partial}{\partial \tau} f(\mathbf{x}, \tau) = -D \sum_{i=1}^{N} \nabla_i^2 \ f(\mathbf{x}, \tau) + D \sum_{i=1}^{N} \nabla_i \cdot [\mathbf{F}_i(\mathbf{x}) f(\mathbf{x}, \tau)] \\ + (E_L(\mathbf{x}) - E_T) f(\mathbf{x}, \tau), \end{aligned}} \tag{4.88}$$

where the *drift* has components

$$\boxed{\mathbf{F}_i(\mathbf{x}): =2\frac{\boldsymbol{\nabla}_i\varphi_T(\mathbf{x})}{\varphi_T(\mathbf{x})} \qquad \text{drift.}} \tag{4.89}$$

For each particle i in a walker, the drift vector $\mathbf{F}_i(\mathbf{x})$ points to the direction of increasing $\varphi_T(\mathbf{x})$. The full drift vector is dN dimensional,

$$\mathbf{F}(\mathbf{x}): =(\mathbf{F}_1(\mathbf{x}), \mathbf{F}_2(\mathbf{x}), \ldots, \mathbf{F}_N(\mathbf{x})). \tag{4.90}$$

The importance sampling evolution can be written in the form

$$\boxed{-\frac{\partial}{\partial\tau}f(\mathbf{x}, \tau) = [\hat{\mathcal{T}} + \hat{\mathcal{D}}(\mathbf{x}) + E_L(\mathbf{x}) - E_T]f(\mathbf{x}, \tau): =(\hat{\mathcal{H}}_F(\mathbf{x}) - E_T)f(\mathbf{x}, \tau),} \tag{4.91}$$

where the drift operator is defined as

$$\boxed{\hat{\mathcal{D}}(\mathbf{x})f(\mathbf{x}, \tau): = \boldsymbol{\nabla} \cdot [F(\mathbf{x})f(\mathbf{x}, \tau)] \qquad \text{drift operator.}} \tag{4.92}$$

The formal solution to the importance sampling evolution problem is

$$f(\mathbf{x}, \tau) = e^{-\tau(\hat{\mathcal{H}}_F(\mathbf{x})-E_T)}f(\mathbf{x}, 0), \tag{4.93}$$

and we recover the importance sampling in equation (4.81). The importance sampling Green's function can be given in two equivalent forms:

$$\boxed{\begin{array}{l} \text{Importance sampling Green's function} \\[4pt] G_F(\mathbf{x}' \leftarrow \mathbf{x}; \tau): =\dfrac{\varphi_T(\mathbf{x}')}{\varphi_T(\mathbf{x})}\langle\mathbf{x}'|e^{-\tau(\hat{\mathcal{H}}-E_T)}|\mathbf{x}\rangle \\[12pt] G_F(\mathbf{x}' \leftarrow \mathbf{x}; \tau): =\langle\mathbf{x}'|e^{-\tau(\hat{\mathcal{H}}_F(\mathbf{x})-E_T)}|\mathbf{x}\rangle = \langle\mathbf{x}'|e^{-\tau(\hat{\mathcal{T}}+\hat{\mathcal{D}}(\mathbf{x})+E_L(\mathbf{x})-E_T)}|\mathbf{x}\rangle. \end{array}} \tag{4.94}$$

Spectral decomposition and why it cannot be done for $\hat{\mathcal{H}}_F$

The standard way to express operators is via spectral decomposition. If the operator $\hat{A}$ satisfies the eigenvalue equation,

$$\hat{A}|\Psi_i\rangle = a_i|\Psi_i\rangle, \tag{4.95}$$

then one can expand

$$\hat{A} = \sum_i a_i|\Psi_i\rangle\langle\Psi_i| \qquad \text{spectral decomposition,} \tag{4.96}$$

and the matrix elements of the exponential operator are

$$\langle\mathbf{x}'|e^{-\tau\hat{A}}|\mathbf{x}\rangle = \sum_i e^{-\tau a_i}\Psi_i(\mathbf{x})\Psi_i^*(\mathbf{x}') \qquad \text{Green's function,} \tag{4.97}$$

where the complex conjugation can be in either function. The significance of the spectral decomposition is that if one can solve the eigenvalues and eigenvectors, one can compute $\langle \mathbf{x}' | e^{-\tau \hat{A}} | \mathbf{x} \rangle$ *at any* τ. However, there is a problem, apart from the trouble of solving the eigenproblem. One can immediately see from equation (4.97) that if a spectral decomposition exists, the Green's function is *symmetric in exchange* $\mathbf{x} \leftrightarrow \mathbf{x}'$,

$$\langle \mathbf{x}' | e^{-\tau \hat{A}} | \mathbf{x} \rangle = \langle \mathbf{x} | e^{-\tau \hat{A}} | \mathbf{x}' \rangle. \tag{4.98}$$

This is not the case for the operator $\hat{\mathcal{H}}_F(\mathbf{x})$, because $G_F(\mathbf{x}', \mathbf{x}; \tau)$ is not symmetric. Mathematically, *the operator* $\hat{\mathcal{H}}_F(\mathbf{x})$ *is a non-self-adjoint operator in Hilbert space.* In general, a non-self-adjoint operator may not have a spectral decomposition.[3]

As a consequence, solving the evolution equation at any imaginary time all at once is not possible. Still, the evolution is governed by a first-order differential equation, which can be integrated from $\tau = 0$ to any imaginary time using *short imaginary-time steps*. In principle, the integration method adopted can be any of the standard numerical methods, such as Euler's (Heun's) method, Runge–Kutta variants, the predictor–corrector method, etc. The order of the integration method dictates how accurate the next point is, assuming the previous point is accurate. The greater the order (accuracy) of the integration method, the more intermediate points are used between the beginning and the end of each step, so higher accuracy comes with a higher computational cost. I will derive a first-order solution to the point where I have a working DMC algorithm and come back to discuss higher-order algorithms, which are also needed in PIMC.

4.4.2.1 Diffusion and drift

Let us first concentrate on diffusion and drift:

$$-\frac{\partial}{\partial \tau} f(\mathbf{x}, \tau) = -D \sum_{i=1}^{N} \nabla_i^2 \, f(\mathbf{x}, \tau) + D \sum_{i=1}^{N} \nabla_i \cdot [\mathbf{F}_i(\mathbf{x}) f(\mathbf{x}, \tau)]. \tag{4.99}$$

Depending on the context, this type of equation is called a **drift–diffusion equation**, a **Fokker–Planck equation**, a **Smoluchowski equation**, or a **convection–diffusion equation**. In QMC, we represent the distribution with M walkers,

$$f(\mathbf{x}, \tau) = \int d\mathbf{x}' f(\mathbf{x}', \tau) \delta(\mathbf{x}' - \mathbf{x}) \approx \frac{1}{M} \sum_{j=1}^{M} \delta(\mathbf{x}_j(\tau) - \mathbf{x}), \tag{4.100}$$

where $\mathbf{x}_j(\tau)$ are sampled from $f(\mathbf{x}', \tau)$. We already solved the diffusion part in section 4.3.2 and found that walkers move about with Gaussian-distributed random steps,

$$\boxed{\mathbf{x}_j(\tau) = \mathbf{x}_j(0) + \sqrt{2D\tau}\,\boldsymbol{\eta}_j \qquad \text{diffusion part.}} \tag{4.101}$$

---

[3] A non-self-adjoint operator may not have real eigenvalues. Also, the set of eigenvectors may be incomplete, and the eigenvectors of distinct eigenvalues are not necessarily orthogonal.

The drift part is a continuity equation for the distribution or density, $f(\mathbf{x}, \tau)$,

$$\underbrace{\frac{\partial}{\partial \tau} f(\mathbf{x}, \tau)}_{\text{density}} + \sum_{i=1}^{N} \nabla_i \cdot \underbrace{[D\mathbf{F}_i(\mathbf{x}) f(\mathbf{x}, \tau)]}_{\text{current}} = 0, \qquad (4.102)$$

so we must be able to write the solution as a flow of a conserved number of walker coordinates $\mathbf{x}(\tau)$.[4] From the factor identified as current, we see that the drift $D\mathbf{F}(\mathbf{x})$ acts as a *velocity field*. Moreover, the flow is deterministic without any randomness in the drift trajectories $\mathbf{x}(\tau)$, which are solutions of the equation

$$\boxed{\frac{d}{d\tau}\mathbf{x}_j(\tau) = D\mathbf{F}(\mathbf{x}_j(\tau)) \qquad \text{drift part.}} \qquad (4.103)$$

The proof of equation (4.103) was given by Vrbik and Rothstein (see [8], appendix A). I will repeat it here using explicit walkers. We insert the walker representation of $f(\mathbf{x}, \tau)$ into the drift–diffusion equation; the factors $1/M$ cancel, and one solution can be found by equating each term in the j summation, term by term. The distribution is unaffected upon relabeling the walkers, so we are content in finding one possible solution. (See also the discussion at end of section 4.3.2.) The resulting drift equation for each walker j is

$$-\frac{d}{d\tau}\delta(\mathbf{x}_j(\tau) - \mathbf{x}) = D\sum_{i=1}^{N} \nabla_i \cdot [\mathbf{F}_i(\mathbf{x})\delta(\mathbf{x}_j(\tau) - \mathbf{x})]. \qquad (4.104)$$

In order to resolve vector components, multiply by the kth coordinate $\mathbf{x}_k$ and integrate over all $\mathbf{x}$:

$$-\frac{d}{d\tau}\int d\mathbf{x}\,\mathbf{x}_k \delta(\mathbf{x}_j(\tau) - \mathbf{x}) = D\sum_{i=1}^{N} \int d\mathbf{x}\,\mathbf{x}_k \nabla_i \cdot [\mathbf{F}_i(\mathbf{x})\delta(\mathbf{x}_j(\tau) - \mathbf{x})] \qquad (4.105)$$

$$\Leftrightarrow -\frac{d}{d\tau}(\mathbf{x}_j(\tau))_k = -D\sum_{i=1}^{N} \int d\mathbf{x}\,\underbrace{(\nabla_i \cdot \mathbf{x}_k)}_{\delta_{ik}}\mathbf{F}_i(\mathbf{x})\delta(\mathbf{x}_j(\tau) - \mathbf{x}) \qquad (4.106)$$

$$\Leftrightarrow \frac{d}{d\tau}(\mathbf{x}_j(\tau))_k = D\int d\mathbf{x}\,\mathbf{F}_k(\mathbf{x})\delta(\mathbf{x}_j(\tau) - \mathbf{x}) \qquad (4.107)$$

$$\Leftrightarrow \frac{d}{d\tau}(\mathbf{x}_j(\tau))_k = D\mathbf{F}_k(\mathbf{x}_j(\tau)), \qquad (4.108)$$

which is equation (4.103) in component form, written for the kth coordinate in walker j. The partial integration on the second line assumes that $\mathbf{F}_i(\mathbf{x})$ goes to zero

---

[4] This is a big asset; in fact, it is one of the reasons why importance sampling is used in DMC. As we shall later see, modified diffusion often fails to give a conservation equation and introduces sinks and sources of walker coordinates.

for large $|\mathbf{x}|$, so that surface terms vanish (this is usually the case). The trial wave function is chosen so that walker coordinates in the far edges of space are not subject to drift.

4.4.3 Combined drift, diffusion, and source

Drift and diffusion do not commute, because you get a different result depending on whether particles first diffuse and then drift or the other way around. As shown by the Baker–Campbell–Hausdorff formula (4.57), drift and diffusion do commute to linear order in τ,

$$\langle \mathbf{x}'|e^{-\tau(\hat{\mathcal{T}}+\hat{\mathcal{D}}(\mathbf{x}))}|\mathbf{x}\rangle = \langle \mathbf{x}'|e^{-\tau\hat{\mathcal{T}}}e^{-\tau\hat{\mathcal{D}}(\mathbf{x})}|\mathbf{x}\rangle + \mathcal{O}(\tau^2), \qquad (4.109)$$

where the error $\mathcal{O}(\tau^2)$ is due to $[\hat{\mathcal{T}}, \hat{\mathcal{D}}(\mathbf{x})] \neq 0$. The same applies to any two operations in the list: drift, diffusion, and source, so the evolution can be approximated by

$$f(\mathbf{x}', \tau) = \int d\mathbf{x}\langle \mathbf{x}'|e^{-\tau(\hat{\mathcal{T}}+\hat{\mathcal{D}}(\mathbf{x})+(E_L(\mathbf{x})-E_T))}|\mathbf{x}\rangle f(\mathbf{x}, 0) \qquad (4.110)$$

$$\approx \int d\mathbf{x}\langle \mathbf{x}'|e^{-\tau\hat{\mathcal{T}}}e^{-\tau\hat{\mathcal{D}}(\mathbf{x})}e^{-\tau(E_L(\mathbf{x})-E_T)}|\mathbf{x}\rangle f(\mathbf{x}, 0). \qquad (4.111)$$

We can now apply the property of a complete basis,

$$\int d\mathbf{x}|\mathbf{x}\rangle\langle \mathbf{x}| = \mathbb{1}\,(\text{unit operator}) \qquad \text{completeness relation.} \qquad (4.112)$$

Every imaginable path can be split into a path from the starting point to any point and another from there to the end point. The list 'any point' is the integral in the completeness relation, and because it can be inserted anywhere in a path without any effect, it acts as a unit operator.

The completeness relation is the workhorse in quantum mechanical changes of basis. For example, the change from basis $\{|i\rangle\}$ to basis $\{|K\rangle\}$ is represented by

$$|i\rangle \equiv \mathbb{1}|i\rangle = \left(\sum_K |K\rangle\langle K|\right)|i\rangle = \sum_K |K\rangle\langle K|i\rangle = \sum_K \langle K|i\rangle|K\rangle, \qquad (4.113)$$

where the factors $\langle K|i\rangle$ are complex numbers. In a continuous space, sums are replaced with integrals.

We now insert a unit operator between the operators,

$$f(\mathbf{x}', \tau) \approx \int d\mathbf{x}d\mathbf{y}d\mathbf{z}\langle \mathbf{x}'|e^{-\tau\hat{\mathcal{T}}}|\mathbf{y}\rangle\langle \mathbf{y}|e^{-\tau\hat{\mathcal{D}}(\mathbf{z})}|\mathbf{z}\rangle\langle \mathbf{z}|e^{-\tau(E_L(\mathbf{x})-E_T)}|\mathbf{x}\rangle f(\mathbf{x}, 0). \quad (4.114)$$

The local energy is a function,

$$\langle \mathbf{z}|e^{-\tau(E_L(\mathbf{x})-E_T)}|\mathbf{x}\rangle = e^{-\tau(E_L(\mathbf{x})-E_T)}\langle \mathbf{z}|\mathbf{x}\rangle = e^{-\tau(E_L(\mathbf{x})-E_T)}\delta(\mathbf{z}-\mathbf{x}), \qquad (4.115)$$

so the **z**-integration is trivial. The evolution is divided into three stages,

$$f(\mathbf{x}', \tau) \approx \underbrace{e^{-\tau(E_L(\mathbf{x})-E_T)}}_{\text{source}} \int d\mathbf{x}d\mathbf{y} \underbrace{\langle \mathbf{x}'|e^{-\tau\hat{T}}|\mathbf{y}\rangle}_{\text{diffusion}} \underbrace{\langle \mathbf{y}|e^{-\tau\hat{D}(\mathbf{z})}|\mathbf{x}\rangle}_{\text{drift}} f(\mathbf{x}, 0) \tag{4.116}$$

in which the matrix elements must be evaluated to first-order accuracy to ensure that the whole expression remains first order. Diffusion,

$$\langle \mathbf{x}'|e^{-\tau\hat{T}}|\mathbf{y}\rangle, \tag{4.117}$$

moves walkers from **y** to **x**', along a random walk trajectory (see equation (4.101))

$$\mathbf{x}' = \mathbf{y} + \sqrt{2D\tau}\,\boldsymbol{\eta}. \tag{4.118}$$

Drift,

$$\langle \mathbf{y}|e^{-\tau\hat{D}(\mathbf{x})}|\mathbf{x}\rangle, \tag{4.119}$$

moves walkers from **x** to **y** along deterministic trajectories, where the velocity is given by equation (4.103), which depends on the drift evaluated at walker positions at τ. The solution trajectory can be found by integrating:

$$\frac{d}{d\tau}\mathbf{x}(\tau) = D\mathbf{F}(\mathbf{x}(\tau)) \Rightarrow \mathbf{x}(\tau) = \mathbf{x}(0) + D\int_0^\tau d\tau'\mathbf{F}(\mathbf{x}(\tau')). \tag{4.120}$$

Numerically, the trajectory can be solved using any differential equation solver. To the first order in τ, we can use Euler's method (a derivative written as a finite difference),

$$\mathbf{x}(\tau) = \mathbf{x}(0) + D\tau\mathbf{F}(\mathbf{x}(0)) + \mathcal{O}(\tau^2), \tag{4.121}$$

which assumes that the drift is approximately constant during the short, τ-long interval. Now the drift is from **x** to **y**:

$$\mathbf{y} = \mathbf{x} + D\tau\mathbf{F}(\mathbf{x}) \qquad \text{first-order drift.} \tag{4.122}$$

The integrals in equation (4.116) are sampled by trajectories, given by the two-stage algorithm

$$\begin{aligned} \mathbf{y} &= \mathbf{x} + D\tau\mathbf{F}(\mathbf{x}) \\ \mathbf{x}' &= \mathbf{y} + \sqrt{2D\tau}\,\boldsymbol{\eta}, \end{aligned} \tag{4.123}$$

which can be combined with a walker coordinate update (applied to each walker),

$$\mathbf{x}' = \mathbf{x} + \sqrt{2D\tau}\,\boldsymbol{\eta} + D\tau\mathbf{F}(\mathbf{x}). \tag{4.124}$$

The downside of this update is that the error in the walker positions is proportional to τ^2, so one needs to take short steps in τ. What makes this update appealing is the fact that only one drift evaluation and no gradients of the drift are required. The first-order drift–diffusion update is often the method of choice.

Finally, the source part in equation (4.116) multiplies the distribution by the factor $e^{-\tau(E_L(\mathbf{x})-E_T)}$. Unlike drift and diffusion, this operation does not preserve the norm of the distribution. Here, we interpret this in terms of branching, where walkers sometimes spawn new walkers or die. We saw this earlier in crude DMC, but now, importance sampling can dramatically reduce branching.

To the first order in τ, the updates are:

> First-order drift–diffusion branching update
>
> (1) $\mathbf{x}' = \mathbf{x} + \sqrt{2D\tau}\,\eta + D\tau\mathbf{F}(\mathbf{x})$
>
> (2) Branch walkers according to factor $e^{-\tau(E_L(\mathbf{x})-E_T)}$. $\qquad$ (4.125)

Before writing down a DMC algorithm, a correction should be put in place to mitigate the adverse effects of using an *approximate* Green's function and the corresponding coordinate update.

4.4.4 Correcting sampling using detailed balance

The evolution sequence of a distribution $f(\mathbf{x}, \tau)$,

$$f(\mathbf{x}, 0) \to f(\mathbf{x}, \tau_1) \to f(\mathbf{x}, \tau_2) \to \cdots \to f(\mathbf{x}, \infty): = f(\mathbf{x}), \qquad (4.126)$$

is a Markov chain because the distribution at each stage depends only on the previous one. As indicated, the importance sampling evolution should converge to a **limiting distribution** $f(\mathbf{x})$. The limiting distribution is also called an asymptotic distribution or an invariant distribution. The very existence of a limiting distribution depends on the condition that once the limiting distribution is reached, one should stay there. If the probability of moving from $\mathbf{x}$ to $\mathbf{x}'$ is $\mathcal{P}(\mathbf{x}' \leftarrow \mathbf{x})$, then the limiting distribution $f(\mathbf{x})$ must satisfy the relation

$$\boxed{\int d\mathbf{x} f(\mathbf{x})\mathcal{P}(\mathbf{x}' \leftarrow \mathbf{x}) = f(\mathbf{x}') = \int d\mathbf{x} f(\mathbf{x}')\mathcal{P}(\mathbf{x} \leftarrow \mathbf{x}').} \qquad (4.127)$$

The left-hand side represents the flux to a point $\mathbf{x}'$ from any point $\mathbf{x}$ sampled from $f(\mathbf{x})$, and it should be the distribution at $\mathbf{x}'$, $f(\mathbf{x}')$. The last integral follows from the fact that the probability of moving to any point from $\mathbf{x}'$ must be unity:

$$\int d\mathbf{x}\mathcal{P}(\mathbf{x} \leftarrow \mathbf{x}') = 1. \qquad (4.128)$$

The condition in equation (4.127) states that the 'dust has settled,' and there is no net flux between points $\mathbf{x}$ and $\mathbf{x}'$. In practice, one does not have to evaluate the integrals in equation (4.127) because the condition is satisfied if the integrands are equal at every $\mathbf{x}$:

$$\boxed{f(\mathbf{x})\mathcal{P}(\mathbf{x}' \leftarrow \mathbf{x}) = f(\mathbf{x}')\mathcal{P}(\mathbf{x} \leftarrow \mathbf{x}') \qquad \text{detailed balance condition.}} \qquad (4.129)$$

The detailed balance condition, also called microscopic reversibility, is a sufficient condition to assure that $f(\mathbf{x})$ is the limiting distribution of the update process $\mathcal{P}$.

The probability of moving, $P(\mathbf{x}' \leftarrow \mathbf{x})$, is made of two decisions (T for 'try' or 'transfer,' and A for 'accept'),[5]

$$\underbrace{P(\mathbf{x}' \leftarrow \mathbf{x})}_{\text{probability of moving}} = \underbrace{T(\mathbf{x}' \leftarrow \mathbf{x})}_{\text{try move}} \underbrace{A(\mathbf{x}' \leftarrow \mathbf{x})}_{\text{acceptance of move}} . \tag{4.130}$$

With explicit try–accept parts, the detailed balance condition gives the ratio

$$\frac{A(\mathbf{x} \leftarrow \mathbf{x}')}{A(\mathbf{x}' \leftarrow \mathbf{x})} = \frac{f(\mathbf{x})T(\mathbf{x}' \leftarrow \mathbf{x})}{f(\mathbf{x}')T(\mathbf{x} \leftarrow \mathbf{x}')} \qquad \text{detailed balance, general tries.} \tag{4.131}$$

If moves are tried asymmetrically, we have to correct the acceptance accordingly. More frequently suggested moves are rejected more often than less frequently suggested ones. One possible solution that satisfies the detailed balance condition is [9]

$$\boxed{\begin{aligned} &A(\mathbf{x}' \leftarrow \mathbf{x}) = \min\left[1, \frac{f(\mathbf{x}')T(\mathbf{x} \leftarrow \mathbf{x}')}{f(\mathbf{x})T(\mathbf{x}' \leftarrow \mathbf{x})}\right] \qquad \text{Metropolis–Hastings algorithm} \\ &\Rightarrow \text{The limiting distribution is} f(\mathbf{x}). \end{aligned}} \tag{4.132}$$

If moves are suggested symmetrically, $T(\mathbf{x}' \leftarrow \mathbf{x}') = T(\mathbf{x} \leftarrow \mathbf{x}')$, they cancel out from equation (4.131), leaving

$$\frac{A(\mathbf{x} \leftarrow \mathbf{x}')}{A(\mathbf{x}' \leftarrow \mathbf{x})} = \frac{f(\mathbf{x})}{f(\mathbf{x}')} \qquad \text{detailed balance, symmetric tries.} \tag{4.133}$$

In VMC, the limiting distribution should be $f(\mathbf{x}) = |\varphi_T(\mathbf{x})|^2$, and moves are symmetric random displacements. The Metropolis algorithm,

$$A(\mathbf{x}' \leftarrow \mathbf{x}) = \min\left[1, \frac{|\varphi_T(\mathbf{x}')|^2}{|\varphi_T(\mathbf{x})|^2}\right], \tag{4.134}$$

is one choice that satisfies the symmetric-tries detailed balance condition because

$$\frac{A(\mathbf{x} \leftarrow \mathbf{x}')}{A(\mathbf{x}' \leftarrow \mathbf{x})} = \frac{\min\left[1, \frac{|\varphi_T(\mathbf{x})|^2}{|\varphi_T(\mathbf{x}')|^2}\right]}{\min\left[1, \frac{|\varphi_T(\mathbf{x}')|^2}{|\varphi_T(\mathbf{x})|^2}\right]} = \frac{|\varphi_T(\mathbf{x})|^2}{|\varphi_T(\mathbf{x}')|^2} \qquad \text{detailed balance in Metropolis VMC.} \tag{4.135}$$

In importance sampling DMC, moves are tried according to the Green's function, $T(\mathbf{x}' \leftarrow \mathbf{x})$ is $G_F(\mathbf{x}' \leftarrow \mathbf{x}; \tau)$. The limiting distribution should be $f(\mathbf{x}) = \varphi_T(\mathbf{x})P(\mathbf{x})$, where P is the actual distribution we are solving. An unknown distribution cannot be used in the detailed balance condition, so the next best thing is to make sure we sample the exact VMC case where $f(\mathbf{x}) = |\varphi_T(\mathbf{x})|^2$. This gives the acceptance [5, 10]

---

[5] The probability of winning a lottery is not the probability of picking the winning ticket out of many. If you do not buy a lottery ticket, your chance of winning is exactly zero, because $T(\text{ticket} \leftarrow \text{no ticket}) = 0$.

$$A(\mathbf{x}' \leftarrow \mathbf{x}) = \min\left[1, \frac{|\varphi_T(\mathbf{x}')|^2 G_F(\mathbf{x} \leftarrow \mathbf{x}'; \tau)}{|\varphi_T(\mathbf{x})|^2 G_F(\mathbf{x}' \leftarrow \mathbf{x}; \tau)}\right]. \tag{4.136}$$

One motivation for choosing this acceptance is that the *exact* importance sampling Green's function satisfies the relation (see equation (4.81))

$$\frac{\varphi_T(\mathbf{x})}{\varphi_T(\mathbf{x}')} G_F(\mathbf{x}' \leftarrow \mathbf{x}; \tau) = G(\mathbf{x}' \leftarrow \mathbf{x}; \tau), \tag{4.137}$$

where the (also exact) G is symmetric in exchange $\mathbf{x}' \leftrightarrow \mathbf{x}$, so

$$\frac{\varphi_T(\mathbf{x})}{\varphi_T(\mathbf{x}')} G_F(\mathbf{x}' \leftarrow \mathbf{x}; \tau) = \frac{\varphi_T(\mathbf{x}')}{\varphi_T(\mathbf{x})} G_F(\mathbf{x} \leftarrow \mathbf{x}'; \tau) \tag{4.138}$$

$$\Leftrightarrow \frac{\varphi_T(\mathbf{x}')^2 G_F(\mathbf{x} \leftarrow \mathbf{x}'; \tau)}{\varphi_T(\mathbf{x})^2 G_F(\mathbf{x}' \leftarrow \mathbf{x}; \tau)} = 1, \tag{4.139}$$

and the acceptance defined in equation (4.136) is $A(\mathbf{x}' \leftarrow \mathbf{x}) = 1$. In practice, we always use an approximate G_F, so the deviation from 100% acceptance is a measure of how inaccurate the approximation is. Using an accept–reject step is known to greatly reduce the finite-τ bias caused by an approximate Green's function, especially the first-order drift–diffusion branching update in equation (4.125). The DMC acceptance should be more than 99% to ensure that the approximate Green's function is not getting too far from the exact one. As a side effect, rejecting some moves slows down the evolution in imaginary time, and the effective time step is slightly shorter than τ.

If the trial wave function is exactly the distribution $P(\mathbf{x})$, for example, the ground-state wave function, then a good algorithm should not move away from it. Another motivation for using the acceptance (4.136) is that the exact case $f(\mathbf{x}) = |\varphi_T(\mathbf{x})|^2$ is a limiting distribution even with an approximate Green's function. If this is the case, then one should have

$$\int d\mathbf{x} |\varphi_T(\mathbf{x})|^2 \mathcal{P}(\mathbf{x}' \leftarrow \mathbf{x}) = |\varphi_T(\mathbf{x}')|^2, \tag{4.140}$$

independent of τ.

Proof: Insert the finite-τ move probability $A(\mathbf{x}' \leftarrow \mathbf{x}) G_F(\mathbf{x}' \leftarrow \mathbf{x}; \tau)$, and use the detailed balance condition equation (4.131) (which follows from acceptance (4.136)),

$$\int d\mathbf{x} |\varphi_T(\mathbf{x})|^2 A(\mathbf{x}' \leftarrow \mathbf{x}) G_F(\mathbf{x}' \leftarrow \mathbf{x}; \tau) \tag{4.141}$$

$$= \int d\mathbf{x} |\varphi_T(\mathbf{x})|^2 \left(A(\mathbf{x} \leftarrow \mathbf{x}') \frac{|\varphi_T(\mathbf{x}')|^2 G_F(\mathbf{x} \leftarrow \mathbf{x}'; \tau)}{|\varphi_T(\mathbf{x})|^2 G_F(\mathbf{x}' \leftarrow \mathbf{x}; \tau)} \right) G_F(\mathbf{x}' \leftarrow \mathbf{x}; \tau) \tag{4.142}$$

$$=|\varphi_T(\mathbf{x}')|^2 \underbrace{\int d\mathbf{x} A(\mathbf{x} \leftarrow \mathbf{x}') G_F(\mathbf{x} \leftarrow \mathbf{x}'; \tau)}_{=1}, \tag{4.143}$$

where the integral is the probability of arriving at any $\mathbf{x}$ from a given $\mathbf{x}'$. The result is independent of τ and valid for an *approximate* G_F.

4.4.5 A first-order DMC algorithm

A first-order DMC algorithm
Drift, diffusion, branching, and an accept–reject step.
Before running, set M_{target}, τ, κ, and E_T.
Choose $\varphi_T(\mathbf{x})$ and, for efficiency, analytically compute the local energy

$$E_L(\mathbf{x}) = \frac{\hat{\mathcal{H}}\varphi_T(\mathbf{x})}{\varphi_T(\mathbf{x})}, \tag{4.144}$$

and the drift $\mathbf{F}(\mathbf{x})$, vector components

$$\mathbf{F}_i(\mathbf{x}) = \frac{2\,\nabla_i \varphi_T(\mathbf{x})}{\varphi_T(\mathbf{x})}. \tag{4.145}$$

1. Generate M_{target} (a large number, $\sim$1000 or more) walkers. During a VMC run, pick coordinates now and then and store them in the walker list.
2. Drift–diffusion: move walkers according to equation (4.124):

$$\mathbf{x}' = \mathbf{x} + \sqrt{2D\tau}\,\eta + D\tau\,\mathbf{F}(\mathbf{x}). \tag{4.146}$$

3. Accept–reject: (Metropolis–Hastings, equation (4.136)). Compute

$$ratio = \frac{|\varphi_T(\mathbf{x}')|^2 G_F(\mathbf{x} \leftarrow \mathbf{x}'; \tau)}{|\varphi_T(\mathbf{x})|^2 G_F(\mathbf{x}' \leftarrow \mathbf{x}; \tau)}. \tag{4.147}$$

If $ratio > 1$, accept the move, else pick $r \in U[0, 1]$, and if $ratio > r$, accept the move, else reject it. Collect acceptance statistics, use smaller τ if acceptance $<$99%.
4. Branching: for each walker, create M^{copies} copies, where

$$M^{\text{copies}} = \text{int}\left(e^{-\tau(\frac{1}{2}(E_L(\mathbf{x})+E_L(\mathbf{x}'))-E_T)} + \xi\right), \ \xi \in U[0, 1]. \tag{4.148}$$

5. Now and then, adjust the trial energy according to

$$E_T = E_0^{\text{guess}} + \kappa \ln\left(\frac{M_{\text{target}}}{M}\right), \tag{4.149}$$

6. (Thermalization: Repeat steps 2–5 until excitations have died out.) Measure quantities of interest.
7. Repeat until statistical errors are small enough.

Details:

- Step 3: the exact Green's function would have 100% acceptance. The approximate Green's function in this algorithm is (d dimensions, N particles)

$$G_F(\mathbf{x}' \leftarrow \mathbf{x}; \tau) = \frac{1}{(4\pi D\tau)^{dN/2}} e^{-\frac{(\mathbf{x}'-\mathbf{x}-D\tau\mathbf{F}(\mathbf{x}))^2}{4D\tau}} e^{-\tau(\frac{1}{2}(E_L(\mathbf{x})+E_L(\mathbf{x}'))-E_T)}. \qquad (4.150)$$

In *ratio*, all $\mathbf{x}' \leftrightarrow \mathbf{x}$ -symmetric factors cancel (a good reason to symmetrize branching in step 4):

$$ratio = \frac{|\varphi_T(\mathbf{x}')|^2}{|\varphi_T(\mathbf{x})|^2} \frac{e^{-\frac{(\mathbf{x}-\mathbf{x}'-D\tau\mathbf{F}(\mathbf{x}'))^2}{4D\tau}}}{e^{-\frac{(\mathbf{x}'-\mathbf{x}-D\tau\mathbf{F}(\mathbf{x}))^2}{4D\tau}}}. \qquad (4.151)$$

In the sample programs, I often use ln(*ratio*) to avoid underflow.

- Step 4: branching simulates the source term factor, which is symmetrized to take the average local energy of the old coordinates $\mathbf{x}$ and the new coordinates $\mathbf{x}'$ before and after the drift–diffusion update,

$$e^{-\tau(E_L(\mathbf{x})-E_T)} \xrightarrow{\text{symmetrize}} e^{-\tau(\frac{1}{2}(E_L(\mathbf{x})+E_L(\mathbf{x}'))-E_T)}. \qquad (4.152)$$

- Step 5: population control. Branching changes the number of walkers, and the walker population fluctuations reflect the fluctuation in the local energy. Too many walkers slow down the computation and may exceed memory limits, while too few walkers cannot represent the distribution. Step 5 controls the walker population. Reducing E_T reduces walker branching in step 4, and vice versa. If $M > M_{\text{target}}$, the logarithm term is negative and E_T is slightly lower than the current best-guess energy E_0^{guess}. Similarly, if $M < M_{\text{target}}$, E_T is slightly higher than E_0^{guess}. The parameter κ should be large enough to let E_T respond to walker population changes: if the population ever drops too low, the statistical significance of the remaining walkers is overemphasized. A overlarge κ forces the population into excessively tight limits and causes unwanted feedback for walkers. The actual value of κ is found using trial and error.

- Step 6: thermalization. DMC needs to run for a while to reach the limiting distribution. The initial walkers represent a distribution that has plenty of excited, high-energy components, and, depending on the time step τ, their amplitudes must be sufficiently reduced before starting measurements. Data collected before DMC has reached the limiting distribution $f(\mathbf{x})$ is biased by the τ-dependent components $f(\mathbf{x}, \tau)$. If an excitation is ΔE above the target-state energy, the excitation amplitude dies out proportionally to

$$e^{-\tau\Delta E}. \qquad (4.153)$$

The high-energy components vanish faster in τ than those close to the target-state energy. Therefore, the number of thermalization steps needed depends on the system and on the time step τ.

4.4.6 Measurements in importance sampling DMC

Importance sampling makes DMC an efficient algorithm, but it comes with a price. Apart from the perfect case $\varphi_T(\mathbf{x}) = \phi_0(\mathbf{x})$,

> In or near the asymptotic limit $\tau \to \infty$, the walker coordinates in importance sampling DMC represent the limiting distribution $f(\mathbf{x}) = \varphi_T(\mathbf{x})P(\mathbf{x})$, so they do not correspond to actual particle coordinates.

Quantities that depend on particle coordinates cannot be evaluated directly as averages over walkers and DMC steps. One such quantity is the expectation value of the potential energy. This should not come as a surprise, considering that *without* importance sampling, the average potential energy of walkers is the total energy.

Mixed estimates

Since importance sampling DMC samples the product $\varphi_T \phi_0$, it is useful to define a *mixed estimate* for an observable $\hat{\mathcal{A}}$,

$$\text{Mixed estimate } \frac{\langle \varphi_T | \hat{\mathcal{A}} | \phi_0 \rangle}{\langle \varphi_T | \phi_0 \rangle}. \tag{4.154}$$

Mixed estimates are directly measured from importance sampling DMC.

Energy and the time step error

The expectation value of the local energy approaches E_0:

$$\frac{\int d\mathbf{x} f(\mathbf{x}, \tau) E_L(\mathbf{x})}{\int d\mathbf{x} f(\mathbf{x}, \tau)} \xrightarrow{\tau \to \infty} \frac{\int d\mathbf{x} \varphi_T(\mathbf{x})\phi_0(\mathbf{x})E_L(\mathbf{x})}{\int d\mathbf{x} \varphi_T(\mathbf{x})\phi_0(\mathbf{x})} = \frac{\int d\mathbf{x} \phi_0(\mathbf{x})\hat{\mathcal{H}}\varphi_T(\mathbf{x})}{\int d\mathbf{x} \varphi_T(\mathbf{x})\phi_0(\mathbf{x})} \tag{4.155}$$

$$= \frac{\int d\mathbf{x}(\hat{\mathcal{H}}\phi_0(\mathbf{x}))\varphi_T(\mathbf{x})}{\int d\mathbf{x} \varphi_T(\mathbf{x})\phi_0(\mathbf{x})} = E_0, \tag{4.156}$$

where $\int d\mathbf{x}\phi_0(\mathbf{x})\hat{\mathcal{H}}\varphi_T(\mathbf{x}) = \int d\mathbf{x}(\hat{\mathcal{H}}\phi_0(\mathbf{x}))\varphi_T(\mathbf{x})$ for a Hermitian $\hat{\mathcal{H}}$. We have shown that

$$\langle E_L \rangle = \langle \hat{\mathcal{H}} \rangle = \frac{\langle \varphi_T | \hat{\mathcal{H}} | \phi_0 \rangle}{\langle \varphi_T | \phi_0 \rangle} = E_0. \tag{4.157}$$

The finite time step τ causes a *time step error* because the evolution of $f(\mathbf{x}, \tau)$ is inaccurate. The first-order DMC algorithm approximates the evolution operator as [11]

$$e^{-\tau\hat{\mathcal{H}}} = e^{-\tau(\hat{\mathcal{T}}+\hat{\mathcal{V}})} \approx e^{-\tau\hat{\mathcal{T}}}e^{-\tau\hat{\mathcal{V}}} =: e^{-\tau\hat{\mathcal{H}}'}, \qquad (4.158)$$

where

$$\hat{\mathcal{H}}' = \hat{\mathcal{H}} - \frac{1}{2}\tau[\hat{\mathcal{H}}, \hat{\mathcal{V}}] + \frac{1}{12}\tau^2[\hat{\mathcal{H}} - 2\hat{\mathcal{V}}, [\hat{\mathcal{H}}, \hat{\mathcal{V}}]] + \mathcal{O}(\tau^3). \qquad (4.159)$$

Let us use the Baker–Campbell–Hausdorff formula given in equation (4.57), repeated here:

$$e^{\hat{A}}e^{\hat{B}} = e^{\hat{A}+\hat{B}+\frac{1}{2}[\hat{A}, \hat{B}]+\frac{1}{12}[\hat{A}-\hat{B}, [\hat{A}, \hat{B}]]+\cdots}, \qquad \text{Baker–Campbell–Hausdorff.} \quad (4.160)$$

In our case,

$$e^{-\tau\hat{\mathcal{T}}}e^{-\tau\hat{\mathcal{V}}} = e^{-\tau(\hat{\mathcal{H}}-\frac{1}{2}\tau[\hat{\mathcal{T}}, \hat{\mathcal{V}}])+\frac{1}{12}\tau^2[\hat{\mathcal{T}}-\hat{\mathcal{V}}, [\hat{\mathcal{T}}, \hat{\mathcal{V}}]]+\mathcal{O}(\tau^3)} =: e^{-\tau\hat{\mathcal{H}}}, \qquad (4.161)$$

from which you can read the operator $\hat{\mathcal{H}}'$. Finally, insert $\hat{\mathcal{T}} = \hat{\mathcal{H}} - \hat{\mathcal{V}}$.

The time step error in energy grows linearly in τ:

$$\langle E_L \rangle := \frac{\langle \varphi_T | \hat{\mathcal{H}} | \phi_0 \rangle}{\langle \varphi_T | \phi_0 \rangle} \approx \frac{\langle \varphi_T | \hat{\mathcal{H}}' | \phi_0 \rangle}{\langle \varphi_T | \phi_0 \rangle} \qquad (4.162)$$

$$= E_0 - \frac{1}{2}\tau \underbrace{\frac{\langle \varphi_T | [\hat{\mathcal{H}}, \hat{\mathcal{V}}] | \phi_0 \rangle}{\langle \varphi_T | \phi_0 \rangle}}_{\neq 0} + \frac{1}{12}\tau^2 \frac{\langle \varphi_T | [\hat{\mathcal{H}} - 2\hat{\mathcal{V}}, [\hat{\mathcal{H}}, \hat{\mathcal{V}}]] | \phi_0 \rangle}{\langle \varphi_T | \phi_0 \rangle} + \mathcal{O}(\tau^3). \qquad (4.163)$$

The linear error term is nonzero due to importance sampling.[6] The ground-state energy E_0 can be found using linear extrapolation to $\tau \to 0$. Figure 4.2 shows two such extrapolations of the He atom energy. As seen in the figure, for a rather good φ_T, the accept–reject step has a minor effect. The linear fit applies to points with $\tau \leqslant 0.02$ Ha^{-1}. Beyond that, the results clearly deviate from linear behavior because of the second-order term in equation (4.163).

Extrapolated estimates

The DMC mixed estimate is not directly useful for an observable $\hat{A}$ that does not commute with the Hamiltonian. The mixed estimate is approximately halfway between the VMC and the ground-state expectation values, so the latter can be found using linear extrapolation [12],

---

[6] Without importance sampling, DMC would sample $\phi_0(\mathbf{x})$ and $\langle \phi_0 | [\hat{\mathcal{H}}, \hat{\mathcal{V}}] | \phi_0 \rangle = \langle \phi_0 | \hat{\mathcal{H}}\hat{\mathcal{V}} - \hat{\mathcal{V}}\hat{\mathcal{H}} | \phi_0 \rangle = 0$. The time step error in energy would be proportional to τ^2.

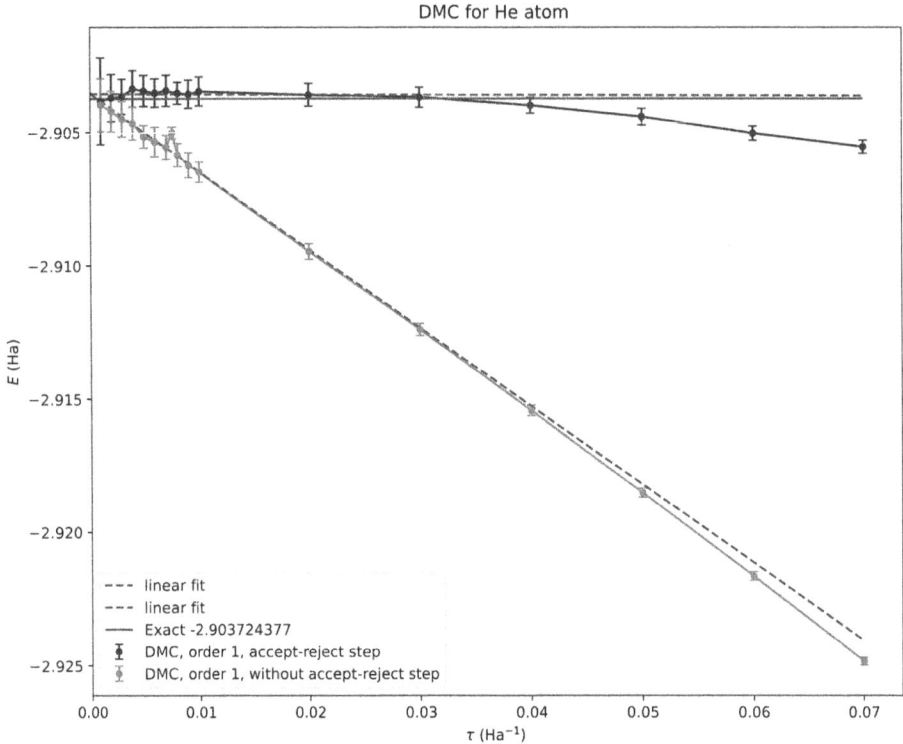

Figure 4.2. He atom energy extrapolation to $\tau \to 0$ using the first-order algorithm, with and without the accept–reject step. The trial wave function is equation (2.37) with $\alpha = 2.0$ and $\beta = 0.1593$. The results were computed using the Julia code `dmc.heatom.jl` with `order=1` and $M = 1000$ walkers.

> **Extrapolated estimate**
> $$\frac{\langle \phi_0 | \hat{\mathcal{A}} | \phi_0 \rangle}{\langle \phi_0 | \phi_0 \rangle} \approx 2 \underbrace{\frac{\langle \phi_0 | \hat{\mathcal{A}} | \varphi_T \rangle}{\langle \phi_0 | \varphi_T \rangle}}_{\text{from DMC}} - \underbrace{\frac{\langle \varphi_T | \hat{\mathcal{A}} | \varphi_T \rangle}{\langle \varphi_T | \varphi_T \rangle}}_{\text{from VMC}}. \tag{4.164}$$

As indicated, one often evaluates the trial state expectation value with VMC and continues from there to DMC and evaluation of the mixed estimate. The accuracy of the extrapolated estimate depends on how 'far' φ_T is from ϕ_0, remembering that each observable has its own measure of distance. The extrapolation is exact in the perfect case $\varphi_T(\mathbf{x}) = \phi_0(\mathbf{x})$; however, for a low-quality trial wave function, the extrapolated estimate is unreliable or completely wrong.

Pure estimates

There is a way to get more accurate ground-state expectation values for quantities that do not commute with $\hat{\mathcal{H}}$, known as **pure estimates**. Ideally, a pure estimate is unbiased, i.e. unaffected by the choice of the trial wave function. The average local energy was already shown to be independent of φ_T, so it is a pure estimate.

Other pure estimates are computed based on information about how long walkers survive and how many descendants they have. The idea is to trace the lineage of the walker family sufficiently far back as to discern the asymptotic offspring. The required length of walker history is found by inspection. Very long walker histories are, however, unstable, and the method has been criticized for its lack of robustness. With care, pure estimates can be used to find the expectation values of coordinate-dependent quantities, such as density, the radial distribution function, and, of course, the kinetic and potential energies. The derivation of the method was given by Liu *et al* in [13], and for a practical implementation, I recommend [14]. Pure estimates are built into many QMC packages, such as QMCPACK.

4.4.7 A second-order DMC algorithm

In some problems, more accurate diffusion–drift trajectories may be beneficial. Assuming one is not willing to compute any gradients of the drift, one can use a numerical differential equation solver. One possibility is the midpoint variant of the second-order Runge–Kutta method (RK2). The drift coordinate update can be written in two stages:

$$\mathbf{y} := \mathbf{x} + D\frac{\tau}{2}\mathbf{F}(\mathbf{x}) \tag{4.165}$$

$$\mathbf{x}' = \mathbf{x} + D\tau\mathbf{F}(\mathbf{y}) + \mathcal{O}(\tau^3) \qquad \text{(RK2 drift trajectory).} \tag{4.166}$$

To check this formula in 1D, Taylor expand the outermost drift to the first order in τ,

$$F\left(x + \frac{D\tau}{2}F(x)\right) = F(x) + \frac{D\tau}{2}F(x)F'(x) + \mathcal{O}(\tau^2), \tag{4.167}$$

where $F'(x): = dF(\xi)/d\xi \mid_{\xi=x}$. Insertion into the RK2 trajectory gives

$$x' = x + D\tau F(x) + \frac{(D\tau)^2}{2}F(x)F'(x) + \mathcal{O}(\tau^3), \tag{4.168}$$

so the velocity is

$$\frac{dx'}{d\tau} = DF(x) + D^2\tau F(x)F'(x) + \mathcal{O}(\tau^2) \tag{4.169}$$

$$= DF(x + D\tau F(x)) + \mathcal{O}(\tau^2) = DF(x') + \mathcal{O}(\tau^2), \tag{4.170}$$

as required. The SymPy code `sympy_drift.py` computes the expansions.

The price is that one needs to evaluate the drift at the halfway point of the predicted trajectory between imaginary times zero and τ. This extra drift evaluation increases the computational burden, which may outweigh the relatively small gain in accuracy.

We would still have to combine diffusion, drift, and the source terms so that the overall accuracy is second order in τ. The RK2 solution is a special case of the more general *operator splitting* technique, which gives us the means to write second- or even higher-order DMC algorithms.

4.4.7.1 Operator splitting

The primitive approximation in equation (4.59) is a first-order operator splitting. Written for $\hat{\mathcal{H}} = \hat{T} + \hat{V}$, it reads

$$e^{-\tau(\hat{\mathcal{H}}-E_T)} \approx e^{-\tau\hat{T}}e^{-\tau(\hat{V}-E_T)} \qquad (\text{error } \mathcal{O}(\tau^2)). \tag{4.171}$$

In coordinate space, the matrix elements of this operator are

$$\langle \mathbf{x}'|e^{-\tau(\hat{\mathcal{H}}-E_T)}|\mathbf{x}\rangle \approx \langle \mathbf{x}'|e^{-\tau\hat{T}}e^{-\tau(\hat{V}-E_T)}|\mathbf{x}\rangle \tag{4.172}$$

$$= \int d\mathbf{x}'' \langle \mathbf{x}'|e^{-\tau\hat{T}}|\mathbf{x}''\rangle \langle \mathbf{x}''|e^{-\tau(\hat{V}-E_T)}|\mathbf{x}\rangle. \tag{4.173}$$

The potential operator is diagonal in coordinate space, meaning the potential is local and there is no potential between two distinct points,

$$\langle \mathbf{x}''|e^{-\tau(\hat{V}-E_T)}|\mathbf{x}\rangle = \langle \mathbf{x}|e^{-\tau(\hat{V}-E_T)}|\mathbf{x}\rangle\delta(\mathbf{x}''-\mathbf{x}) = e^{-\tau(\hat{V}-E_T)}\delta(\mathbf{x}''-\mathbf{x}). \tag{4.174}$$

This forces the paths to pass through the point $\mathbf{x}'' = \mathbf{x}$, and the matrix elements in equation (4.173) become

$$\langle \mathbf{x}'|e^{-\tau(\hat{\mathcal{H}}-E_T)}|\mathbf{x}\rangle \approx \langle \mathbf{x}'|e^{-\tau\hat{T}}|\mathbf{x}\rangle e^{-\tau(V(\mathbf{x})-E_T)}, \tag{4.175}$$

which gives the evolution equation

$$P(\mathbf{x}',\tau) \approx \int d\mathbf{x}\,\underbrace{\langle \mathbf{x}'|e^{-\tau\hat{T}}|\mathbf{x}\rangle}_{\text{diffusion}}\,\underbrace{e^{-\tau(V(\mathbf{x})-E_T)}}_{\text{source}}P(\mathbf{x},0) \qquad \text{crude DMC}, \tag{4.176}$$

and we have recovered the first-order crude DMC algorithm. Looking back, we now see why the energy shift E_T was kept with the potential. From now on, I will leave out E_T but add it back to the final expression.

Higher-order approximations bring in operator combinations such as $\hat{T}\hat{V}$ and $\hat{V}\hat{T}$,

$$e^{-\tau\hat{\mathcal{H}}} = \sum_{n=0}^{\infty}\frac{(-\tau)^n}{n!}\hat{\mathcal{H}}^n \tag{4.177}$$

$$= 1 - \tau(\hat{T}+\hat{V}) + \frac{\tau^2}{2!}(\hat{T}+\hat{V})(\hat{T}+\hat{V})$$
$$- \frac{\tau^3}{3!}(\hat{T}+\hat{V})(\hat{T}+\hat{V})(\hat{T}+\hat{V}) + ..., \tag{4.178}$$

which contains terms such as

$$(\hat{T}+\hat{V})(\hat{T}+\hat{V}) = \hat{T}^2 + \hat{T}\hat{V} + \hat{V}\hat{T} + \hat{V}^2. \tag{4.179}$$

A second-order splitting of $\hat{\mathcal{H}} = \hat{T} + \hat{V}$ can be written in two forms,

$$e^{-\tau\hat{\mathcal{H}}} \approx e^{-\frac{1}{2}\tau\hat{T}} e^{-\tau\hat{V}} e^{-\frac{1}{2}\tau\hat{T}} \tag{4.180}$$

$$e^{-\tau\hat{\mathcal{H}}} \approx e^{-\frac{1}{2}\tau\hat{V}} e^{-\tau\hat{T}} e^{-\frac{1}{2}\tau\hat{V}}. \tag{4.181}$$

The choice $\hat{\mathcal{H}} = \hat{T} + \hat{V}$ is just an example; in general,

$$\boxed{\begin{array}{l} \text{second-order } 1/2\text{–}1\text{–}1/2 \text{ splitting of } \hat{\mathcal{H}} = \hat{A} + \hat{B} \\ e^{-\tau\hat{\mathcal{H}}} \approx e^{-\frac{1}{2}\tau\hat{A}} e^{-\tau\hat{B}} e^{-\frac{1}{2}\tau\hat{A}} \quad (\text{error } \mathcal{O}(\tau^3)). \end{array}} \tag{4.182}$$

The exponent operator expansions can be done with pen and paper, but they are most easily checked using any symbolic calculus package that can use operators. The SymPy code expH_sympy.py expands $e^{-\tau\hat{\mathcal{H}}}$ up to the third order in τ, and the two expansions given above are clearly seen to agree up to the second order in τ. The code prints out the warning message 'Warning: SymPy expands exp(-tau*T) to series without checking its convergence'; the reason for this is explained in the next remark.

Warning: do not expand $e^{-\tau\hat{T}}$ using the series expansion $e^A = 1 + A + \frac{1}{2}A^2 + \frac{1}{3!}A^3 + \dots$ without hesitation. Consider the 1D evolution,

$$P(x', \tau) = \int_{-\infty}^{\infty} dx \langle x' | e^{-\tau\hat{T}} | x \rangle P(x, 0), \tag{4.183}$$

which represents diffusion from x to x'. On the other hand, the evolution described by the series expansion would be

$$P_{\text{strange}}(x', \tau) = \int_{-\infty}^{\infty} dx \langle x' | 1 - \tau\hat{T} + \frac{1}{2}\tau^2\hat{T}^2 - \frac{\tau^3}{3!}\hat{T}^3 + \dots | x \rangle P(x, 0). \tag{4.184}$$

Remember that $\hat{T}$ is, apart from a constant factor, the second derivative d^2/dx^2. Like all operators, $\hat{T}$ commutes with itself, so this is not a problem. The problem is *domains* and the fact that the kinetic energy eigenvalues are not bounded, making $\hat{T}$ an *unbounded operator*. To see how the series expansion can fail, consider the two evolutions starting from the distribution

$$P(x, 0) = \begin{cases} 1, & \text{for } 0 < x < 1 \\ 0, & \text{elsewhere} \end{cases}. \tag{4.185}$$

The correct diffusion evolution spreads this around to finally cover all space. The series expansion, however, has $\hat{T}P(x, 0) \propto d^2/dx^2 P(x, 0) = 0$ outside the domain $0 < x < 1$, and so all $\hat{T}^n P(x, 0) = 0$ outside that domain. This means that the series expansion evolution never lets the distribution $P_{\text{strange}}(x', \tau)$ spread out of the original domain $0 < x < 1$; it tries to, and hitting the boundaries produces an interesting fractal pattern.

The latter form is especially convenient because the matrix elements are

$$\langle \mathbf{x}'|e^{-\tau\hat{\mathcal{H}}}|\mathbf{x}\rangle \approx e^{-\frac{1}{2}\tau V(\mathbf{x}')}\langle \mathbf{x}'|e^{-\tau\hat{\mathcal{T}}}|\mathbf{x}\rangle e^{-\frac{1}{2}\tau V(\mathbf{x})} = e^{-\tau(\frac{1}{2}(V(\mathbf{x}')+V(\mathbf{x})))}\langle \mathbf{x}'|e^{-\tau\hat{\mathcal{T}}}|\mathbf{x}\rangle. \quad (4.186)$$

This still leads to a crude DMC algorithm, albeit one that is second-order accurate. Moving on to importance sampling, the drift–diffusion source matrix elements are

$$\langle \mathbf{x}'|e^{-\tau(\hat{\mathcal{H}}_F - E_T)}|\mathbf{x}\rangle = \langle \mathbf{x}'|e^{-\tau(\hat{\mathcal{T}}+\hat{\mathcal{D}}(\mathbf{x})+E_L(\mathbf{x})-E_T)}|\mathbf{x}\rangle \quad (4.187)$$

$$\approx \langle \mathbf{x}'|e^{-\frac{1}{2}\tau(E_L(\mathbf{x})-E_T)}e^{-\tau(\hat{\mathcal{T}}+\hat{\mathcal{D}}(\mathbf{x}))}e^{-\frac{1}{2}\tau(E_L(\mathbf{x})-E_T)}|\mathbf{x}\rangle \quad (4.188)$$

$$= \int d\mathbf{y}d\mathbf{z}\langle \mathbf{x}'|e^{-\frac{1}{2}\tau(E_L(\mathbf{x_1})-E_T)}|\mathbf{y}\rangle\langle \mathbf{y}|e^{-\tau(\hat{\mathcal{T}}+\hat{\mathcal{D}}(\mathbf{x_2}))}|\mathbf{z}\rangle\langle \mathbf{z}|e^{-\frac{1}{2}\tau(E_L(\mathbf{x})-E_T)}|\mathbf{x}\rangle, \quad (4.189)$$

where I used the second-order 1/2–1–1/2 -splitting to obtain

$$\hat{\mathcal{H}}_F - E_T = (\hat{\mathcal{T}} + \hat{\mathcal{D}}) \overset{\text{split here}}{+} (E_L - E_T). \quad (4.190)$$

Notice how the position-dependent factors always pick up the 'current' coordinate on their right. The local energy is diagonal in coordinate space,

$$\langle \mathbf{x}'|e^{-\frac{1}{2}\tau(E_L(\mathbf{x_1})-E_T)}|\mathbf{y}\rangle = e^{-\frac{1}{2}\tau(E_L(\mathbf{x}')-E_T)}\delta(\mathbf{x}' - \mathbf{y}), \quad (4.191)$$

so the matrix elements are

$$\langle \mathbf{x}'|e^{-\tau(\hat{\mathcal{H}}_F - E_T)}|\mathbf{x}\rangle \approx e^{-\tau(\frac{1}{2}(E_L(\mathbf{x}')+E_L(\mathbf{x}))-E_T)}\langle \mathbf{x}'|e^{-\tau(\hat{\mathcal{T}}+\hat{\mathcal{D}}(\mathbf{x}))}|\mathbf{x}\rangle. \quad (4.192)$$

Next, separate the operators $\hat{\mathcal{T}}$ and $\hat{\mathcal{D}}$. Considering drift is more expensive to compute than diffusion, split them as half diffusion–full drift–half diffusion:

$$\langle \mathbf{x}'|e^{-\tau(\hat{\mathcal{H}}_F - E_T)}|\mathbf{x}\rangle \approx e^{-\tau(\frac{1}{2}(E_L(\mathbf{x}')+E_L(\mathbf{x}))-E_T)}\langle \mathbf{x}'|e^{-\frac{1}{2}\tau\hat{\mathcal{T}}}e^{-\tau\hat{\mathcal{D}}(\mathbf{x})}e^{-\frac{1}{2}\tau\hat{\mathcal{T}}}|\mathbf{x}\rangle \quad (4.193)$$

$$= \underbrace{e^{-\tau(\frac{1}{2}(E_L(\mathbf{x}')+E_L(\mathbf{x}))-E_T)}}_{\text{branching}} \int d\mathbf{y}d\mathbf{z}\underbrace{\langle \mathbf{x}'|e^{-\frac{1}{2}\tau\hat{\mathcal{T}}}|\mathbf{z}\rangle}_{\tau/2-\text{diffusion}}\underbrace{\langle \mathbf{z}|e^{-\tau\hat{\mathcal{D}}(\mathbf{y})}|\mathbf{y}\rangle}_{\tau-\text{drift}}\underbrace{\langle \mathbf{y}|e^{-\frac{1}{2}\tau\hat{\mathcal{T}}}|\mathbf{x}\rangle}_{\tau/2-\text{diffusion}} \quad (4.194)$$

The factor marked τ-drift *is not* sampled by the trajectory $\mathbf{z} = \mathbf{y} + D\tau\mathbf{F}(\mathbf{y})$, because this would contaminate the algorithm, making it just first-order accurate. We need to solve also the drift trajectory to the second order in τ using, for example, the RK2 drift trajectory in equation (4.166). Reading equation (4.194) from right to left and using the second-order drift solution gives the following steps:

$$
\begin{aligned}
\mathbf{y}: &= \mathbf{x} + \sqrt{D\tau}\,\eta & &\tau/2-\text{diffusion} \\
\mathbf{z}': &= \mathbf{y} + D\frac{\tau}{2}\mathbf{F}(\mathbf{y}) & &\tau-\text{drift, first part} \\
\mathbf{z}: &= \mathbf{y} + D\tau\mathbf{F}(\mathbf{z}') & &\tau-\text{drift, last part} \\
\mathbf{x}' &= \mathbf{z} + \sqrt{D\tau}\,\eta & &\tau/2 - \text{diffusion}.
\end{aligned}
\qquad (4.195)
$$

Combine the last two steps to get a slightly more compact algorithm:

Second-order drift–diffusion branching update for each walker

(1) $\mathbf{y} := \mathbf{x} + \sqrt{D\tau}\,\boldsymbol{\eta}$

(2) $\mathbf{z} := \mathbf{y} + D\dfrac{\tau}{2}\mathbf{F}(\mathbf{y})$

(3) $\mathbf{x}' = \mathbf{y} + D\tau\mathbf{F}(\mathbf{z}) + \sqrt{D\tau}\,\boldsymbol{\eta}$

(4) Branch with factor $e^{-\tau\left(\frac{1}{2}(E_L(\mathbf{x}')+E_L(\mathbf{x}))-E_T\right)}$.

(4.196)

The Gaussian random vectors $\boldsymbol{\eta}$ are two distinct dN-dimensional vectors, and the update contains two computationally expensive drift evaluations. The second-order update can be used in the DMC algorithm in section 4.4.5, except that no accept–reject step is applied.[7] Figure 4.3 shows how the second-order algorithm performs

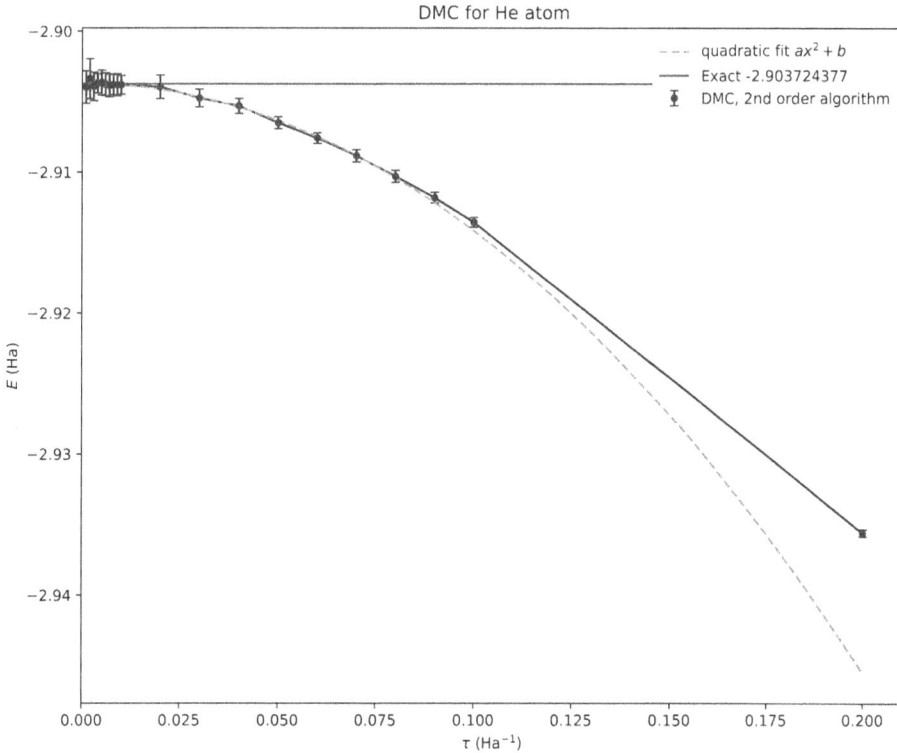

Figure 4.3. He atom energy extrapolation to $\tau \to 0$ using the second-order algorithm (4.196). The trial wave function is equation (2.37) with $\alpha = 2.0$ and $\beta = 0.1593$. The results were computed using the Julia code dmc. heatom.jl with order=2 and $M = 1000$ walkers. The extrapolated energy is $E = -2.903\,712$ Ha.

---

[7] The approximate second-order Green's function contains an integral, whose evaluation at every step would be impractical. I have not encountered any adoption of accept–reject steps in second-order DMC in the literature.

for the He atom. The second-order algorithm derived above is equation (25) in reference [11], where many possible second-order algorithms are compared.

The 1/2–1–1/2 -splitting also works for a sum of three operators and is used for operator pairs:

$$e^{-\tau(\hat{A}+\hat{B}+\hat{C})} \approx e^{-\frac{1}{2}\tau\hat{A}}e^{-\tau(\hat{B}+\hat{C})}e^{-\frac{1}{2}\tau\hat{A}} \approx e^{-\frac{1}{2}\tau\hat{A}}e^{-\frac{1}{2}\tau\hat{B}}e^{-\tau\hat{C}}e^{-\frac{1}{2}\tau\hat{B}}e^{-\frac{1}{2}\tau\hat{A}}. \tag{4.197}$$

The SymPy code `triple_sympy.py` shows this is valid to the second order in τ.

4.4.8 About fourth-order DMC algorithms

The first- and second-order operator splittings we have been using are

$$e^{-\tau(\hat{A}+\hat{B})} \approx e^{-\tau\hat{A}}e^{-\tau\hat{B}} \qquad \text{first order, error } \mathcal{O}(\tau^2) \tag{4.198}$$

$$e^{-\tau(\hat{A}+\hat{B})} \approx e^{-\frac{1}{2}\tau\hat{A}}e^{-\tau\hat{B}}e^{-\frac{1}{2}\tau\hat{A}} \qquad \text{second order, error } \mathcal{O}(\tau^3), \tag{4.199}$$

so a logical step would be to find coefficients a_i and b_i so that the expansion

$$e^{-\tau(\hat{A}+\hat{B})} \approx \prod_{i=1}^{N_i} e^{-a_i\tau\hat{A}}e^{-b_i\tau\hat{B}}, \tag{4.200}$$

is fourth order. Indeed, one such expansion is that of Ruth and Forest [15], found earlier (in 1983) by Ruth and rediscovered several times around the year 1990:

> **Forest–Ruth fourth-order splitting**
> $$e^{-\tau(\hat{A}+\hat{B})} \approx e^{-a_1\tau\hat{A}}e^{-b_1\tau\hat{B}}e^{-a_2\tau\hat{A}}e^{-b_2\tau\hat{B}}e^{-a_2\tau\hat{A}}e^{-b_1\tau\hat{B}}e^{-a_1\tau\hat{A}}$$
> where
> $$s = 2^{1/3}(\approx 1.259\,92)$$
> $$a_1 = \frac{1}{2}\frac{1}{2-s}, \quad a_2 = -\frac{1}{2}\frac{s-1}{2-s}$$
> $$b_1 = \frac{1}{2-s}, \quad b_2 = -\frac{s}{2-s}. \tag{4.201}$$

The **coefficients a_2 and b_2 are negative**, so if $\hat{T}$ is among the operators, then the corresponding matrix element is

$$\langle\mathbf{x}_1|e^{\text{(positive constant)}\tau\hat{T}}|\mathbf{x}_2\rangle. \tag{4.202}$$

Such *reverse diffusion* cannot be simulated, so the Forest–Ruth splitting is unsuitable for DMC. Furthermore, the Sheng–Suzuki theorem [16, 17] states that no fourth-order splitting (4.200) can have all coefficients positive.

A new element has to be added that prevents reverse diffusion from appearing. The Baker–Campbell–Hausdorff formula contains commutators, such as $[\hat{A}, \hat{B}]$ and $[\hat{A} - \hat{B}, [\hat{A}, \hat{B}]]$, so it could be helpful to replace a bare operator somewhere with a commutator. Suzuki [18] suggested that one should make use of a particularly convenient commutator, which happens to be a local function,

$$[\hat{\mathcal{V}}, [\hat{\mathcal{T}}, \hat{\mathcal{V}}]] = \frac{\hbar^2}{m} |\, \boldsymbol{\nabla}_\mathbf{x} V(\mathbf{x})|^2 = \frac{\hbar^2}{m} \sum_{i=1}^{N} |\mathbf{f}_i(\mathbf{x})|^2, \text{ where } \mathbf{f}_i: = -\boldsymbol{\nabla}_i V. \quad (4.203)$$

In this case, $\mathbf{f}_i$ is the classical force on particle i.

Proof: $\hat{\mathcal{T}} = -D\nabla_\mathbf{x}^2, V: = V(\mathbf{x}), D: = \frac{\hbar^2}{m},$

$$[\hat{\mathcal{V}}, [\hat{\mathcal{T}}, \hat{\mathcal{V}}]]\phi(\mathbf{x}) = [-V^2\hat{\mathcal{T}} - \hat{\mathcal{T}}V^2 + 2V\hat{\mathcal{T}}V]\phi = D[V^2\nabla_\mathbf{x}^2\phi + \nabla_\mathbf{x}^2(V^2\phi) - 2V\nabla_\mathbf{x}^2(V\phi)] \quad (4.204)$$

$$=D\left[V^2\nabla_\mathbf{x}^2\phi + \boldsymbol{\nabla}_\mathbf{x} \cdot [2V(\boldsymbol{\nabla}_\mathbf{x}V)\phi + V^2\boldsymbol{\nabla}_\mathbf{x}\phi] - 2V\boldsymbol{\nabla} \cdot [(\boldsymbol{\nabla}_\mathbf{x}V)\phi + V\boldsymbol{\nabla}_\mathbf{x}\phi] \right] \quad (4.205)$$

$$= D\Big[V^2\nabla_\mathbf{x}^2\phi + \underline{2(\boldsymbol{\nabla}_\mathbf{x}V)^2\phi} + 2(V\nabla_\mathbf{x}^2 V)\phi + 2V(\boldsymbol{\nabla}_\mathbf{x}V) \cdot \boldsymbol{\nabla}_\mathbf{x}\phi$$
$$+ 2V\boldsymbol{\nabla}_\mathbf{x}V \cdot \boldsymbol{\nabla}_\mathbf{x}\phi + V^2\nabla_\mathbf{x}^2\phi \quad (4.206)$$
$$- 2V\nabla_\mathbf{x}^2 V\phi - 2V\boldsymbol{\nabla}V \cdot \boldsymbol{\nabla}_\mathbf{x}\phi - 2V\boldsymbol{\nabla}_\mathbf{x}V \cdot \boldsymbol{\nabla}_\mathbf{x}\phi - 2V^2\nabla_\mathbf{x}^2\phi \Big]$$

All colored terms cancel, so only the underlined term survives:

$$=2D(\boldsymbol{\nabla}_\mathbf{x}V)^2\phi. \quad (4.207)$$

The result is valid for any function of positions, for example

$$[E_L(\mathbf{x}), [\hat{\mathcal{T}}, E_L(\mathbf{x})]] = 2D|\boldsymbol{\nabla}_\mathbf{x}E_L(\mathbf{x})|^2. \quad (4.208)$$

Takahashi and Imada [19], and Li and Broughton [20], suggested the propagator

> Takahashi–Imada second-order splitting ('almost fourth order')
> $$e^{-\tau(\hat{\mathcal{T}}+\hat{\mathcal{V}})} \approx e^{-\frac{1}{2}\tau\hat{\mathcal{T}}}e^{-\frac{1}{2}\tau\hat{\mathcal{V}}}e^{-\frac{1}{24}\tau^3[\hat{\mathcal{V}}, [\hat{\mathcal{T}}, \hat{\mathcal{V}}]]}e^{-\frac{1}{2}\tau\hat{\mathcal{V}}}e^{-\frac{1}{2}\tau\hat{\mathcal{T}}} + \mathcal{O}(\tau^3) \quad (4.209)$$
> $$=e^{-\frac{1}{2}\tau\hat{\mathcal{T}}}e^{-\tau\hat{\mathcal{V}}-\frac{1}{24}\tau^3[\hat{\mathcal{V}}, [\hat{\mathcal{T}}, \hat{\mathcal{V}}]]}e^{-\frac{1}{2}\tau\hat{\mathcal{T}}} + \mathcal{O}(\tau^3).$$

This is actually the familiar second-order 1/2–1–1/2 splitting,

$$e^{-\tau(\hat{\mathcal{T}}+\hat{\mathcal{V}})} \approx e^{-\frac{1}{2}\tau\hat{\mathcal{T}}}e^{-\tau\hat{\mathcal{W}}}e^{-\frac{1}{2}\tau\hat{\mathcal{T}}} + \mathcal{O}(\tau^3), \quad (4.210)$$

with an τ-dependent effective potential,

$$\hat{\mathcal{W}}: =\hat{\mathcal{V}} + \frac{1}{24}\tau^2[\hat{\mathcal{V}}, [\hat{\mathcal{T}}, \hat{\mathcal{V}}]]. \quad (4.211)$$

The propagator is only second order, therefore it is not applied in DMC. However, it is frequently used in PIMC calculations of the partition function,

$$Z = \mathrm{Tr}\,e^{-\beta\hat{\mathcal{H}}}, \quad (4.212)$$

because this trace turns out to be fourth order. The formula is often said to be 'almost fourth order,' and it was, in fact, designed for accurate trace calculation (for a detailed discussion, see [21]). We will use the Takahashi–Imada or Li–Broughton Green's function in the PIMC section.

The search for an all-positive fourth-order expansion continued, and after a systematic screening process, Chin [22] found several high-order splittings. One of them is

Chin's fourth-order splitting

$$e^{-\tau(\hat{T}+\hat{V})} \approx e^{-v_0 \tau \hat{V}} e^{-t_1 \tau \hat{T}} e^{-v_1 \tau \hat{W}} e^{-t_2 \tau \hat{T}} e^{-v_1 \tau \hat{W}} e^{-t_1 \tau \hat{T}} e^{-v_0 \tau \hat{V}}$$

$$\hat{W} := \hat{V} + \frac{u_0}{v_1}\tau^2[\hat{V}, [\hat{T}, \hat{V}]]$$

(4.213)

$$v_0 = \frac{6t_1(t_1 - 1) + 1}{12(t_1 - 1)t_1}, \quad t_2 = 1 - 2t_1$$

$$v_1 = \frac{1}{2} - v_0, \quad u_0 = \frac{1}{48}\left(\frac{1}{6t_1(1 - t_1)^2} - 1\right).$$

(4.214)

The parameter t_1 is tunable within the limits $0 \leqslant t_1 \leqslant 1 - 1/\sqrt{3}$. In some systems, even the fifth-order errors cancel with the choice $t_1 = 0.350\,23$. All coefficients are positive, so in principle this splitting is readily applicable to DMC.

Importance sampling causes extra trouble because we have used diffusion, drift, and branching, with the corresponding operators $\hat{T}$, $\hat{D}$ and the function $E_L - E_T$. That makes two fourth-order splittings and two new commutators. Schematically,

$$\hat{T} + \hat{D}|E_L - E_T \Rightarrow [E_L, [\hat{T} + \hat{D}, E_L]] \tag{4.215}$$

$$\hat{T}|\hat{D} \Rightarrow [\hat{D}, [\hat{T}, \hat{D}]]. \tag{4.216}$$

The drift trajectory has to be solved accurately using, for example, a fourth-order Runge–Kutta algorithm, but the need for higher derivatives of the local energy and drift makes this approach rather cumbersome. Nevertheless, a fourth-order DMC algorithm was developed and applied in [23].

4.5 Case study: importance sampling DMC for a particle in a box

A single particle in a box is an exactly solvable system, which can be used to elucidate some features of the first-order DMC algorithm from section 4.4.5:

- What are the drift trajectories $\mathbf{x}(\tau)$? What is the role of diffusion?
- How important is the accept–reject step in the first-order DMC algorithm?
- Does the initialization of the walkers affect the results?
- What are the so-called *persistent configurations*?

The exact ground-state wave function of a single particle in a unit box in three dimensions is ($\mathbf{x} := (x, y, z)$):

$$\phi_0(\mathbf{x}) = \left(\frac{\pi}{2}\right)^3 \sin(\pi x)\sin(\pi y)\sin(\pi z), \text{ if } x, y, z \in [0, 1], \tag{4.217}$$

and zero elsewhere. I shall use this eigenstate as the trial wave function,

$$\varphi_T(\mathbf{x}) = \phi_0(\mathbf{x}), \; E_T = E_L(\mathbf{x}) = E_0, \tag{4.218}$$

and since the local energy is constant, *the imaginary-time evolution has no source term* (and it is not in the Python code `particle_in_a_box.py`).

4.5.1 Drift trajectories

The exact drift is

$$\mathbf{F}(\mathbf{x}) = 2\frac{\nabla\phi_0(\mathbf{x})}{\phi_0(\mathbf{x})} = 2\pi(\frac{\cos(\pi x)}{\sin(\pi x)}, \frac{\cos(\pi y)}{\sin(\pi y)}, \frac{\cos(\pi z)}{\sin(\pi z)}), \text{if } x, y, z \in [0, 1], \tag{4.219}$$

and zero elsewhere. Since φ_T is positive everywhere, it can be represented using M walkers, whose drift trajectories satisfy the differential equation (4.103)

$$\frac{d}{d\tau}\mathbf{x}(\tau) = D\mathbf{F}(\mathbf{x}). \tag{4.220}$$

The exact solutions are

$$\mathbf{x}(\tau) = \frac{1}{\pi}\arccos(\cos(\pi\mathbf{x}(0))e^{-2\pi^2 D\tau}). \tag{4.221}$$

In the limit $\tau \to \infty$, the drift trajectories converge to the point (0.5, 0.5, 0, 5), which is in the middle of the box, where φ_T is at a maximum. Figure 4.4 shows a few of the trajectories. Simultaneously, diffusion random walk expands the walkers further from each other, so diffusion counteracts drift.

In the first-order importance sampling DMC algorithm, the drift trajectories $\mathbf{x}(\tau)$ are solved approximately, and the walker positions are updated according to

$$\mathbf{x}' = \mathbf{x} + D\tau F(\mathbf{x}) + \sqrt{2D\tau}\,\eta. \tag{4.222}$$

The walkers now represent the probability distribution

$$f(\mathbf{x}, \tau) := \varphi_T(\mathbf{x})\phi(\mathbf{x}, \tau) = \phi_0(\mathbf{x})\phi(\mathbf{x}, \tau), \tag{4.223}$$

so the expected limiting distribution is

$$f(\mathbf{x}, \tau \to \infty) = |\phi_0(\mathbf{x})|^2. \tag{4.224}$$

In the first-order algorithm, the accept–reject step is based on the quantity

$$ratio = \frac{|\varphi_T(\mathbf{x}')|^2 G_F(\mathbf{x} \leftarrow \mathbf{x}'; \tau)}{|\varphi_T(\mathbf{x})|^2 G_F(\mathbf{x}' \leftarrow \mathbf{x}; \tau)} = \frac{|\varphi_T(\mathbf{x}')|^2 \exp\left(-\dfrac{(\mathbf{x} - \mathbf{x}' - D\tau F(\mathbf{x}'))^2}{4D\tau}\right)}{|\varphi_T(\mathbf{x})|^2 \exp\left(-\dfrac{(\mathbf{x}' - \mathbf{x} - D\tau F(\mathbf{x}))^2}{4D\tau}\right)}. \tag{4.225}$$

The DMC solution is computed using the Python code `particle_in_a_box.py`. Usually, walkers are initialized during a VMC run, which initializes walkers

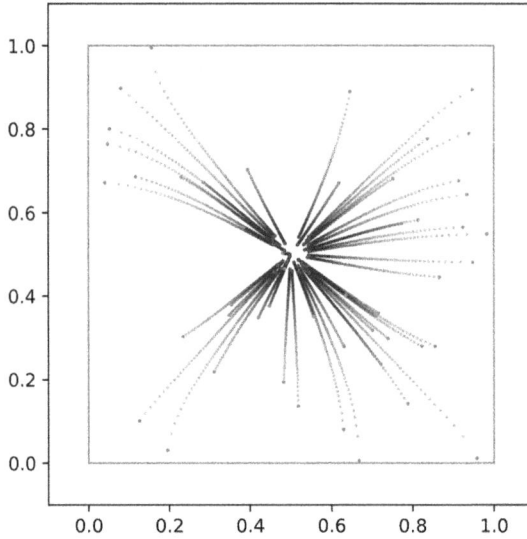

Figure 4.4. A few exact drift trajectories $\mathbf{x}(\tau)$, shown for plotting purposes in a two-dimensional box (blue square). The original positions $\mathbf{x}(0)$ are marked with red points, and their locations after time $\tau = 0.2$ are indicated by blue points. The points on a trajectory are equidistant in τ and their spatial distance is proportional to the drift velocity, showing that the drift is largest near the box edges. See the code `particle_in_a_box_drift_trajectories.py`.

directly to the exact limiting distribution $|\phi_0(\mathbf{x})|^2$. To show how the walker density evolves, the Python code has two initializations: (1) all walkers in the middle of the box, and (2) walkers scattered randomly in the box.

Figure 4.5 compares the limiting distributions of $M = 5000$ walkers with the exact solution $|\phi_0(\mathbf{x})|^2$ using two different walker initializations. Initializing all walkers to the middle of the box gives the correct limiting distribution.

4.5.2 Persistent configurations (aka trapped walkers)

The DMC run in figure 4.6 happens to have a few walkers initially close to the box edges, and they do not move at all.

The cause of the persistent configurations is drift leading to an impossible walker position, in this case, movement out of the box. The accept–reject step always rejects such moves because

$$ratio \propto |\varphi_T(\mathbf{x}')|^2 = 0, \; \text{if} \, \mathbf{x}' \text{is outside the box.} \tag{4.226}$$

Consider a walker close to the edge, at $\mathbf{x} = (0.001, 0.5, 0.5)$, in the code `particle_in_a_box.py`. The trial wave function has the value

$$\varphi_T(\mathbf{x}) = \sin(\pi x)\sin(\pi y)\sin(\pi z) \approx 0.003, \tag{4.227}$$

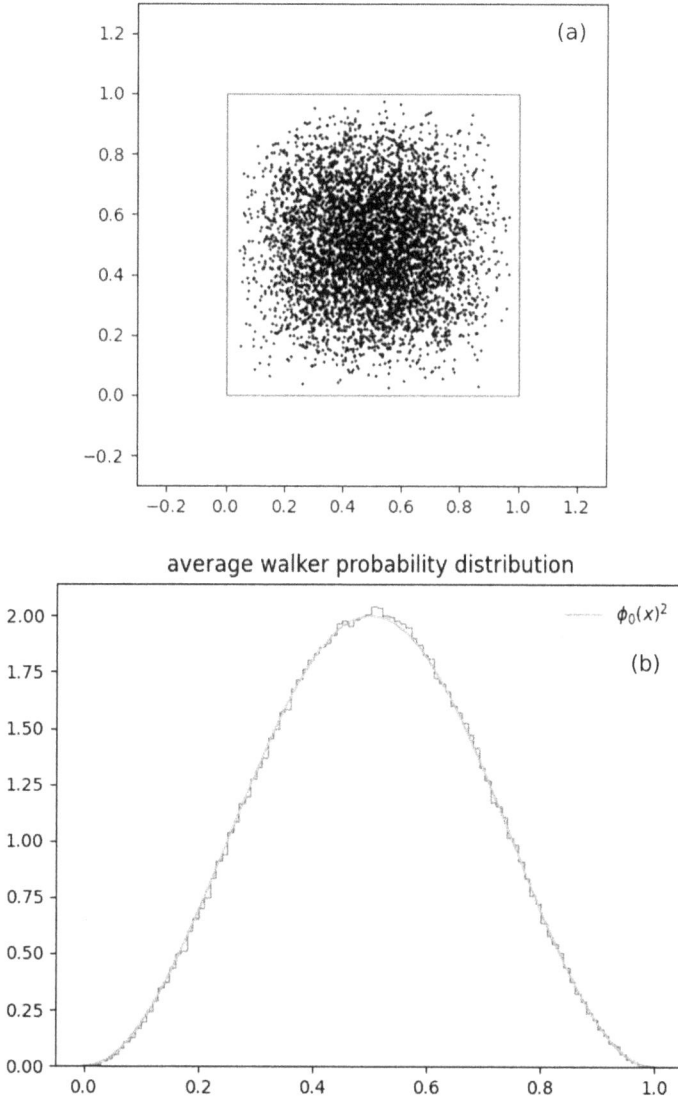

Figure 4.5. Initial walker distribution with all walkers in the middle of the box soon develops into the spherically symmetric limiting distribution. A snapshot of walker positions is shown in (a). In (b), the average distribution after a short run is compared with the exact result $|\phi_0(x)|^2$.

so the probability of finding a walker at $\mathbf{x}$ is small but nonzero. The drift vector points toward the center of the box,

$$\mathbf{F}(\mathbf{x}) \approx (2000.0,\ 0.0,\ 0.0), \tag{4.228}$$

so the drift update $\mathbf{x}' = \mathbf{x} + D\tau F(\mathbf{x})$ overshoots the box center unless

$$\tau \leqslant 0.001. \tag{4.229}$$

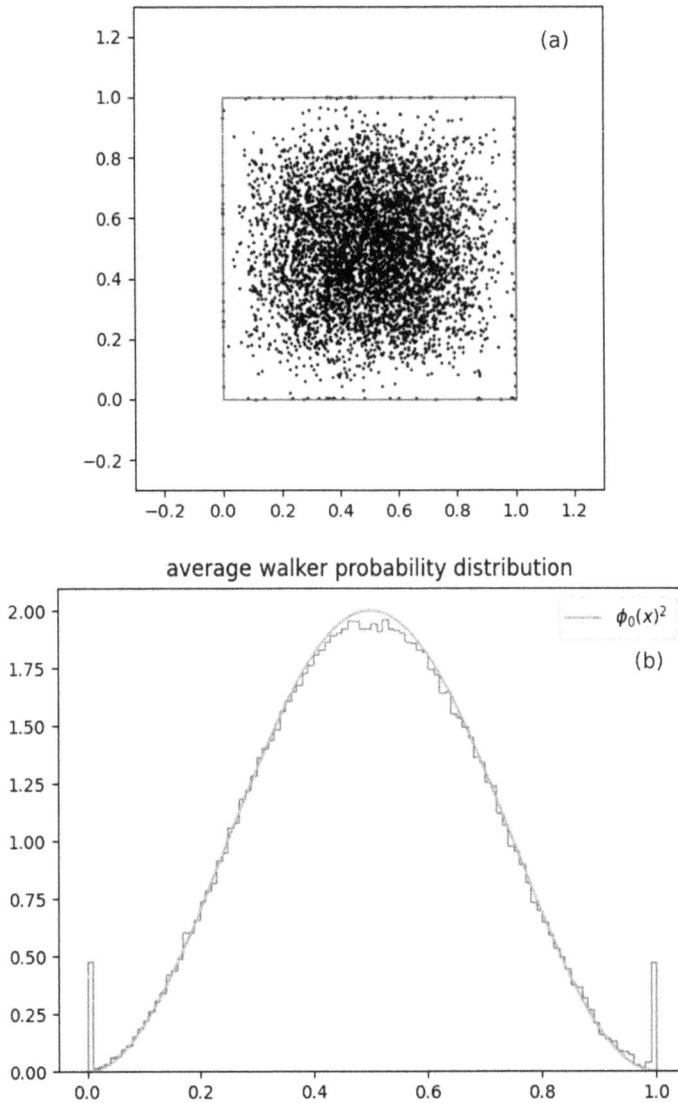

Figure 4.6. Walkers are initially randomly located in the box. As panel (a) shows, some walkers are stuck near the box edges and do not move at all. As a result, the walker distribution in (b) is badly distorted, with extra weight at the edges.

Diffusion does not change the picture that much. Figure 4.7 shows the probability that the suggested new walker positions $\mathbf{x}'$ are inside the box after a drift–diffusion update $\mathbf{x}' = \mathbf{x} + D\tau \mathbf{F}(\mathbf{x}) + \sqrt{2D\tau}\,\eta$. The walker is originally at $\mathbf{x} = (0.0001, 0.5, 0.5)$, and the probability is estimated using 10^6 move attempts. As shown in the inset, the drift is toward the middle of the box, but the drift move badly overshoots the midpoint if τ is too large.

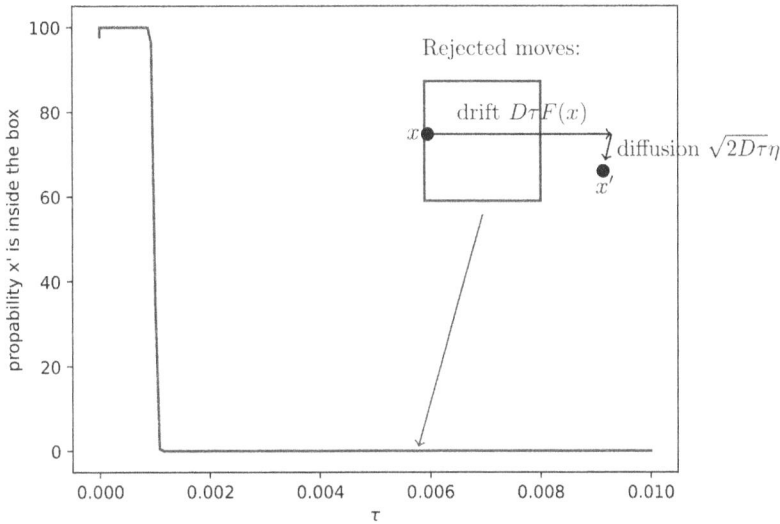

Figure 4.7. After one million attempts to drift–diffuse a walker near the edge at $\mathbf{x} = (0.001, 0.5, 0.5)$, the probability that $\mathbf{x}'$ is inside the box is approximately zero for $\tau \geqslant 0.001$. Moves out of the box are always rejected in the accept–reject step, and the walker remains trapped near the edge. The random vector η may pick values from the tail of the normal distribution, and the resulting long diffusion moves out of the box cause a small dip in the probability near $\tau = 0$.

Summary

An overlong imaginary-time step τ causes the drift to overshoot a walker to a location where it should not be. Using the code `particle_in_a_box.py`, it is easy to demonstrate that shorter values of τ reduce the chances of persistent configurations. The first-order importance sampling update $\mathbf{x}' = \mathbf{x} + D\tau\mathbf{F}(\mathbf{x}) + \sqrt{2D\tau}\,\eta$ is inaccurate and may try to sample impossible walker configurations.

Persistent configurations are often detected as sudden jumps in the local energy. Fermion nodes, where $\varphi_T(\mathbf{x}) = 0$, are complicated multidimensional surfaces, and plotting walker coordinates as shown in figure 4.6 does not provide much information.

The particle-in-a-box problem is a toy model, but persistent configurations are also possible in real systems. In fixed-node DMC, the node boundaries of the fermion trial wave function have essentially the same properties as the box edges.

4.5.3 Dealing with drift and and local energy divergence near boundaries

The trial wave function $\varphi_T(\mathbf{x})$ is almost zero near any no-cross boundary, which can be an impenetrable wall, nodal surface, or, for an electron, the location of the nucleus. As we saw, this may cause systematic errors in the DMC results, but there are ways to make DMC more stable.

The drift $\mathbf{F}(\mathbf{x}) = 2\nabla\varphi_T(\mathbf{x})/\varphi_T(\mathbf{x})$ gets too large if $\varphi_T(\mathbf{x}) \approx 0$. This happens near any of the no-cross boundaries listed above. A frequently used remedy was suggested by Umrigar, Nightingale, and Runge (UNR) in [24]:

Stabilized drift using UNR scaling:
For each particle i, compute (the drift velocity is $\mathbf{v}_i := \boldsymbol{\nabla}_i \varphi_T / \varphi_T$, so $\mathbf{v}_i := \mathbf{F}_i/2$)

$$v_i^2 := |\mathbf{v}_i|^2 \tag{4.230}$$

$$s_i = \frac{-1 + \sqrt{1 + 2av_i^2\tau}}{av_i^2\tau} \qquad \text{scaling factor} \tag{4.231}$$

$$\tilde{\mathbf{F}}_i := s_i \mathbf{F}_i, \tag{4.232}$$

with an adjustable parameter a (usually a scalar value $a \in [0, 1]$, or a position-dependent $a(\mathbf{x})$).

The scaled drift update $D\tau\tilde{\mathbf{F}}_i(\mathbf{x})$ is at most

$$D\sqrt{2\tau/a}\,\frac{1}{2}\frac{\mathbf{F}_i(\mathbf{x})}{|\mathbf{F}_i(\mathbf{x})|}, \tag{4.233}$$

a constant-length vector. This is reasonable, considering that the first-order drift–diffusion update is based on the assumption that the drift is approximately constant during the move $\mathbf{x} \to \mathbf{x}'$. Meanwhile, the small-$\tau$ limit of the scaled drift update is just $D\tau\mathbf{F}_i(\mathbf{x})$, so scaling the drift improves stability while keeping the extrapolated estimates at $\tau \to 0$ unimpaired.

In addition to the drift, the local energy also blows up near boundaries, causing the branching factor,

$$e^{-\tau(E_L(\mathbf{x})-E_T)}, \tag{4.234}$$

to blow up. We are at liberty to manipulate this expression as we please, as long as it reduces to the original form in the limit $\tau \to 0$. One option is to let E_L vanish near the boundary by using the UNR-scaled drift:

$$e^{-\tau(E_{\text{best}} - E_T + (E_L(\mathbf{x})-E_{\text{best}})\frac{|\tilde{\mathbf{F}}(\mathbf{x})|}{|\mathbf{F}(\mathbf{x})|})}, \tag{4.235}$$

where E_{best} is the current best guess of the ground-state energy. It is easy to see that in the $\tau \to 0$ limit, the branching reduces to that in equation (4.234). There is one snag, pointed out in [25], namely that the $\tau > 0$ DMC energy is not size consistent, meaning the energy does not scale linearly with size. This underlines the fact that although the $\tau \to 0$ limit scales correctly, actual calculations are always performed using finite τ, and *energy differences* may come out wrong.

As discussed in [25], the binding energies of molecules A and B, $E_{\text{bind}} = E_{A+B} - (E_A + E_B)$, may suffer from size inconsistency if the free molecule energies E_A and E_B are evaluated in two independent DMC runs. Meanwhile, E_{bind} has no size inconsistency if the energy $E_A + E_B$ is evaluated in the same DMC run while keeping molecules A and B far away from each other.

The authors suggest a size-consistent branching factor:

Size-consistent, stabilized branching factor (see [25]):

$$e^{-\tau(\tilde{E}_L(\mathbf{x}) - E_T)},\qquad(4.236)$$

where

$$\tilde{E}_L(\mathbf{x}) := E_{\text{best}} + \text{sign}(E_L(\mathbf{x}) - E_{\text{best}}) \times \min[E_{\text{cut}}, |E_L(\mathbf{x}) - E_{\text{best}}|],\qquad(4.237)$$

$E_{\text{cut}} := a\sqrt{N/\tau}$ for N electrons, and $a \approx 0.2$.

For very small τ, one has $\min[E_{\text{cut}}, |E_L(\mathbf{x}) - E_{\text{best}}|] = |E_L(\mathbf{x}) - E_{\text{best}}|$, and because $sign(z)|z| \equiv z$, one has

$$\text{sign}[E_L(\mathbf{x}) - E_{\text{best}}] \times \min[E_{\text{cut}}, |E_L(\mathbf{x}) - E_{\text{best}}|] = E_L(\mathbf{x}) - E_{\text{best}},\qquad(4.238)$$

and we recover

$$\tilde{E}_L(\mathbf{x}) = E_L(\mathbf{x}) \qquad \text{small-}\tau \text{ limit},\qquad(4.239)$$

as required. Near the boundaries, $E_L(\mathbf{x})$ strays far from E_{best}, so that $\min[E_{\text{cut}}, |E_L(\mathbf{x}) - E_{\text{best}}|] = E_{\text{cut}}$. This leads to

$$\tilde{E}_L(\mathbf{x}) = E_{\text{best}} + \text{sign}(E_L(\mathbf{x}) - E_{\text{best}})E_{\text{cut}},\qquad(4.240)$$

so $\tilde{E}_L(\mathbf{x})$ is pushed away from E_{best} in the *direction* of $E_L(\mathbf{x})$, but only by the amount E_{cut}.[8]

4.6 Case study: DMC for a particle in a box without importance sampling

Can the particle-in-a-box problem be solved without importance sampling, using the crude DMC algorithm? The exact ground-state wave function is not used at all and we are on our own, with only the box potential well and the kinetic operator. To the first order in τ, the imaginary-time evolution is

$$\Psi(\mathbf{x}', \tau) = \int d\mathbf{x}\langle\mathbf{x}'|e^{-\tau\hat{\mathcal{H}}}|\mathbf{x}\rangle\Psi(\mathbf{x}, 0) \approx \int d\mathbf{x}\langle\mathbf{x}'|e^{-\tau\hat{T}}e^{-\tau\hat{V}}|\mathbf{x}\rangle\Psi(\mathbf{x}, 0)\qquad(4.241)$$

$$= \int d\mathbf{x}\langle\mathbf{x}'|e^{-\tau\hat{T}}|\mathbf{x}\rangle e^{-\tau V(\mathbf{x})}\Psi(\mathbf{x}, 0).\qquad(4.242)$$

---

[8] This is the reason why $E_{\text{cut}} \propto \sqrt{N}$ is related to energy variance; for further details, see [25]).

The potential factor is

$$e^{-\tau V(\mathbf{x})} = \begin{cases} 1, \text{ if } \mathbf{x} \text{ is inside the box} \\ 0, \text{ if } \mathbf{x} \text{ is outside the box} \end{cases}. \tag{4.243}$$

In three-dimensional free space, the matrix elements for $N = 1$ are

$$\langle \mathbf{x}' | e^{-\tau \hat{T}} | \mathbf{x} \rangle = \frac{1}{(4\pi D\tau)^{3/2}} e^{-\frac{(\mathbf{x}'-\mathbf{x})^2}{4D\tau}}, \tag{4.244}$$

sampled as $\mathbf{x}' = \mathbf{x} + \sqrt{2D\tau}\,\eta$. The evolution according to the crude DMC algorithm in section 4.3.4 gives the results shown in figure 4.8. Both the energy and the walker probability density converge to the exact ground-state results in the limit of a zero imaginary-time step. The particle-in-a-box problem can be solved without importance sampling, and the crude DMC algorithm is stable, albeit inefficient.

I would like to draw your attention to the peculiar way DMC finds the ground-state energy. The Hamiltonian is known, so there must exist energy estimators, and the potential $V(\mathbf{x}) = 0$ inside the box defines the zero energy level. In section 4.3.4, it was mentioned that for a self-bound potential, $E_0 \approx \langle V \rangle_{\text{walkers}}$, but now walkers are always inside the box with potential $V(\mathbf{x}) = 0$. The other obvious energy measure is the potential outside the box, but that is infinite. From this perspective, it is a small miracle that we are able to find that $E_0 \approx 14.8$ a.u., close to the exact value of 14.804 4066 a.u.

If you scrutinize the code, you find that the branching is (for random numbers $\xi \in U[0, 1]$)

$$M_i^{\text{copies}} = \text{int}\,(e^{-\tau(V(\mathbf{x})-E_T)} + \xi) = \begin{cases} \text{int}\,(e^{\tau E_T} + \xi), \text{ inside} \\ 0, \text{ outside}, \end{cases} \tag{4.245}$$

and E_T is adjusted using, as usual,

$$E_T = E_0^{\text{guess}} + \kappa \ln\left(\frac{M_{\text{target}}}{M}\right). \tag{4.246}$$

In the beginning, I summarily set $E_0^{\text{guess}} = 10$ and update it to the average E_T during the calculation. In other words, E_T is bootstrapped to maintain M_{target} walkers on average.

Without importance sampling, the DMC ground-state energy of a particle in a box is the average trial energy E_T that stabilizes the number of walkers.

As walkers diffuse out of the box, they are deleted; meanwhile, new walkers spawn inside the box at the rate $e^{\tau E_T}$. These two rates are in balance when $E_T = E_0$.

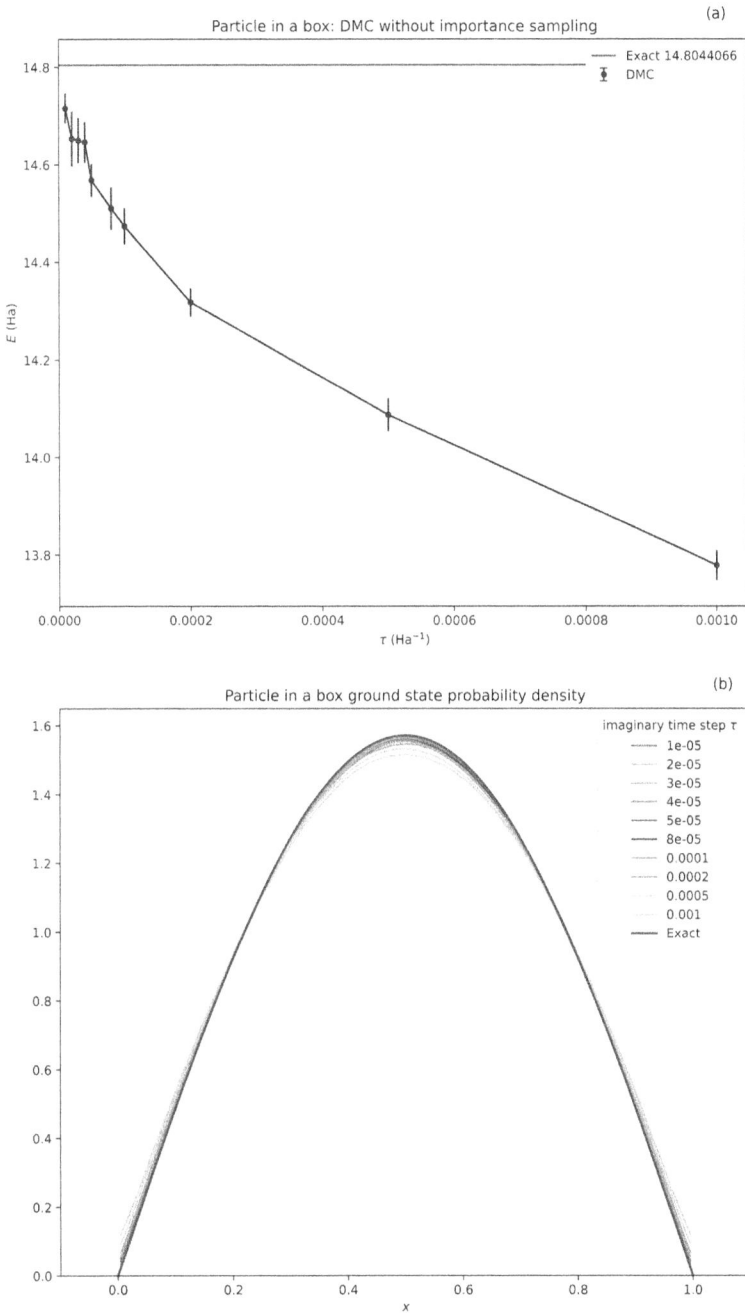

Figure 4.8. Ground-state energy as a function of the imaginary-time step τ (a), and (b) the spherically symmetric probability density profile, compared with the exact result $\pi/2\sin(\pi x)$. The results were computed using the Python code `particle_in_a_box_no_importance_sampling.py` with 5000 walkers.

This is a rather convoluted way of obtaining the energy, which explains why the variance is large.

The algorithm used in the code `particle_in_a_box_no_importance_-sampling.py` is accurate only in the limit $\tau \rightarrow 0$ because we used the first-order approximation $e^{-\tau(\hat{\mathcal{H}} - E_T)} \approx e^{-\tau \hat{\mathcal{T}}} e^{-\tau \hat{\mathcal{V}}} e^{\tau E_T}$. But what does this actually mean in this case? The evolution equation can be written in the exact form

$$\Psi(\mathbf{x}', \tau) = \int d\mathbf{x} \langle \mathbf{x}' | e^{-\tau \hat{\mathcal{T}}} | \mathbf{x} \rangle e^{\tau E_T} \Psi(\mathbf{x}, 0) | \text{inside box}, \tag{4.247}$$

with the infinite potential well converted to a boundary condition. *The non-commutation of $\hat{\mathcal{T}}$ and the potential energy has turned into a problem of diffusion in a bound region.* In a bound region, the free diffusion in equation (4.244) is valid only in the $\tau \rightarrow 0$ limit. For any finite τ, walkers can diffuse out of the box and need to be respawned inside.

Can we do better with diffusion tailored to the box geometry? The eigenfunction expansion gives (see appendix B, unit box)

$$\langle \mathbf{x}' | e^{-\tau \hat{\mathcal{T}}} | \mathbf{x} \rangle = \sum_{\mathbf{q}} e^{-i \mathbf{q} \cdot (\mathbf{x}' - \mathbf{x})} e^{-\tau D q^2}, \tag{4.248}$$

and by choosing the wave vectors so that the plane waves vanish at the box walls, we obtain

$$\langle \mathbf{x}' | e^{-\tau \hat{\mathcal{T}}} | \mathbf{x} \rangle = \sum_{\mathbf{q}} \sin(q_x x') \sin(q_y y') \sin(q_z z') \sin(q_x x) \sin(q_y y) \sin(q_z z) e^{-\tau D q^2}, \tag{4.249}$$

with $\mathbf{q} = \pi \mathbf{n}$: $= \pi(n_1, n_2, n_3)$, $n_i = 1, 2, 3, \dots$. The exact evolution equation is

$$\Psi(\mathbf{x}', \tau) = \int d\mathbf{x} \sum_{\mathbf{q}} \sin(q_x x') \sin(q_y y') \sin(q_z z') \sin(q_x x) \sin(q_y y) \sin(q_z z) e^{-\tau D q^2} e^{\tau E_T} \Psi(\mathbf{x}, 0), \tag{4.250}$$

which is a 'modified' diffusion that respects box limits for any τ. The τ-dependence is trivial, and a stable limiting solution can be found if we choose

$$E_T = D q^2 = D \pi^2 |\mathbf{n}|^2, \text{ for some } \mathbf{n} = (n_1, n_2, n_3), \tag{4.251}$$

so the lowest E_T gives the ground-state energy,

$$E_T = D \pi^2 3 = \frac{1}{2} \pi^2 3 \, \text{a.u.} \approx 14.804\,406\,601 \, \text{a.u.} \tag{4.252}$$

With this choice, the evolution equation is

$$\Psi(\mathbf{x}', \tau) = \int d\mathbf{x} \sum_{\mathbf{n}} \sin(\pi n_1 x') \sin(\pi n_2 y') \sin(\pi n_3 z') \sin(\pi n_1 x) \sin(\pi n_2 y) \sin(\pi n_3 z) \tag{4.253}$$
$$\times e^{-\tau D \pi^2 (n^2 - 1)} \Psi(\mathbf{x}, 0)$$

$$\xrightarrow{\tau \rightarrow \infty} \sin(\pi x') \sin(\pi y') \sin(\pi z') \int d\mathbf{x} \, \sin(\pi x) \sin(\pi y) \sin(\pi z) \Psi(\mathbf{x}, 0). \tag{4.254}$$

Unless $\Psi(\mathbf{x}, 0)$ is orthogonal to the ground state, the integral gives a constant, and we recover the ground-state wave function. The calculation above demonstrates that free diffusion and branching can be replaced with modified diffusion in the simple particle-in-a-box problem.

4.7 DMC for bosons and fermions

The nonrelativistic Hamiltonian contains all assumptions about the system, with one exception: the wave function of identical particles in 3D is either symmetric or antisymmetric. As a consequence, the imaginary-time Schrödinger equation (4.16),

$$-\frac{\partial}{\partial \tau}\Psi(\mathbf{x}, \tau) = (\hat{\mathcal{H}} - E_T)\Psi(\mathbf{x}, \tau) = -D\sum_{i=1}^{N} \nabla_i^2 \ \Psi(\mathbf{x}, \tau) + (V(\mathbf{x}) - E_T)\Psi(\mathbf{x}, \tau), \quad (4.255)$$

has to be supplemented with a symmetry constraint. The overwhelming majority of published DMC calculations do this in terms of **importance sampling**, writing

$$\Psi(\mathbf{x}, \tau) = \varphi_T(\mathbf{x})f(\mathbf{x}, \tau), \quad\quad\quad (4.256)$$

where the trial wave function $\varphi_T(\mathbf{x})$ is symmetric for bosons and antisymmetric for fermions, and all that evolves is the symmetric function $f(\mathbf{x}, \tau)$. For bosons, this is an excellent solution. For fermions, it leads to the so-called *fixed-node DMC* (FN-DMC), first suggested by J B Anderson [26]. The nodes are fixed because, where $\varphi_T(\mathbf{x}) = 0$, there $\Psi(\mathbf{x}, \tau) = 0$. If the nodes are slightly wrong—which they almost always are—we find *a* fermion ground state, but not quite the fermion ground state of our $\hat{\mathcal{H}}$. The good news is that the variational principle is valid, and the FN-DMC energy is an upper bound to the exact ground-state energy [27]. Not surprisingly, fermion DMC calculations revolve around finding better and better trial wave functions.

Before discussing bosons and fermions in terms of importance sampling, it is instructive to take a look at what we are up against when dealing with identical particles in QMC.

Consider two identical particles. Their imaginary-time evolution is

$$\Psi(\mathbf{r}', \mathbf{r}', \tau) = \int d\mathbf{x}\langle \mathbf{r}'_1, \mathbf{r}'_2 | e^{-\tau(\hat{\mathcal{H}} - E_T)} | \mathbf{r}_1, \mathbf{r}_2\rangle\Psi(\mathbf{r}_1, \mathbf{r}_2, 0), \quad\quad (4.257)$$

where $\int d\mathbf{x} := \int d\mathbf{r}_1 d\mathbf{r}_2$. There is nothing wrong with this expression, but we cannot immediately see whether this evolution relates to two identical bosons or two identical fermions. Labeling identical particles with 1 and 2 is suboptimal but necessary for bookkeeping purposes. In practice, fixed labeling is forced upon us by the limitations of (non-quantum) computers. In what follows, $\langle \mathbf{r}'_1, \mathbf{r}'_2 | e^{-\tau(\hat{\mathcal{H}} - E_T)} | \mathbf{r}_1, \mathbf{r}_2\rangle$ denotes a propagator that takes particles $\mathbf{r}_1 \rightarrow \mathbf{r}'_1$ and $\mathbf{r}_2 \rightarrow \mathbf{r}'_2$.

The simplest approach is to force global symmetry from the outset by setting

$$\Psi(\mathbf{r}_1, \mathbf{r}_2, 0) = \pm\Psi(\mathbf{r}_2, \mathbf{r}_1, 0), \quad\quad\quad (4.258)$$

where the upper sign is for bosons and the lower for fermions. Hence, equation (4.257) can be written in the form

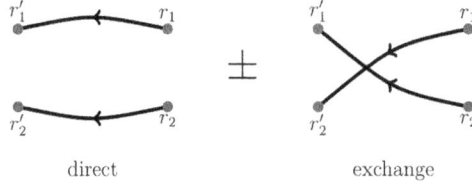

Figure 4.9. The direct and exchange quantum processes for two identical bosons (plus sign) or fermions (minus sign).

$$\Psi(\mathbf{r'}_1, \mathbf{r'}_2, \tau) = \int d\mathbf{x} \langle \mathbf{r'}_1, \mathbf{r'}_2 | e^{-\tau(\hat{\mathcal{H}}-E_T)} | \mathbf{r}_1, \mathbf{r}_2 \rangle (\pm\Psi(\mathbf{r}_2, \mathbf{r}_1, 0)) \tag{4.259}$$

$$= \pm \int d\mathbf{x} \langle \mathbf{r'}_1, \mathbf{r'}_2 | e^{-\tau(\hat{\mathcal{H}}-E_T)} | \mathbf{r}_2, \mathbf{r}_1 \rangle \Psi(\mathbf{r}_1, \mathbf{r}_2, 0), \tag{4.260}$$

where the first equality follows from the (anti)symmetry, and the second follows from swapping the integration variables. Add equations (4.257) and (4.260), and divide by two to get a manifestly bosonic (fermionic) evolution:

$$\Psi(\mathbf{r'}_1, \mathbf{r'}_2, \tau) = \int d\mathbf{x} \frac{1}{2} [\underbrace{\langle \mathbf{r'}_1, \mathbf{r'}_2 | e^{-\tau(\hat{\mathcal{H}}-E_T)} | \mathbf{r}_1, \mathbf{r}_2 \rangle}_{\text{direct}} \pm \underbrace{\langle \mathbf{r'}_1, \mathbf{r'}_2 | e^{-\tau(\hat{\mathcal{H}}-E_T)} | \mathbf{r}_2, \mathbf{r}_1 \rangle}_{\text{exchange}}]$$
$$\underbrace{}_{\text{two-boson(fermion) Green's function}}$$
$$\times \Psi(\mathbf{r}_1, \mathbf{r}_2, 0). \tag{4.261}$$

The terms marked 'direct' and 'exchange' describe the two possible propagators, shown schematically in figure 4.9. The tags 'direct' and 'exchange' refer to our fictitious coordinate labels 1 and 2, introduced for bookkeeping purposes. You could swap 'direct' and 'exchange' with no physical effect whatsoever, because from a quantum point of view they are just two possibilities that cannot be told apart.

4.7.1 Two identical particles in a box

An interparticle potential does not add anything to boson/fermion symmetry, so we consider the particle-in-a-box problem for two ideal bosons or fermions. In section 4.6, we found two ways to solve the evolution without importance sampling for one particle: (i) either use the first-order algorithm with free diffusion and branching or (ii) modify diffusion to stay in the box. Let us start from the latter approach, without actually using either diffusion or Monte Carlo!

Solving imaginary-time evolution on paper

The kinetic operator is $\hat{\mathcal{T}} = \hat{\mathcal{T}}_1 + \hat{\mathcal{T}}_2$, where $\hat{\mathcal{T}}_1$ operates on particle 1 and $\hat{\mathcal{T}}_2$ on particle 2. Furthermore,

$$\hat{\mathcal{T}}_1 \phi_{\mathbf{q}}(\mathbf{r}_1) = Dq^2 \phi_{\mathbf{q}}(\mathbf{r}_1), \tag{4.262}$$

and similarly for $\hat{\mathcal{T}}_2$. The single-particle states $\phi_{\mathbf{q}}(\mathbf{r})$ are plane waves confined to the box,

$$\phi_{\mathbf{q}}(\mathbf{r}): = 2^{3/2} \sin(q_x x)\sin(q_y y)\sin(q_z z), \tag{4.263}$$

with $\mathbf{q} = \pi(n_1, n_2, n_3)$, where $n_i = 1, 2, 3. \ldots$ The single-particle Green's function is

$$\langle \mathbf{r}'|e^{-\tau\hat{T}_1}|\mathbf{r}_1\rangle = \sum_{\mathbf{q}}\phi_{\mathbf{q}}(\mathbf{r}')\phi_{\mathbf{q}}(\mathbf{r}_1)e^{-\tau Dq^2}. \tag{4.264}$$

The two-particle evolution is

$$\Psi(\mathbf{r}', \mathbf{r}', \tau) = \int d\mathbf{x}\frac{1}{2}\left[\sum_{\mathbf{q}}\phi_{\mathbf{q}}(\mathbf{r}')\phi_{\mathbf{q}}(\mathbf{r}_1)e^{-\tau Dq^2}\sum_{\mathbf{p}}\phi_{\mathbf{p}}(\mathbf{r}')\phi_{\mathbf{p}}(\mathbf{r}_2)e^{-\tau Dp^2}\right.$$

$$\left. \pm \sum_{\mathbf{q}}\phi_{\mathbf{q}}(\mathbf{r}')\phi_{\mathbf{q}}(\mathbf{r}_1)e^{-\tau Dq^2}\sum_{\mathbf{p}}\phi_{\mathbf{p}}(\mathbf{r}')\phi_{\mathbf{p}}(\mathbf{r}_2)e^{-\tau Dp^2}\right]e^{\tau E_T}\Psi(\mathbf{r}_1, \mathbf{r}_2, 0). \tag{4.265}$$

For bosons, a stable $\tau \to \infty$ limit requires E_T to cancel the τ-dependence of the term $\mathbf{p} = \mathbf{q} = \pi(1, 1, 1)$: $=$"1", and in units of $D = 1/2$, we find

Two non-interacting bosons in a box:

$E_0 = E_T = 3\pi^2$ ground state energy $\qquad\qquad$ (4.266)

$\Phi_0(\mathbf{x}) = \text{const.} \times \phi_1(\mathbf{r}_1)\phi_1(\mathbf{r}_2)$ ground state wave function

$$\phi_1(\mathbf{r}) = \sin(\pi x)\sin(\pi y)\sin(\pi z). \tag{4.267}$$

Notice that we did not specify $\Psi(\mathbf{r}_1, \mathbf{r}_2, 0)$, only that it is not orthogonal to the ground state,

$$\text{const.} = \int d\mathbf{x}\phi_1(\mathbf{r}_1)\phi_1(\mathbf{r}_2)\Psi(\mathbf{r}_1, \mathbf{r}_2, 0) \neq 0, \tag{4.268}$$

which implies that $\Psi(\mathbf{r}_1, \mathbf{r}_2, 0)$ must be symmetric.

For fermions, the same choice would give $\Psi(\mathbf{r}'_1, \mathbf{r}'_2, \tau) = 0$, which contradicts the assumption of two fermions in a box. If we choose E_T to have the second-lowest value, the wave vectors are "1": $=\pi(1, 1, 1)$ and "2": $=\pi(2, 1, 1)$ (degenerate with $\pi(1, 2, 1)$ and $\pi(1, 1, 2)$),

Two non-interacting fermions in a box:

$E_0 = E_T = \dfrac{9}{2}\pi^2$ ground-state energy

$\Phi_0(\mathbf{x}) = \text{const.} \times \begin{vmatrix} \phi_1(\mathbf{r}_1) & \phi_2(\mathbf{r}_1) \\ \phi_1(\mathbf{r}_2) & \phi_2(\mathbf{r}_2) \end{vmatrix}$ ground-state wave function $\qquad$ (4.269)

$\phi_1(\mathbf{r}) = \sin(\pi x)\sin(\pi y)\sin(\pi z)$

$\phi_2(\mathbf{r}) = \sin(2\pi x)\sin(\pi y)\sin(\pi z),$

where ϕ_2 can be any of the three degenerate cases. Again, the constant multiplying the ground state must be nonzero,

$$\text{const.} = \int d\mathbf{x}[\phi_1(\mathbf{r}_1)\phi_2(\mathbf{r}_2) - \phi_2(\mathbf{r}_1)\phi_1(\mathbf{r}_2)]\Psi(\mathbf{r}_1, \mathbf{r}_2, 0) \neq 0, \tag{4.270}$$

which implies that $\Psi(\mathbf{r}_1, \mathbf{r}_2, 0)$ must be antisymmetric and non-orthogonal with the exact ground-state wave function.

Swapping the integration variables, we find that

$$\text{const.} = \frac{1}{2} \int d\mathbf{x}\Big[[\phi_1(\mathbf{r}_1)\phi_2(\mathbf{r}_2) - \phi_2(\mathbf{r}_1)\phi_1(\mathbf{r}_2)]\Psi(\mathbf{r}_1, \mathbf{r}_2, 0)$$
$$+ [\phi_1(\mathbf{r}_2)\phi_2(\mathbf{r}_1) - \phi_2(\mathbf{r}_2)\phi_1(\mathbf{r}_1)]\Psi(\mathbf{r}_2, \mathbf{r}_1, 0)] \tag{4.271}$$

$$= \frac{1}{2} \int d\mathbf{x}[\phi_1(\mathbf{r}_1)\phi_2(\mathbf{r}_2) - \phi_2(\mathbf{r}_1)\phi_1(\mathbf{r}_2)][\Psi(\mathbf{r}_1, \mathbf{r}_2, 0) - \Psi(\mathbf{r}_2, \mathbf{r}_1, 0)], \tag{4.272}$$

and upon comparing this with the original integral (4.270), we see that $\Psi(\mathbf{r}_1, \mathbf{r}_2, 0) = -\Psi(\mathbf{r}_2, \mathbf{r}_1, 0)$.

The exact ground state has nodes at

$$\Phi_0(\mathbf{x}) = \phi_1(\mathbf{r}_1)\phi_2(\mathbf{r}_2) - \phi_2(\mathbf{r}_1)\phi_1(\mathbf{r}_2) = 0 \tag{4.273}$$

$$\Leftrightarrow \sin(\pi x_1)\sin(2\pi x_2) - \sin(\pi x_2)\sin(2\pi x_1) = 0 \tag{4.274}$$

$$\Leftrightarrow x_1 = x_2. \tag{4.275}$$

Unlike in 1D, in 3D this does not mean the fermions coincide in space, because their y and z coordinates can take any value between zero and one. The real wave function of N identical fermions in d dimensions has $dN = 6$ variables. One nodal condition takes the nodal surface down to $dN - 1 = 5$ dimensions. In contrast, coincidence in space has $d = 3$ conditions (now $x_1 = x_2$, $y_1 = y_2$, $z_1 = z_2$), which brings the coincidence plane down to $dN - d = 3$ dimensions. 'In more than one dimension these coincidence planes are only a scaffolding through which the nodes pass' [28].

Antisymmetry in $\Psi(\mathbf{r}_1, \mathbf{r}_2, 0)$ does not guarantee that the overlap integral (4.270) is nonzero. Take, for example, the antisymmetric function

$$\Psi(\mathbf{r}_1, \mathbf{r}_2, 0) = \phi_1(\mathbf{r}_1)\phi_3(\mathbf{r}_2) - \phi_3(\mathbf{r}_1)\phi_1(\mathbf{r}_2), \tag{4.276}$$

with ϕ_1 as defined in equation (4.269) and $\phi_3(\mathbf{r}) = \sin(3\pi x)\sin(\pi x)\sin(\pi x)$. The function (4.276) has nodes at $x_1 = x_2$, just like the ground state; however, it also has them at $x_2 = 1 - x_1$, which is orthogonal to the ground state. In this case, the imaginary time evolution does not converge to the fermion ground state but to a fermion state with slightly higher energy.

Solving the imaginary time evolution of two fermions using DMC

The Julia code `two_fermions_in_a_box_no_importance_sampling.jl` uses the exact nodes of the ground-state wave function, or, alternatively, those of (4.276). Notice that the code really uses only the nodes, and simply returns `true` if the walker is still in the positive cell where it started, or `false` if not. In the latter case, the walker is deleted, to be balanced by the $\epsilon^{\tau E_T}$ birth rate—this is the only reason why the $\tau \to 0$ limit is necessary. The DMC results are shown in figure 4.10. Here, the nodal surface acts as a multidimensional no-cross barrier and is handled in exactly the same way as box walls.

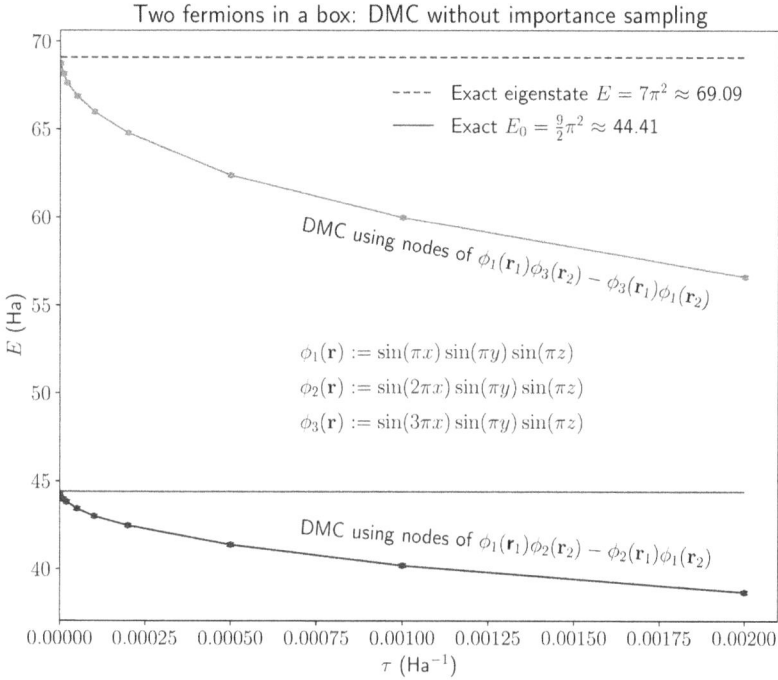

Figure 4.10. Using the exact ground-state nodes allows DMC to find the fermion ground-state energy E_0, while the other choice of nodes forces DMC to find the corresponding fermion excited-state energy. No importance sampling was used, only nodes. The DMC results were computed using the Julia code `two_fermions_in_a_box_no_importance_sampling.jl`.

> DMC imaginary-time evolution finds the fermion ground state only if the nodes of $\Psi(\mathbf{r}_1, \mathbf{r}_2, 0)$ are those of the exact ground state.

To this day, the $(dN - 1)$-dimensional manifold of the exact ground-state nodes has resisted accurate solutions, and there is no reason why interacting fermion nodes should be those of noninteracting fermions.

Solving imaginary-time evolution using DMC without importance sampling

DMC without importance sampling solves the evolution equation

$$
\Psi(\mathbf{r}'_1, \mathbf{r}'_2, \tau) = \int d\mathbf{x} [\langle \mathbf{r}'_1 | e^{-\tau \hat{T}_1} | \mathbf{r}_1 \rangle \langle \mathbf{r}'_2 | e^{-\tau \hat{T}_2} | \mathbf{r}_2 \rangle \pm \langle \mathbf{r}'_2 | e^{-\tau \hat{T}_1} | \mathbf{r}_1 \rangle \langle \mathbf{r}'_1 | e^{-\tau \hat{T}_2} | \mathbf{r}_2 \rangle] \\
\times e^{\tau E_T} \Psi(\mathbf{r}_1, \mathbf{r}_2, 0).
\tag{4.277}
$$

As we saw, the only restriction on $\Psi(\mathbf{r}_1, \mathbf{r}_2, 0)$ is that it must be (anti)symmetric.

Two bosons

Consider M walkers, each with two coordinates picked from the box at random. They represent the initial wave function

$$\Psi(\mathbf{r}_1, \mathbf{r}_2, 0) = \begin{cases} 1, & \text{if } \mathbf{r}_1, \mathbf{r}_2 \text{ in the box} \\ 0, & \text{elsewhere} \end{cases}. \tag{4.278}$$

This is clearly symmetric, so the imaginary-time evolution converges to the boson ground state (see figure 4.11). When free diffusion is used in equation (4.277), walkers spill out of the box; therefore, only the $\tau \to 0$ limit is correct.

Two fermions

The symmetric boson ground state is the 'natural' limiting distribution of the DMC algorithm. The question is how to force the evolution to converge to something fermionic. We need to express the fermion wave function as distributions that can subsequently be represented as walkers. Only then do we have walkers that can meaningfully diffuse.

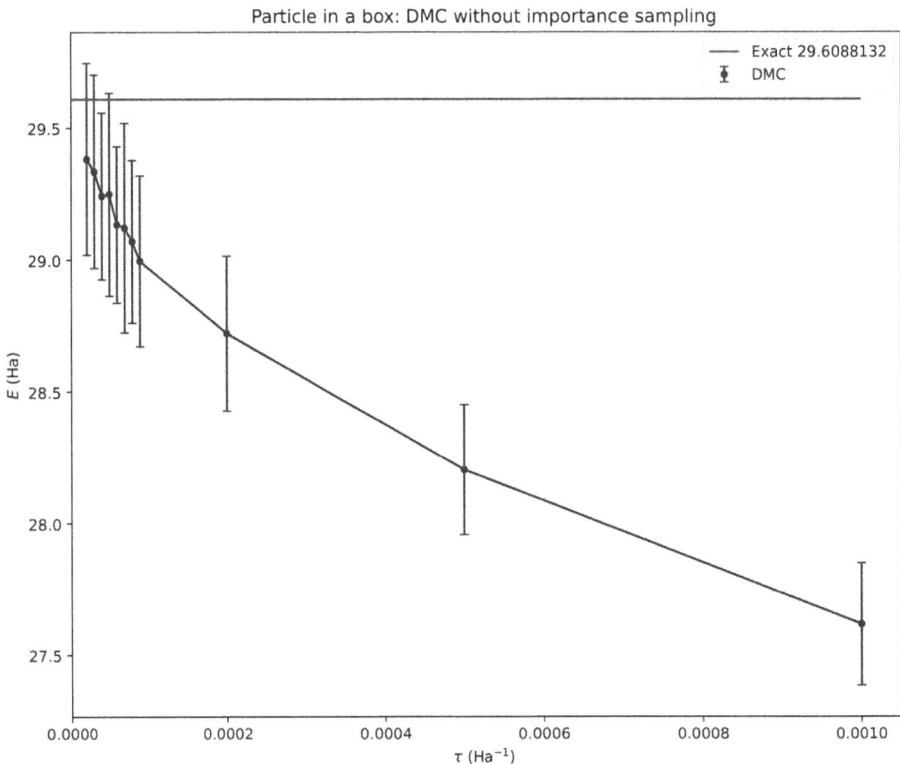

Figure 4.11. The two-boson ground-state energy as a function of the imaginary-time step τ. The results were computed using the Python code `two_bosons_in_a_box_no_importance_sampling.py` with 5000 walkers. The code is essentially the $N = 2$ version of `particle_in_a_box_no_importance_sampling.py`.

Let us first isolate positive and negative domains of the wave function by writing it as the difference of two positive functions,

$$\Psi(\mathbf{r}_1, \mathbf{r}_2) = \underbrace{\Psi^+(\mathbf{r}_1, \mathbf{r}_2)}_{\geqslant 0} - \underbrace{\Psi^-(\mathbf{r}_1, \mathbf{r}_2)}_{\geqslant 0}. \tag{4.279}$$

The two functions are $\Psi^\pm = \frac{1}{2}(|\Psi| \pm \Psi)$, that is,

$$\Psi^+(\mathbf{r}_1, \mathbf{r}_2) = \begin{cases} \Psi(\mathbf{r}_1, \mathbf{r}_2), & \text{if } \Psi(\mathbf{r}_1, \mathbf{r}_2) > 0, \quad \text{positive cell} \\ 0, & \text{else} \end{cases} \tag{4.280}$$

$$\Psi^-(\mathbf{r}_1, \mathbf{r}_2) = \begin{cases} -\Psi(\mathbf{r}_1, \mathbf{r}_2), & \text{if } \Psi(\mathbf{r}_1, \mathbf{r}_2) < 0, \text{ negative cell} \\ 0, \text{else} \end{cases}. \tag{4.281}$$

Figure 4.12 is a schematic view of the functions Ψ, Ψ^+, and Ψ^-.

The wave functions in the positive and negative cells are related by particle exchange,

$$\Psi(\mathbf{r}_1, \mathbf{r}_2) = -\Psi(\mathbf{r}_2, \mathbf{r}_1) \Leftrightarrow \Psi^-(\mathbf{r}_1, \mathbf{r}_2) = \Psi^+(\mathbf{r}_2, \mathbf{r}_1), \tag{4.282}$$

and we can sample the cells using 'positive walkers' and 'negative walkers,' respectively. However, sampling *both* cells is wasteful, because they hold the same information, so let us sample, say, Ψ^+,

$$\Psi^+(\mathbf{r}_1, \mathbf{r}_2) \approx \frac{1}{M}\sum_{j=1}^{M}\delta((\mathbf{r}_j)_1^+ - \mathbf{r}_1)\delta((\mathbf{r}_j)_2^+ - \mathbf{r}_2). \tag{4.283}$$

The imaginary-time evolution is

$$\Psi^+(\mathbf{r}'_1, \mathbf{r}'_2, \tau) = \int d\mathbf{x} \langle \mathbf{r}'_1 | e^{-\tau \hat{T}_1} | \mathbf{r}_1 \rangle \langle \mathbf{r}'_2 | e^{-\tau \hat{T}_2} | \mathbf{r}_2 \rangle e^{\tau E_T} \Psi^+(\mathbf{r}_1, \mathbf{r}_2, 0). \tag{4.284}$$

Notice that there is no exchange diffusion term—that would shift walkers to the negative cell. This simplifies the task, as positive walkers have fixed labels. It would seem that all we need is to pick the initial positive walkers and then let them diffuse and branch.

At this point, we hit a serious problem: How can we choose $\Psi^+(\mathbf{r}_1, \mathbf{r}_2, 0)$ or positive walkers that sample it? Choosing $\Psi^+(\mathbf{r}_1, \mathbf{r}_2, 0)$ is equivalent to choosing

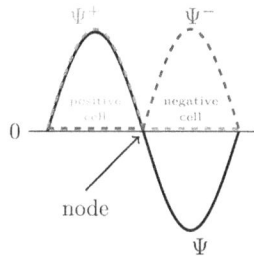

Figure 4.12. Schematic showing how a real wave function Ψ with a node is sampled using two positive functions, Ψ^+ and Ψ^-.

$\Psi(\mathbf{r}_1, \mathbf{r}_2, 0)$, but an antisymmetric function has nodes, so no matter how we solve the imaginary-time evolution, those nodes are also in $\Psi(\mathbf{r}_1, \mathbf{r}_2, \tau)$ at any τ. In other words, we are doing *fixed-node DMC without importance sampling*. The nodes in $\Psi(\mathbf{r}_1, \mathbf{r}_2, 0)$ are the borders of the positive cell, which affect our DMC simulation just like the box walls, except that the positive cell borders are a very complicated multidimensional surface.

Suppose we just throw in random walkers into the box, boldly define them as positive walkers, and sample an unspecified $\Psi^+(\mathbf{r}_1, \mathbf{r}_2, 0)$? After a small-$\tau$ free diffusion, some walkers may step out of the box and be deleted. Furthermore, some walkers may step out of the positive cell and should be deleted, and here lies the problem: we did not specify $\Psi^+(\mathbf{r}_1, \mathbf{r}_2, 0)$, so we have no idea when walkers leave the positive cell.

One could start with positive and negative walkers and concoct a rule that states how they react to each other. This would be something unheard of because, in our case, a walker consists of two coordinates, and correlating walkers means something like 'two fermions feel the presence of another two fermions'. (I am trying to avoid identifying walker coordinates as particle positions.) For example, postulating that there is only one positive and one negative cell, one could imagine that same-sign walkers form a connected region of space. The connection should be local, so that a positive walker needs to be within a certain distance of another positive walker. This should create a positive amoeba and a negative amoeba.

4.7.2 More diffusion near walls or nodes

In the case of noninteracting particles, the rather poor τ-behavior of the crude DMC algorithm stems from walkers crossing actual no-cross boundaries, either leaving the box (bosons and fermions) or crossing node surfaces (fermions). The frequent crossing is caused by the chosen free diffusion, so we would like to use a diffusion Green's function that obeys the appropriate boundaries from the outset. That is, we would like to find eigenstates of the single-particle kinetic operator T_1 with the boundary conditions

$$\hat{T}_1\phi_{\mathbf{q}}(\mathbf{r}_1) = \epsilon(\mathbf{q})\phi_{\mathbf{q}}(\mathbf{r}_1) \tag{4.285}$$

and use them in the spectral expansion, similar to what we did in equation (4.264),

$$\langle\mathbf{r}'_1|e^{-\tau\hat{T}_1}|\mathbf{r}\rangle = \sum_{\mathbf{q}}\phi_{\mathbf{q}}(\mathbf{r}'_1)\phi_{\mathbf{q}}(\mathbf{r}_1)e^{-\tau\epsilon(\mathbf{q})}. \tag{4.286}$$

We already solved the diffusion-in-a-box problem for the unit box $x, y, z \in [0, 1]$ in equations (4.263) and (4.264). This leads to the sum

$$\begin{aligned}\langle\mathbf{r}'|e^{-\tau\hat{T}_1}|\mathbf{r}\rangle = 2\sum_{n_1=1}^{\infty}&\sin(n_1\pi x')\sin(n_1\pi x)e^{-\tau D\pi^2 n_1^2}\\ \times 2\sum_{n_2=1}^{\infty}&\sin(n_2\pi y')\sin(n_2\pi y)e^{-\tau D\pi^2 n_2^2}\\ \times 2\sum_{n_3=1}^{\infty}&\sin(n_3\pi z')\sin(n_3\pi z)e^{-\tau D\pi^2 n_3^2}.\end{aligned} \tag{4.287}$$

The 1D diffusion evolution is

$$P(x', \tau) = e^{\tau E_T} \int_0^1 dx \langle x'|e^{-\tau \hat{T}_1}|x \rangle P(x, 0), \tag{4.288}$$

with the 1D box Green's function

$$\boxed{\langle x'|e^{-\tau \hat{T}_1}|x \rangle = 2\sum_{n=1}^{\infty} \sin(n\pi x')\sin(n\pi x)e^{-\tau D\pi^2 n^2}.} \tag{4.289}$$

As before, a stable $\tau \to \infty$ limit in the noninteracting evolution requires $n = 1$ and $E_T = D\pi^2$, and we can sample points x from the single-particle ground state $\sin(\pi x)$ using, for example, the rejection method (see appendix C). However, we are not always solving the noninteracting problem, and we often need to evaluate the sum in full. In other words, the propagation in an interacting system involves higher eigenstates of $\hat{T}_1$.

In most cases, the kinetic Green's function $\langle x'|e^{-\tau \hat{T}_1}|x \rangle$ is just one factor in a bigger picture where we have been forced to assume that τ is small. In this scenario, the infinite sum converges really slowly, but it can be rewritten in a fast-converging form (see the discussion below)

$$\boxed{\langle x'|e^{-\tau \hat{T}_1}|x \rangle = \frac{1}{\sqrt{4\pi D\tau}} \sum_{n=-\infty}^{\infty} \left[e^{-\frac{(x'-x+2n)^2}{4D\tau}} - e^{-\frac{(x'+x+2n)^2}{4D\tau}} \right].} \tag{4.290}$$

In this sum, the index n no longer enumerates the eigenstates of $\hat{T}_1$ but the number of Gaussians. The two ways of evaluating the n-sums are compared in figure 4.13. The same-n terms cancel each other at the boundary $x = 0$, while at the boundary $x = 1$, the positive term with index n cancels the negative one with index $n - 1$.

The partition function of a particle in an infinite potential well of unit length (1D box), namely

$$Z = \sum_{n=1}^{\infty} e^{-\beta E_n} = \sum_{n=1}^{\infty} e^{-\beta n^2 D\pi^2}, \quad \beta: = \frac{1}{k_B T}, \tag{4.291}$$

can be written in terms of the *Jacobi theta function* $\vartheta(x)$,

$$\vartheta(x): = \sum_{n=-\infty}^{\infty} e^{-n^2\pi^2 x} \Leftrightarrow \sum_{n=1}^{\infty} e^{-n^2\pi^2 x} = \frac{1}{2}(\vartheta(x) - 1). \tag{4.292}$$

Remarkably, $\sqrt{x}\,\vartheta(x) = \vartheta(1/x)$, and since, in our case, we have $\vartheta(D\tau)$, this property of the Jacobi theta function is exactly what we need to turn the small-τ sum that slowly converges with $D\tau$ into one that converges fast with $1/(D\tau)$. This fact was exploited in the field of biochemistry in [29], in a Monte Carlo study of the motion and binding of ligand molecules between two membranes. First, use the identity $\sin(a)\sin(b) = [\cos(a - b) - \cos(a + b)]/2$ to get

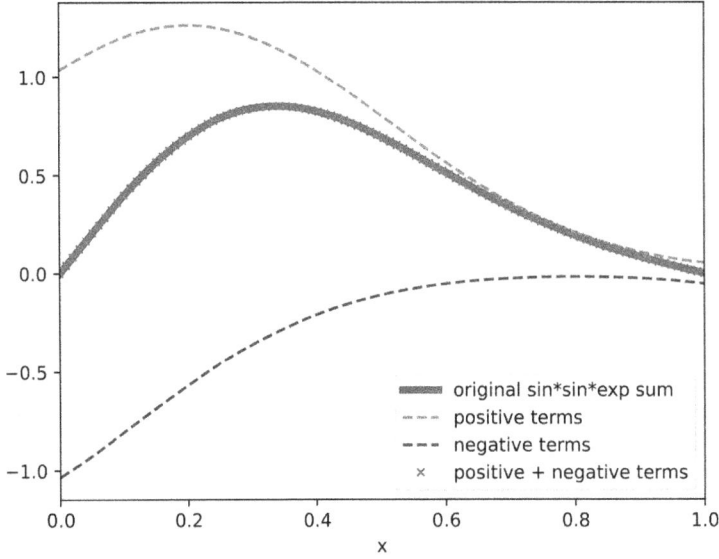

Figure 4.13. The thick curve shows $\langle x'|e^{-\tau \hat{T}_1}|x \rangle$ for fixed $x = 0.2$ and $\tau = 0.1$ as a function of x', computed using the sum in equation (4.289). The dashed lines show the positive and negative Gaussian term sums in equation (4.290). The figure was generated using the Python code plot_box_gaussians.py.

$$\langle x'|e^{-\tau \hat{T}_1}|x \rangle = 2\sum_{n=1}^{\infty} \sin(n\pi x')\sin(n\pi x)e^{-\tau D\pi^2 n^2} \tag{4.293}$$

$$= \sum_{n=1}^{\infty}[\cos(n\pi(x' - x))e^{-\tau D\pi^2 n^2} - \cos(n\pi(x' + x))e^{-\tau D\pi^2 n^2}] \tag{4.294}$$

$$= \frac{1}{2} \sum_{n=-\infty}^{\infty} [\cos(n\pi(x' - x))e^{-\tau D\pi^2 n^2} - \cos(n\pi(x' + x))e^{-\tau D\pi^2 n^2}], \tag{4.295}$$

because the terms are even functions of n, and $n = 0$ gives a zero term. Next, use the Jacobi theta function identity,

$$\sum_{n=-\infty}^{\infty} \cos(2\pi n y)e^{-n^2\pi^2 x} = \frac{1}{\sqrt{\pi x}} \sum_{n=-\infty}^{\infty} e^{-\frac{(n+y)^2}{x}}, \text{ for } y \in \mathbb{R}, \, x > 0. \tag{4.296}$$

Replace $y \to (x' \pm x)/2$ and let $x \to D\tau$ to get equation (4.290).

The diffusion algorithm devised in [29] uses reflecting boundary conditions common in classical diffusion:

$$\frac{\partial}{\partial x} density(x, t) = 0, \text{ at boundaries;} \tag{4.297}$$

however, quantum particles have a vanishing wave function at the boundaries. The effect is that in equation (4.293), classical diffusion has $\cos(n\pi x')\cos(n\pi x)$ instead of the quantum diffusion $\sin(n\pi x')\sin(n\pi x')$. Both choices lead to an equation like

(4.290), but because $\cos(a)\cos(b) = [\cos(a - b) + \cos(a + b)]/2$, the classical result has a plus sign between the summed terms, while the quantum case has a minus sign.

The terms in equation (4.290) are Gaussians centered at $x - 2n$ and $-x + 2n$, so they can be interpreted as the free diffusion of 'positive walkers,' with evolution $x' = x - 2n + \sqrt{2D\tau}\,\eta$, and the free diffusion of 'negative walkers,' with evolution $x' = -x - 2n + \sqrt{2D\tau}\,\eta$, respectively. Here, η are normally distributed random numbers. If τ is so small that the Gaussians have standard deviation, $\sqrt{2D\tau} \ll 1$, then the Gaussians reaching the cell $x' \in [0, 1]$ are just those with centers in same cell with $n = 0$, and those with centers two cells to the left with $n = 1$ or those with centers two cells to the right with $n = -1$. These give the approximation

$$\langle x'|e^{-\tau\hat{T}_1}|x\rangle \approx \frac{1}{\sqrt{4\pi D\tau}}\left[e^{-\frac{(x'-x)^2}{4D\tau}} - e^{-\frac{(x'+x)^2}{4D\tau}} - e^{-\frac{(x'+x-2)^2}{4D\tau}} \right], \qquad \text{small } \tau. \quad (4.298)$$

The evolution is

$$P(x', \tau) \approx \int_0^1 dx \int d\eta \frac{1}{\sqrt{2\pi}} e^{-\eta^2/2}\left(1 - e^{-\frac{x^2}{D\tau} - \frac{x\eta}{\sqrt{D\tau/2}}} - e^{-\frac{(x-1)^2}{D\tau} - \frac{(x-1)\eta}{\sqrt{D\tau/2}}} \right)$$
$$\times \delta(x' - (x + \sqrt{4D\tau}\,\eta))P(x, 0), \qquad (4.299)$$

which is free diffusion $x' = x + \sqrt{4D\tau}\,\eta$ with an extra weight given in parentheses. Because τ is small, the x-dependent weight is always positive and can be added using an accept–reject step.

The three terms in equation (4.298) could also be interpreted as representing the free diffusion of 'positive walkers' in the cell $[0, 1]$, the counter-diffusion of 'negative walkers' from the left of $x = 0$, and the counter-diffusion of 'negative walkers' from the right of $x = 1$. Remembering that $\langle x'|e^{-\tau\hat{T}_1}|x\rangle$ as a whole is positive, we come to the following conclusion:

The division into 'positive' and 'negative' walkers is an artifact of expressing the problem of diffusion between two walls in terms of freely diffusing walkers.

Even though negative walkers make the free diffusion approach cumbersome, the sum expression in equation (4.290) is valid and can be sampled directly.

Sampling 1D diffusion between walls directly (ineffective but educative)

Is the rejection method applicable to our diffusion problem? The matrix elements do have an upper limit,

$$\langle x'|e^{-\tau\hat{T}_1}|x\rangle = 2\sum_{n=1}^{\infty} \sin(n\pi x')\sin(n\pi x)e^{-\tau D\pi^2 n^2} \qquad (4.300)$$

$$\leqslant 2\sum_{n=1}^{\infty} e^{-\tau D \pi^2 n^2}: = \mathcal{M}, \tag{4.301}$$

so $\mathcal{M}$ is (twice) the single-particle partition function $Z_1(\beta)$ at a fixed inverse temperature $\beta = \tau$. Keeping x and τ as parameters, we could tentatively use the rejection method with the weight

$$w(x') \stackrel{\text{try}}{=} \langle x' | e^{-\tau \hat{\mathcal{T}}} | x \rangle. \tag{4.302}$$

However, the matrix element is not a density,

$$\int_0^1 dx' w(x') = 4 \sum_{n=1;\,\text{odd}}^{\infty} \frac{\sin(n\pi x)}{n\pi} e^{-\tau D \pi^2 n^2} \neq \text{constant}. \tag{4.303}$$

If τ is small, we can use

$$\int_0^1 dx' w(x') \approx \frac{1}{\sqrt{4\pi D\tau}} \int_0^1 dx' \left[e^{-\frac{(x'-x)^2}{4D\tau}} - e^{-\frac{(x'+x)^2}{4D\tau}} - e^{-\frac{(x'+x-2)^2}{4D\tau}} \right] \tag{4.304}$$

$$= \text{erf}\left(\frac{x}{\sqrt{4D\tau}}\right) - \text{erf}\left(\frac{x-1}{\sqrt{4D\tau}}\right) + \frac{1}{2}\text{erf}\left(\frac{x-2}{\sqrt{4D\tau}}\right) - \frac{1}{2}\text{erf}\left(\frac{x+1}{\sqrt{4D\tau}}\right). \tag{4.305}$$

A 'walker-friendly' evolution is now

$$P(x', \tau) = e^{\tau E_T} \int_0^1 dx \frac{\langle x' | e^{-\tau \hat{\mathcal{T}}_1} | x \rangle}{\int_0^1 dx' w(x')} P(x, 0). \tag{4.306}$$

Summary for two noninteracting, identical particles in a box:
- The single-particle states $\phi_i(\mathbf{r})$ (originally labeled as $\phi_\mathbf{q}(\mathbf{r})$) first appeared as the eigenfunctions of the single-particle kinetic operators $\hat{\mathcal{T}}_1$ and $\hat{\mathcal{T}}_2$, and from there in the eigenfunction decomposition of the kinetic matrix elements. The fact that the kinetic operator is a sum of single-particle operators, $\hat{\mathcal{T}} = \hat{\mathcal{T}}_1 + \hat{\mathcal{T}}_2$, splits the formulas into single-particle contributions.
- The imaginary-time evolution did not need any branching term, because the potential only keeps particles in the box, and we could adjust the single-particle kinetic operator eigenstates to do the exact same task.
- The fermion ground state has higher energy than the boson ground state.
- No fermion sign problem: the minus sign in the fermion Green's function caused no problems because no Monte Carlo randomness was needed.
- The fermion ground state has a degeneracy of three.
- The fermion ground state has **two nodal cells**, one where $\Phi_0 > 0$ and one where $\Phi_0 < 0$, stemming from the sign change in $\sin(2\pi x)$ in the single-particle state $\phi_2(\mathbf{r})$ (or the degenerate $\sin(2\pi y)$ and $\sin(2\pi z)$).
- The imaginary-time evolution converges to the ground state, provided the initial $\Psi(\mathbf{r}_1, \mathbf{r}_2, 0)$ is symmetric for bosons. For fermions, fixed-node DMC

gives the ground state as the limiting distribution only if $\Psi(\mathbf{r}_1, \mathbf{r}_2, 0)$ is both antisymmetric and has the exact ground-state nodes.

- To the first order in τ, the boson (fermion) evolution is

$$\Psi(\mathbf{r}'_1, \mathbf{r}'_2, \tau) \approx \int d\mathbf{r}_1 d\mathbf{r}_2 \frac{1}{2}[\langle \mathbf{r}'_1, \mathbf{r}'_2|e^{-\tau\hat{T}}|\mathbf{r}_1, \mathbf{r}_2\rangle \pm \langle \mathbf{r}'_2, \mathbf{r}'_1|e^{-\tau\hat{T}}|\mathbf{r}_1, \mathbf{r}_2\rangle]$$
$$\times e^{-\tau(V(\mathbf{r}_1)+V(\mathbf{r}_2))}e^{\tau E_T}\Psi(\mathbf{r}_1, \mathbf{r}_2, 0). \tag{4.307}$$

Use the single-particle free diffusion Green's function,

$$\langle \mathbf{r}'_1|e^{-\tau\hat{T}}|\mathbf{r}_1\rangle = \frac{1}{(4\pi D\tau)^3}e^{-\frac{(\mathbf{r}'_1-\mathbf{r}_1)^2}{4D\tau}}, \tag{4.308}$$

together with branching, so that walkers moving out of the box are respawned inside. This was shown to give the correct result in the $\tau \to 0$ limit. The fermion Green's function for free diffusion can be written as a determinant,

$$\frac{1}{2}[\langle \mathbf{r}'_1, \mathbf{r}'_2|e^{-\tau\hat{T}}|\mathbf{r}_1, \mathbf{r}_2\rangle - \langle \mathbf{r}'_2, \mathbf{r}'_1|e^{-\tau\hat{T}}|\mathbf{r}_1, \mathbf{r}_2\rangle] = \frac{1}{2}\frac{1}{(4\pi D\tau)^6}\begin{vmatrix} e^{-\frac{(\mathbf{r}'_1-\mathbf{r}_1)^2}{4D\tau}} & e^{-\frac{(\mathbf{r}'_1-\mathbf{r}_2)^2}{4D\tau}} \\ e^{-\frac{(\mathbf{r}'_2-\mathbf{r}_1)^2}{4D\tau}} & e^{-\frac{(\mathbf{r}'_2-\mathbf{r}_2)^2}{4D\tau}} \end{vmatrix} \tag{4.309}$$

where the minus sign causes the so-called **fermion sign problem**.

- Modify diffusion so that walkers do not leave the box. The box potential is present only as a boundary condition.

Free diffusion moves walkers out of the box to infinite potential; therefore, we have two choices: either use $\Psi(\mathbf{r}_1, \mathbf{r}_2, 0)$ to keep particles inside the box or modify diffusion so that particles cannot move through walls.

The initial wave function $\Psi(\mathbf{x}, 0): = \Psi(\mathbf{r}_1, \mathbf{r}_2, 0)$ can be represented by M walkers,

$$\Psi(\mathbf{x}, 0) \approx \frac{1}{M}\sum_{j=1}^{M}\delta(\mathbf{x} - \mathbf{x}_j(0)), \tag{4.310}$$

where $\mathbf{x}_j(0)$ are pairs of random coordinates in the box.

4.7.2.1 Fermions

In the absence of magnetic fields, the Hamiltonian is real, and the fermion ground-state wave function $\Phi_0(\mathbf{x})$ can be chosen to be real-valued. Because of the Pauli exclusion principle, the fermion wave functions have nodes, but a function that changes sign cannot be interpreted using one distribution for every $\mathbf{x}$. However, we can isolate the positive and negative domains,

$$\Phi_0(\mathbf{x}) = \int d\mathbf{y}\Phi_0(\mathbf{y})\delta(\mathbf{y} - \mathbf{x})$$
$$= \int d\mathbf{y} \underbrace{[\Phi_0(\mathbf{y})_+]}_{\text{distribution } (+)} \delta(\mathbf{y} - \mathbf{x}) - \int d\mathbf{y} \underbrace{[\Phi_0(\mathbf{y})_-]}_{\text{distribution } (-)} \delta(\mathbf{y} - \mathbf{x}), \tag{4.311}$$

and sample the positive walkers $\{(\mathbf{x}_i)_+\}$ and the negative walkers $\{(\mathbf{x}_i)_-\}$ from the two distributions,

$$\Phi_0(\mathbf{x}) \approx \frac{1}{M}\sum_{i=1}^{M}\delta((\mathbf{x}_i)_+ - \mathbf{x}) - \frac{1}{M}\sum_{i=1}^{M}\delta((\mathbf{x}_i)_- - \mathbf{x}). \tag{4.312}$$

Again, we would like to see the antisymmetry of the wave function explicitly. Following the steps we used to derive the two-boson result for two fermions, and using $\Phi_0(\mathbf{r}'_1, \mathbf{r}'_2) = -\Phi_0(\mathbf{r}'_2, \mathbf{r}'_1)$, we arrive at the expression

$$\Phi_0(\mathbf{r}'_1, \mathbf{r}'_2) = \int d\mathbf{r}_1 d\mathbf{r}_2 \frac{1}{2}[\underbrace{\delta(\mathbf{r}_1 - \mathbf{r}'_1)\delta(\mathbf{r}_2 - \mathbf{r}'_2)}_{\text{direct}} - \underbrace{\delta(\mathbf{r}_2 - \mathbf{r}'_1)\delta(\mathbf{r}_1 - \mathbf{r}'_2)}_{\text{exchange}}]\Phi_0(\mathbf{r}_1, \mathbf{r}_2). \tag{4.313}$$

Next, we separate the positive and the negative domains of the wave function,

$$\Phi_0(\mathbf{r}_1, \mathbf{r}_2) = \underbrace{\Phi_0^+(\mathbf{r}_1, \mathbf{r}_2)}_{\geqslant 0} - \underbrace{\Phi_0^-(\mathbf{r}_1, \mathbf{r}_2)}_{\geqslant 0}. \tag{4.314}$$

The parts Φ_0^+ and Φ_0^- are related by particle exchange,

$$\Phi_0(\mathbf{r}_2, \mathbf{r}_1) = -\Phi_0(\mathbf{r}_1, \mathbf{r}_2) \Leftrightarrow \Phi_0^-(\mathbf{r}_1, \mathbf{r}_2) = \Phi_0^+(\mathbf{r}_2, \mathbf{r}_1), \tag{4.315}$$

and Φ_0^+ satisfies the integral equations

$$\boxed{\Phi_0^+(\mathbf{r}'_1, \mathbf{r}'_2) = \int d\mathbf{r}_1 d\mathbf{r}_2 \delta(\mathbf{r}_1 - \mathbf{r}'_1)\delta(\mathbf{r}_2 - \mathbf{r}'_2)\Phi_0^+(\mathbf{r}_1, \mathbf{r}_2).} \tag{4.316}$$

For positive functions a, b, c, and d, one has

$$a(x) - b(x) = c(x) - d(x), \forall x \Leftrightarrow a(x) = c(x) + \alpha \text{ and } b(x) = d(x) + \alpha, \forall x, \tag{4.317}$$

where α is a constant. Therefore,

$$\Phi_0(\mathbf{r}_2, \mathbf{r}_1) = -\Phi_0(\mathbf{r}_1, \mathbf{r}_2) \Leftrightarrow \Phi_0^+(\mathbf{r}_2, \mathbf{r}_1) - \Phi_0^-(\mathbf{r}_2, \mathbf{r}_1) = \Phi_0^-(\mathbf{r}_1, \mathbf{r}_2) - \Phi_0^+(\mathbf{r}_1, \mathbf{r}_2) \tag{4.318}$$

$$\Leftrightarrow \Phi_0^-(\mathbf{r}_1, \mathbf{r}_2) = \Phi_0^+(\mathbf{r}_2, \mathbf{r}_1) + \alpha. \tag{4.319}$$

Now $\Phi_0(\mathbf{r}_1, \mathbf{r}_2)$ is a normalizable wave function, so $\alpha = 0$.

Insert $\Phi_0 = \Phi_0^+ - \Phi_0^-$ into equation (4.313):

$$\begin{aligned}
\Phi_0^+(\mathbf{r}'_1, \mathbf{r}'_2) &- \Phi_0^-(\mathbf{r}'_1, \mathbf{r}'_2) \\
&= \int d\mathbf{r}_1 d\mathbf{r}_2 \frac{1}{2}[\delta(\mathbf{r}_1 - \mathbf{r}'_1)\delta(\mathbf{r}_2 - \mathbf{r}'_2) - \delta(\mathbf{r}_2 - \mathbf{r}'_1)\delta(\mathbf{r}_1 - \mathbf{r}'_2)] \\
&\times [\Phi_0^+(\mathbf{r}_1, \mathbf{r}_2) - \Phi_0^-(\mathbf{r}_1, \mathbf{r}_2)].
\end{aligned} \tag{4.320}$$

The positive and negative terms give two coupled integral equations,

$$\Phi_0^+(\mathbf{r}'_1, \mathbf{r}'_2) = \int d\mathbf{r}_1 d\mathbf{r}_2 \frac{1}{2}\Big[\delta(\mathbf{r}_1 - \mathbf{r}'_1)\delta(\mathbf{r}_2 - \mathbf{r}'_2)\Phi_0^+(\mathbf{r}_1, \mathbf{r}_2) + \delta(\mathbf{r}_2 - \mathbf{r}'_1)\delta(\mathbf{r}_1 - \mathbf{r}'_2)\Phi_0^-(\mathbf{r}_1, \mathbf{r}_2)\Big] \tag{4.321}$$

$$\Phi_0^-(\mathbf{r'}_1, \mathbf{r'}_2) = \int d\mathbf{r}_1 d\mathbf{r}_2 \frac{1}{2} \Big[\delta(\mathbf{r}_2 - \mathbf{r'}_1)\delta(\mathbf{r}_1 - \mathbf{r'}_2)\Phi_0^+(\mathbf{r}_1, \mathbf{r}_2) + \delta(\mathbf{r}_1 - \mathbf{r'}_1)\delta(\mathbf{r}_2 - \mathbf{r'}_2)\Phi_0^-(\mathbf{r}_1, \mathbf{r}_2) \Big]. \quad (4.322)$$

The equations decouple with the aid of the fact that Φ_0^+ and Φ_0^- are related by particle exchange, equation (4.315). The equation for Φ_0^+ is

$$\Phi_0^+(\mathbf{r'}_1, \mathbf{r'}_2) = \int d\mathbf{r}_1 d\mathbf{r}_2 \frac{1}{2} \Big[\delta(\mathbf{r}_1 - \mathbf{r'}_1)\delta(\mathbf{r}_2 - \mathbf{r'}_2)\Phi_0^+(\mathbf{r}_1, \mathbf{r}_2) + \delta(\mathbf{r}_2 - \mathbf{r'}_1)\delta(\mathbf{r}_1 - \mathbf{r'}_2)\Phi_0^+(\mathbf{r}_2, \mathbf{r}_1) \Big], \quad (4.323)$$

and swapping the integration variables in the latter term gives equation (4.316).

We can now represent Ψ_0^+ with 'positive walkers':

Representation of the two-fermion ground-state wave function using walkers

$$\Phi_0^+(\mathbf{r}_1, \mathbf{r}_2) \approx \frac{1}{M} \sum_{j=1}^{M} \delta(\mathbf{r'}_1 - \mathbf{r}_1)\delta(\mathbf{r'}_2 - \mathbf{r}_2); \quad (4.324)$$

walker coordinates $\mathbf{r'}_1, \mathbf{r'}_2$ are sampled from $\Phi_0^+(\mathbf{r'}_1, \mathbf{r'}_2)$.

Similarly, we can represent Φ_0^- with 'negative walkers,' and it is enough to find a way to sample either kind of walker. This is ensured by the so-called *tiling property*.

4.7.3 Tiling property

The N-fermion wave function in 3D changes sign on *nodal surfaces* that have dimension $3N - 1$. The surfaces are boundaries of regions called *nodal pockets* or *nodal cells*. The positive and negative walkers are in different nodal cells, but the two kinds of walkers are related by particle permutations. Exchanging an odd number of fermions changes the sign of the wave function,

$$\hat{\mathcal{P}}\Phi_0(\mathbf{x}) = (-1)^P \Phi_0(\mathbf{x}), \quad (4.325)$$

where the permutation operator $\hat{\mathcal{P}}$ represents P pairwise fermion exchanges. Both $\Phi_0(\mathbf{x})$ and $\hat{\mathcal{P}}\Phi_0(\mathbf{x})$ are fermion ground states, and we are not able to tell them apart. Thanks to the tiling property, we can concentrate on solving the fermion ground state in one nodal cell [28]:

Tiling property:
   Pick any reference point. All points that can be reached from this point by following a continuous path without crossing a nodal surface is called a nodal cell. All nodal cells are equivalent and related to each other by particle permutations.

The proof is based on the principle that an added node increases energy. Heuristically, a node means certain configurations are impossible; hence, you know more about the system. The 'added node increases energy' principle is valid for the Hamiltonian

$$\hat{\mathcal{H}} = -D\sum_{i=1}^{N} \nabla_i^2 + V(\mathbf{x}), \tag{4.326}$$

and it also implies that the bosonic (symmetric) ground state is always energetically below the fermionic (antisymmetric) one.

The conclusion that the symmetric ground-state energy is energetically below the antisymmetric ground-state energy is not valid for all Hamiltonians. $\hat{\mathcal{H}}$ is what we assume about the system, and nothing prevents us from adding new assumptions. For example, in statistical physics, the grand canonical (GC) ensemble uses $\hat{\mathcal{H}}_{GC} = \hat{\mathcal{H}} - \mu\hat{\mathcal{N}}$, where μ is the chemical potential and $\hat{\mathcal{N}}$ is the number operator. Similarly, we can add a large energy penalty for symmetric states [30],

$$\hat{\mathcal{H}}_a = \hat{\mathcal{H}} + \mathcal{A}(1 + \hat{\pi}), \tag{4.327}$$

where the parity operator operates on $\Psi(\mathbf{x})$ such that $\hat{\pi}\Psi(\mathbf{x}) = \Psi(-\mathbf{x})$ and $\mathcal{A}$ is a large positive constant, large enough to lift the symmetric state energies well above the antisymmetric ones. This would make the fermion ground state the lowest-energy state, obviously a very useful benefit for DMC. The price is that the modified Hamiltonian is no longer local. Apart from such modifications to the Hamiltonian, we can safely assume that adding a node increases energy and move on to discussing the tiling property.

The reasoning behind the tiling property goes like this [31]. Let us say the fermion ground state is $\Phi_0(\mathbf{x})$. Pick any nodal cell and all cells that are copies of it with particles relabeled, and color them blue. Now assume there is a nodal cell in $\Phi_0(\mathbf{x})$ that is not blue. Color that cell red, and find its permutation copies and color them red as well. If there are still uncolored cells, continue the coloring process with new colors. Somewhere, there is a blue and a non-blue cell touching, with a nodal surface in between. If you eliminate the nodal surface and attach the cells, you can again antisymmetrize the wave function and get a new, normalizable $\Phi'(\mathbf{x})$. However, the energy was lowered by removing the nodal surface, so $\Phi'(\mathbf{x})$ has a lower energy than $\Phi_0(\mathbf{x})$, and consequently $\Phi_0(\mathbf{x})$ could not have been the fermion ground state. This is contrary to what was assumed; hence, all nodal cells must be reachable from one cell via particle permutations.

Foulkes *et al* [31] point out that fermion nodal surfaces are not the same as planes mapped by fermion coordinate coincidence, because the two may have different numbers of dimensions. On the coincident hyperplanes, $\Psi(\mathbf{x}) = 0$ *because* two or more fermions are at the same location. In d-dimensional space, the coincident planes have $dN - d$ dimensions, while nodal surfaces have $dN - 1$ dimensions. As it happens, in 1D systems, the coincident planes exhaust the set of nodal surfaces, which makes solving 1D fermion problems easy with a simple 'no-crossing rule': prevent fermions from hopping past each other, and that is all that is required in 1D.

In two or more dimensions, it is not enough to keep identical fermions from overlapping, because there are also other configurations $\mathbf{x}$ such that $\Psi(\mathbf{x}) = 0$.

4.8 Fixed-node DMC of atoms and small molecules

Computational quantum chemistry is a healthy competition between density functional theory (DFT) and QMC. The coupled cluster (CC) theory and configuration interaction (CI) calculations provide benchmark results, but due to their computational cost, they can only be applied to small molecules. DFT wins hands down in speed, but QMC achieves higher accuracy in total energies and in electron–electron correlation energies, so it is preferable in studies of strongly correlated materials, such as transition-metal oxides and heavy fermion systems. The reliability of DFT results relies on how well the chosen energy functional, such as the local density approximation (LDA) or generalized gradient approximation (GGA), suits the studied system. Each choice of a DFT functional has to be validated against experimental data or CC, CI, or QMC results. Compared to DFT, the scaling of QMC to large systems is excellent because the Monte Carlo algorithms are inherently parallelizable.

To put up a good fight, QMC of atoms and molecules should set a few goals:

- Lowest level: beat Kohn–Sham DFT in accuracy. If we do not, we are just wasting resources. Best to let DFT win here.
- Acceptable level: reach the so-called *chemical accuracy*.

> The chemical accuracy is the accuracy required to make realistic chemical predictions and is generally considered to be 1 kcal mol^{-1} $\approx$ 4 kJ mol^{-1} $\approx$ 43 meV $\approx$ 0.001 59 Ha mol^{-1}.

- Excuse: we may be excused for not achieving the above accuracy goals if we gain a better understanding of the system compared to the story told by the DFT electron density.

The standard Kohn–Sham DFT accuracy is typically 2–3 kcal mol^{-1} with currently available functionals, but to be fair, the accuracy was recently pushed below the chemical accuracy limit using machine learning [32]. To ensure the first goal is achieved, one customarily starts from self-consistent DFT results, generally the DFT single-particle orbitals, and improves the correlations using QMC. However, all is not lost, because although inferior to current DFT, the venerable Hartree–Fock (HF) method has also proved to be a good starting point for QMC. The popularity of HF is based on the simplicity of the wave function and on the fact that, unlike DFT, HF is a self-consistent, non-parametric *ab initio* method. DFT has correlations that may interfere with QMC optimization, and using HF reduces the chances of overparameterization. QMC often uses the Jastrow factor to add explicit electron–electron correlations, and with HF, one can clearly tell which correlation comes from the Jastrow factor and which comes from the orbitals.

HF codes are abundant; in fact, every quantum chemistry package has one. The Python-based Simulations of Chemistry Framework (PYSCF) uses a Python interface on top of core routines written in C. The code JuliaChem (JuliaChem @github) is written in Julia [33]. Another quantum chemistry package coded in Julia is the Fermi quantum chemistry module `Fermi.jl` described in [34] (link to GitHub FermiQC page), which can also perform Møller–Plesset perturbation theory and CC calculations. If you are even less DFT-oriented than I am, you can download optimized basis sets from www.basissetexchange.org to get started.

I will start from the ground state of atoms, which have single-particle orbitals filled in the order given by the *Aufbau principle*:

Aufbau principle
Madelung's $n + l$-rules are:
- Orbitals are filled in order of increasing $n + l$.
- For orbitals with the same $n + l$, the one with lower n is filled first,

---

which gives the filling order 1s, 2s, 2p, 3s, 3p, 4s, 3d, 4p, 5s, 4d, 5p, 6s, 4f, 5d, 6p and so on.

If needed, apply Hund's rule: electrons fill degenerate orbitals, such as $2p_x$, $2p_y$, and $2p_z$, so that an electron enters each degenerate orbital until they all have one electron with parallel spins; after that, the next electrons start to pair up in the orbitals.

4.8.1 Beryllium atom ground state

The Be atom ground state is ^1S in the spectroscopic notation ^{2S+1}L, with the four electrons paired to give a total spin of $S = 0$ and a total angular momentum of $L = 0$. The HF method assumes the ground state to be a single determinant of spin orbitals $\chi_i(x_j)$ (see section 2.2, The multielectron wave function). Using the spin orbitals χ_1 and χ_2, the HF ground-state wave function is the Slater determinant

$$\Psi_T^{HF}(x) = \frac{1}{\sqrt{4!}} \begin{vmatrix} \chi_1(x_1) & \chi_2(x_1) & \chi_1(x_1) & \chi_2(x_1) \\ \chi_1(x_2) & \chi_2(x_2) & \chi_1(x_2) & \chi_2(x_2) \\ \chi_1(x_3) & \chi_2(x_3) & \chi_1(x_3) & \chi_2(x_3) \\ \chi_1(x_4) & \chi_2(x_4) & \chi_1(x_4) & \chi_2(x_4) \end{vmatrix}. \tag{4.328}$$

The T in Ψ_T^{HF} reminds us that this is a ground-state trial wave function, not the exact ground state. Next, separate the spin orbitals into spatial orbitals and spin. The lowest-energy spatial orbitals are $\phi_{1s}(\mathbf{r})$ and $\phi_{2s}(\mathbf{r})$, so the ground-state configuration is $1s^2 2s^2$. What makes this simple is that the s-orbitals are spherically symmetric, so the electron distances (r_1, r_2, r_3, or r_4) from the nucleus are the only arguments. Using spin functions α and β, the determinant is

$$\Psi_T^{\mathrm{HF}}(\boldsymbol{x}) = \frac{1}{\sqrt{4!}} \begin{vmatrix} \phi_{1s}(r_1)\alpha(\sigma_1) & \phi_{2s}(r_1)\alpha(\sigma_1) & \phi_{1s}(r_1)\beta(\sigma_1) & \phi_{2s}(r_1)\beta(\sigma_1) \\ \phi_{1s}(r_2)\alpha(\sigma_2) & \phi_{2s}(r_2)\alpha(\sigma_2) & \phi_{1s}(r_2)\beta(\sigma_2) & \phi_{2s}(r_2)\beta(\sigma_2) \\ \phi_{1s}(r_3)\alpha(\sigma_3) & \phi_{2s}(r_3)\alpha(\sigma_3) & \phi_{1s}(r_3)\beta(\sigma_3) & \phi_{2s}(r_3)\beta(\sigma_3) \\ \phi_{1s}(r_4)\alpha(\sigma_4) & \phi_{2s}(r_4)\alpha(\sigma_4) & \phi_{1s}(r_4)\beta(\sigma_4) & \phi_{2s}(r_4)\beta(\sigma_4) \end{vmatrix} . \quad (4.329)$$

The **restricted Hartree–Fock** (RHF) forces electrons with different spins to occupy the same orbitals, and the resulting wave function is an eigenfunction of the total and projected spins. The ground state is a product of spin α and β determinants (spins up and down, or vice versa),

$$\varphi_T^{\mathrm{HF}}(\mathbf{x}) = \frac{1}{\sqrt{4!}} \begin{vmatrix} \phi_{1s}(r_1)\alpha(\sigma_1) & \phi_{2s}(r_1)\alpha(\sigma_1) & 0 & 0 \\ \phi_{1s}(r_2)\alpha(\sigma_2) & \phi_{2s}(r_2)\alpha(\sigma_2) & 0 & 0 \\ 0 & 0 & \phi_{1s}(r_3)\beta(\sigma_3) & \phi_{2s}(r_3)\beta(\sigma_3) \\ 0 & 0 & \phi_{1s}(r_4)\beta(\sigma_4) & \phi_{2s}(r_4)\beta(\sigma_4) \end{vmatrix} \quad (4.330)$$

$$= \frac{1}{\sqrt{4!}} \begin{vmatrix} \phi_{1s}(r_1) & \phi_{2s}(r_1) \\ \phi_{1s}(r_2) & \phi_{2s}(r_2) \end{vmatrix}_\alpha \begin{vmatrix} \phi_{1s}(r_1) & \phi_{2s}(r_1) \\ \phi_{1s}(r_2) & \phi_{2s}(r_2) \end{vmatrix}_\beta . \quad (4.331)$$

The filling of single-particle orbitals is shown in figure 4.14, with the choice of spin-up electrons 1 and 2 and spin-down electrons 3 and 4.

As mentioned earlier, $\varphi_T^{\mathrm{HF}}(\mathbf{x})$ is not the exact ground state, which is obvious since it has **no dependence on electron–electron distances** and therefore nothing that accounts for the correlations induced by the electron–electron Coulomb interactions. This is true for all HF many-electron wave functions; therefore, we define the correlation energy as the energy missing from HF:

HF energy + correlation energy ≡ exact energy.

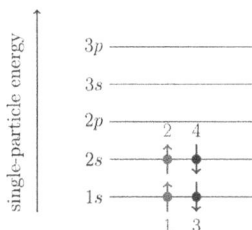

Figure 4.14. The $1s^2 2s^2$ configuration of the Be atom.

Correlation energies typically described by various methods are:

Hartree–Fock   0% by definition

optimized Slater–Jastrow $\sim 85\%$

FN-DMC $\sim 95\%$ or more

Exact   100%.

Table 4.1 shows how much correlation energy is missing and also why the Be atom is such a fruitful case for DMC.

4.8.1.1 Be atom nodal cells and the tiling property

The two HF determinants are zero if either $r_1 = r_2$ or $r_3 = r_4$, so these coincidence conditions define *some* of the nodes, but not all of them. All nodes can be solved using the condition

$$(\mathbf{r}_1 - \mathbf{r}_2) \cdot (\mathbf{r}_3 - \mathbf{r}_4) = 0. \tag{4.332}$$

The orbitals are positive, so we can separate positive and negative domains,

$$\varphi_T^{HF}(\mathbf{x}) = [\underbrace{\phi_{1s}(r_1)\phi_{2s}(r_2)\phi_{1s}(r_3)\phi_{2s}(r_4)}_{\text{`1234'}} + \underbrace{\phi_{1s}(r_2)\phi_{2s}(r_1)\phi_{1s}(r_4)\phi_{2s}(r_3)}_{\text{`2143'}}]$$
$$- [\underbrace{\phi_{1s}(r_1)\phi_{2s}(r_2)\phi_{1s}(r_4)\phi_{2s}(r_3)}_{\text{`1243'}} + \underbrace{\phi_{1s}(r_2)\phi_{2s}(r_1)\phi_{1s}(r_3)\phi_{2s}(r_4)}_{\text{`2134'}}]. \tag{4.333}$$

There are now *four nodal cells*, two positive and two negative, related by electron permutations. A schematic illustration of this is shown in figure 4.15. If you start, say, from the positive second term, coordinate indices 2143, and swap 1 and 2, you

Table 4.1. The HF and exact ground-state energies for selected few-electron atoms. As indicated, the Be atom has a fairly large correlation energy.

Atom	Ground state	HF	Exp.	Correlation
He	$1s^2$	$-2.861\ 679\ 993$	$-2.903\ 386\ 83$	$-0.041\ 706$
Li	$1s^2 2s^1$	$-7.432\ 726\ 924$	$-7.478\ 060\ 323$	$-0.045\ 333$
Be	$1s^2 2s^2$	$-14.573\ 023\ 13$	$-14.667\ 3564$	$-0.094\ 333$
B	$1s^2 2s^2 2p^1$	$-24.529\ 060\ 69$	$-24.541\ 246$	$-0.012\ 185$

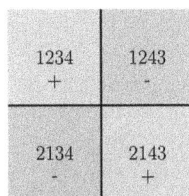

1234 +	1243 -
2134 -	2143 +

Figure 4.15. Schematic illustration of the four nodal cells of the $1s^2 2s^2$ wave function and the nodal surfaces (thick lines).

enter 1243, which is the negative third term. If you then swap 4 and 3, you come to 1234, the positive first term. Finally, swap 1 and 2 to get to the negative fourth term 2134. Hence, the tiling property is satisfied, and this is a viable trial wave function.

Instead of four, the exact Be atom ground state is *known to have only two nodal cells*. The tiling property does not reveal the number of nodal cells, but it has been conjectured that many-fermion systems have two nodal cells and a single nodal hypersurface [35, 36].

The HF $1s^2 2s^2$ configuration has four nodal cells, and the only way to reduce the number is to extend to post-HF states and include more orbitals and more Slater determinants. This improvement essentially removes nodal surfaces between positive cells and between negative cells, respectively, as shown in figure 4.16. Presumably, the true ground state is mostly made of low-energy HF orbitals, so the 2p orbital should be a good addition,

$$\varphi_T = (1s^2 2s^2) + c(1s^2 2p^2). \tag{4.334}$$

As shown in [37], this tiny CI expansion with $c \neq 0$ closes the gap between same-sign nodal cells, leaving only two, and the whole nodal cell topology changes.

The three 2p orbitals, $2p_x$, $2p_y$, and $2p_z$, are not occupied in the HF ground state, so they are, in that sense, **virtual orbitals**. The ^1S symmetry of the Be atom dictates that the coefficients of all three 2p orbitals are the same; otherwise, the cigar-shaped 2p orbitals would not add up to a symmetric combination. The **configuration state functions** (CSFs) mentioned in section 2.2.3 are wave function constructions that incorporate the symmetries of the system explicitly. A CSF of a Be atom would now be

$$\varphi_T^{CSF} = (1s^2 2s^2) + c\left(1s^2 2p_x^2 + 1s^2 2p_y^2 + 1s^2 2p_z^2\right). \tag{4.335}$$

Adding electron–electron correlations via a Jastrow factor gives the Slater–Jastrow trial wave function; for example, the configuration $1s^2 2s^2$ gives

$$\varphi_T(\mathbf{x}) = e^{J(\mathbf{x})} \begin{vmatrix} \phi_{1s}(r_1) & \phi_{2s}(r_1) \\ \phi_{1s}(r_2) & \phi_{2s}(r_2) \end{vmatrix}_\alpha \begin{vmatrix} \phi_{1s}(r_1) & \phi_{2s}(r_1) \\ \phi_{1s}(r_2) & \phi_{2s}(r_2) \end{vmatrix}_\beta \tag{4.336}$$

$$= e^{J(\mathbf{x})} D^\uparrow(\mathbf{r}_1, \mathbf{r}_2) D^\downarrow(\mathbf{r}_3, \mathbf{r}_4). \tag{4.337}$$

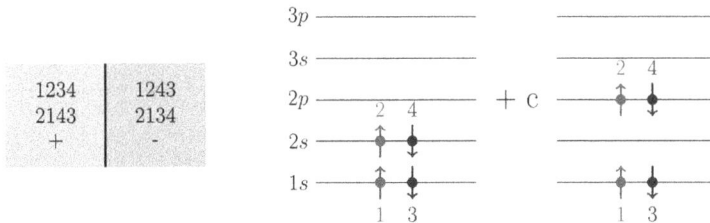

Figure 4.16. Schematic illustration of the two nodal cells of the $(1s^2 2s^2) + c(1s^2 2p^2)$ wave function and the filling of single-particle orbitals.

To use a trial wave function in importance sampling requires computation of the gradients for the drift $\mathbf{F}(\mathbf{x}) = \nabla_{\mathbf{x}} \varphi_T(\mathbf{x})$ and the Laplacian for the local energy $E_L = (\hat{\mathcal{H}} \varphi_T(\mathbf{x}))/\varphi_T(\mathbf{x})$. We need a fast way to compute the gradients and Laplacians of determinants D, preferably as updates to existing values after each walker move.

4.8.1.2 Be atom results

Table 4.2 shows the Slater-type orbitals (STOs) with one or two exponents $c \exp(-\zeta r)$ for each orbital. The parameters shown in table 4.2 were optimized using the code `Atom_Slater_Jastrow_optimization.jl`.

Figure 4.17 shows DMC results as a function of the time step. Recall that the $\tau \to 0$ limit has to be applied for just one reason: the Green's function approximation in the drift–diffusion update of walker coordinates is inaccurate. Nodal surfaces are no-cross boundaries, and if walkers have no idea where they are before they have already crossed a boundary, then it is hardly surprising that only the $\tau \to 0$ limit is correct.

Table 4.2. The Be atom trial wave function parameters used to compute the results shown in figure 4.17. For STO, the linear combination of orbitals was $(1s^2 2s^2) - 0.236(1s^2 2p^2)$, while for STO-$2\zeta$ it was $(1s^2 2s^2) - 0.239(1s^2 2p^2)$.

Orbital parameters				
	STO		STO-2ζ	
Orbital	ζ	c	ζ	c
1s	3.85	1.000	3.47	1.090
			5.18	0.356
2s	1.11	0.361	1.11	0.428
			4.91	-0.055
2p	1.09	0.510	1.09	0.510
			3.56	0.079

Jastrow parameters		
	STO	STO-2ζ
α	0.010	-0.039
β_1	0.839	1.153
β_2	0.740	0.936
α_{12}	1.000	1.000

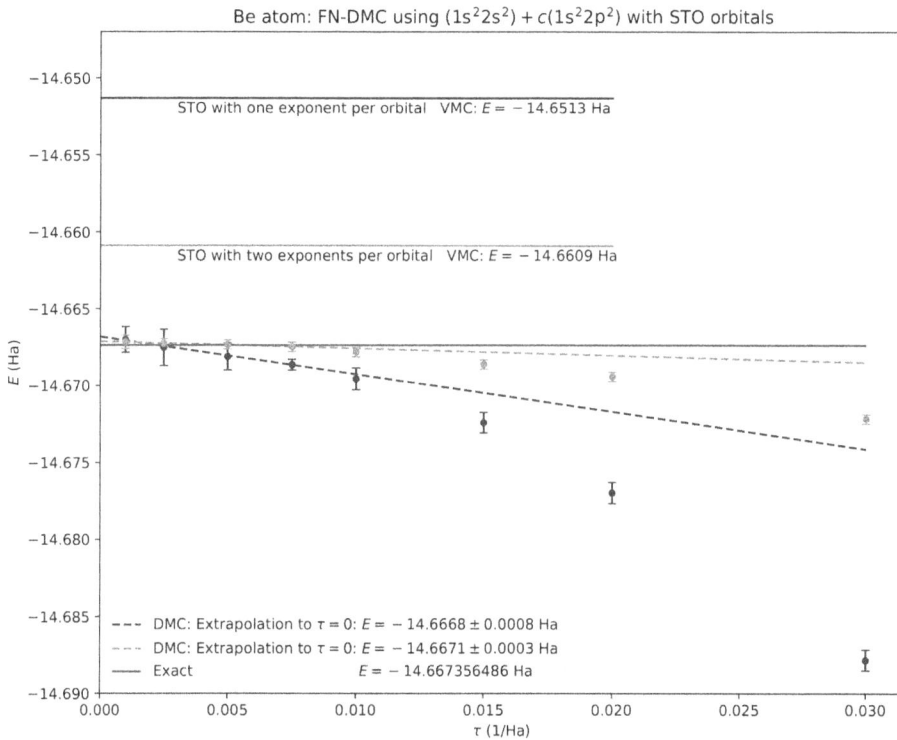

Figure 4.17. VMC and fixed-node DMC energies computed using the two Slater–Jastrow parametrizations given in table 4.2. Only the four small-τ data points were used in the linear extrapolation; the remaining data points demonstrate how $E(\tau)$ ceases to be linear as τ increases. The data was computed using the code `Atom_Slater_Jastrow.jl`.

About the Julia code `Atom_Slater_Jastrow_optimization.jl`
- Derivatives w.r.t. parameters are computed using automatic differentiation (AD). AD is slower than functions with analytical derivatives, but you do not need to add much new code to the existing VMC code base, just a function that invokes AD to derive the trial wave function and the local energy. That code is now

```
function G_i_AD(R ::MMatrix{dim, N, Float64},
               wf_params ::WfParams, G ::Function)
```

which calculates the gradients of *any* function G that can be called with

```
function G(R, wf_params)
```

Here, the argument R contains the electron coordinates of a single walker, and `wf_params` is a structure that contains the trial wave function parameters. The trial wave function is φ_T(R, wf_params) and the local energy is EL(R, wf_params), so G_i_AD can differentiate either of them. For AD, the code uses the Julia ForwardDiff package method gradient, and AD requires that the differentiated arguments of the functions must have types that can accommodate both floating-point numbers and AD dual numbers (a dual number is basically the value of a parameter and its derivative). For example, the structure of atomic orbitals (AOs) is

```
struct AtomicOrbital{A<:AbstractString, T<:Number, Z<:Number}
    atom          ::A
    orbital_type  ::A
    n             ::Int
    l             ::Int
    exponents     ::Vector{T}
    coefficients  ::Vector{Z}
    spin          ::A
end
```

Here, Vector{T} is a vector of type T, and $T <: \text{Number}$ means it is any type derived from the Number type, so AD dual numbers can be stored as exponents. Exponents and coefficients are derived separately; hence, their types are separated into $T <: \text{Number}$ and $Z <: \text{Number}$. The structure AtomicOrbital has some redundancy, as the quantum numbers n, l already define the orbital_type, such as 1s or 2s.

- Parameter updates $\Delta\alpha$ are computed using the linear optimization method described in section 3.3.5. Averages over a finite set of walkers are replaced with covariances whenever appropriate in order to reduce statistical fluctuations and improve the accuracy of the gradients. If the energy is not decreasing, the parameter updates are turned in the direction of steepest descent by adding a constant a_opt to the $\bar{H}$ matrix diagonals. This process is open to fine-tuning. The sample code compares the energies given by three values of a_opt and tries to guess the best value.

- Be sure to test any optimization code against a known case. The code is flawed if it cannot find the H atom ground state $\Psi_0 = e^{-r}$ starting from $\varphi_T = e^{-1.1\,r}$. Try running the code with
  'julia   Atom_Slater_Jastrow_optimization.jl   atom=H basis_set=STO.
  ' This uses the input data file atom_data/H_STO, where the exponent is deliberately set to 1.1.

- The Jastrow factor uses two β parameters because parallel and antiparallel spin electron correlations may decay differently.

- The serial code is not very fast, and you may run out of patience after $\sim 10\,000$ walkers. Many tasks are easy to parallelize, though.

4.8.2 Molecules

Quantum chemistry is an old discipline with decades of experience with electrons in molecules. Most readers should choose one of the available QMC codes that are well suited for the computation of a selected set of properties of just about any atom or molecule, given enough computational resources. Widely used QMC packages for electronic structure calculations of atomic and molecular systems include (in no particular order; many of these are open source):

- CASINO [38]
- CHAMP (link to CHAMP on GitHub)
- QWALK [39]
- QMCPACK [40]

Some of the newcomers include

- CSIRO
- QMCTorch (link to QMCTorch on GitHub) is a Python package based on PyTorch. Install it from source or use the Python package manager: 'pip install qmctorch.'

If you need code for HF, Møller–Plesset perturbation theory, or CC, the package Fermi.jl is a recent addition that utilizes Julia's multiple dispatch, metaprogramming, and interactive usage. It is also modular, so in the future, QMC may also be part of it [34].

The QMC packages mentioned above are not quite black-box routines; you also need some quantum chemistry knowledge to fully benefit from their capabilities. The code performance is fine-tuned, and the documentation is adequate to get started. I am not trying to compete with these multipurpose QMC packages, but it is good to know some principles of molecular QMC and the so-called molecular orbital theory, which superseded the valence bond theory in the early 1930s.

The molecular orbital theory has many facets. Those who dislike it may have learned about the shortcomings of the theory at the HF level and recall that, for example, the H_2 molecule bond dissociation is poorly described and predicts unphysical fractional charges.[9] The HF orbital energies are often too high, so the predicted HOMO–LUMO gaps and ionization energies can be systematically wrong.[10] DFT works with electron density; nevertheless, the Kohn–Sham DFT uses single-electron orbitals, and the quality of the results depends on the energy functional; in particular, LDA results can indeed be unreliable. However, the molecular orbital theory *is not* the same as the HF theory. The other side of the coin is post-HF methods and CI. In particular, full configuration interaction (FCI) **gives quantitatively accurate results**. In addition, QMC with multideterminant wave

---

[9] As we shall see, the electron density is unevenly divided between the separated atoms, so the size-consistency requirement is not satisfied.

[10] HOMO = highest occupied molecular orbital, LUMO=lowest unoccupied molecular orbital.

functions can predict accurate dissociation curves, ionization energies, and HOMO–LUMO gaps.

A set of non-orthonormal AOs $\phi_i(\mathbf{r})$ can be combined to form **molecular orbitals** (MOs). Such MOs go under the name 'linear combination of atomic orbitals' (LCAOs). To tell them apart, AOs are sometimes enumerated with Greek letters and MOs with roman letters. Each MO $\Phi_i(\mathbf{r})$ is a linear combination of AOs,

$$\Phi_i(\mathbf{r}) = \sum_\mu C_{\mu i}\phi_\mu(\mathbf{r}).$$
(4.338)

AOs typically have coefficients and exponents (c's and ζ's in STOs) as parameters, and the coefficients C are added as new, optimizable parameters. Antisymmetry is ensured by collecting these orthonormal, single-electron MOs in a Slater determinant, supplemented by a Jastrow factor.

The restriction to single-electron MOs is relaxed in quantum chemistry in the so-called multireference methods, which are used in cases where the molecular wave function cannot be adequately described using a single determinant of MOs. In systems with periodic boundary conditions, such as solids, a plane wave basis is very effective. Unlike MOs, which are delocalized over the whole molecule, the so-called natural bond orbitals (NBOs) are localized to specific bonds, antibonds, or lone pairs (nonbonding valence electrons); NBOs are designed to give a compact description of the chemical bond.

4.8.2.1 Dynamic and static correlations
Some ways of obtaining AOs and MOs

It is beneficial to start with orbitals computed using a self-consistent calculation, possibly HF or DFT. For example, QMCTorch takes its AOs and MOs from the codes pyscf and ADF (Amsterdam density functional). Some input orbitals satisfy the cusp conditions from the onset, but Gaussian orbitals do not, and QMC is more stable if run with Gaussian orbitals fitted to single-exponent STOs. In his Pisa lecture notes, Mike Towler reveals a common problem:

> 'Often people find that their HF/DFT code of choice is not supported, so they give up before they start.'

Thankfully, many QMC codes are flexible and understand inputs in many formats.

After obtaining a set of MOs and their energies, you can move on. Only some MOs are 'active': those near the top of the occupied MOs and those near the lowest unoccupied MOs. That is, orbitals around the HOMO and the LUMO. The rest are fully occupied core MOs and empty virtual orbitals. The set of all active MOs is called the **complete active space (CAS)**. The notation $[n, m]$-CAS refers to a multiconfigurational wave function of n active electrons in m active MOs. Restricting the configuration space to the CAS keeps the calculations feasible because the number of determinants grows as the factorial of the number of MOs. Serious QMC molecular calculations must still deal with tens of thousands of determinants, and much effort has been dedicated to algorithmic improvements.

The algorithms are quite involved, and I suggest you rely on the specialized QMC software packages mentioned earlier rather than coding them yourself.

Quantum chemists talk about multiconfiguration self-consistent field (MCSCF), an iterative optimization of the coefficients and the orbital parameters. The QMC community often uses a variant of MCSCF called the complete active space self-consistent field (**CASSCF**) [41] process, where the MO coefficients (the Cs in equation (4.338), i.e. how the AOs combine to form MOs) and the CI coefficients (weights of different electron configurations, that is, Slater determinants) are optimized. During optimization, the MO energies are updated, and some reordering takes place. Experience has shown that the optimization of MOs is most effective when done simultaneously with the Jastrow factor because it then leads to a substantial improvement in the nodal surface. After applying the CASSCF, you have a pretty accurate set of MOs, whose occupations give the HOMO and the LUMO.

4.8.2.2 H_2 molecule

The hydrogen molecule H_2 shows why the HF description is inadequate and how the wave function could be improved. The AOs in two hydrogen atoms A and B are 1s orbitals, and a simple MO is a linear combination of them:

$$\Phi(\mathbf{r}) = c_A \phi_{1s;A}(\mathbf{r}) + c_B \phi_{1s;B}(\mathbf{r}). \tag{4.339}$$

This MO is a single-electron function that describes how the electron is distributed among atoms A and B. The ground state of H_2 is symmetric; ignoring normalization,

$$\sigma_g(\mathbf{r}) = \phi_{1s;A}(\mathbf{r}) + \phi_{1s;B}(\mathbf{r}) \qquad \text{in-phase, bonding MO}, \tag{4.340}$$

while the other possibility is

$$\sigma_u(\mathbf{r}) = \phi_{1s;A}(\mathbf{r}) - \phi_{1s;B}(\mathbf{r}) \qquad \text{out-of-phase, antibonding MO}. \tag{4.341}$$

The bonding and antibonding orbitals are shown in figure 4.18. The bonding σ_g has large electron density between the atoms, indicating they share an electron. The antibonding σ_u has low electron density between the atoms, even zero in the middle, indicating that the atoms do not share electrons. In the ground state, σ_g is the important one, but as we shall see, σ_g does not adequately describe dissociation into two hydrogen atoms,

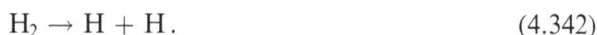

$$H_2 \rightarrow H + H. \tag{4.342}$$

The RHF (see section 4.8.1) ground-state trial wave function of electrons 1 and 2 is [42, 43]

$$\Psi^{\text{RHF}}(1, 2) = \frac{1}{\sqrt{2}} \begin{vmatrix} \sigma_g(1)\alpha(1) & \sigma_g(1)\beta(1) \\ \sigma_g(2)\alpha(2) & \sigma_g(2)\beta(2) \end{vmatrix} \tag{4.343}$$

$$= \underbrace{\sigma_g(1)\sigma_g(2)}_{\text{symmetric spatial part}} \underbrace{\frac{1}{\sqrt{2}}(\alpha(1)\beta(2) - \beta(1)\alpha(2))}_{\text{spin singlet, antisymmetric}}. \tag{4.344}$$

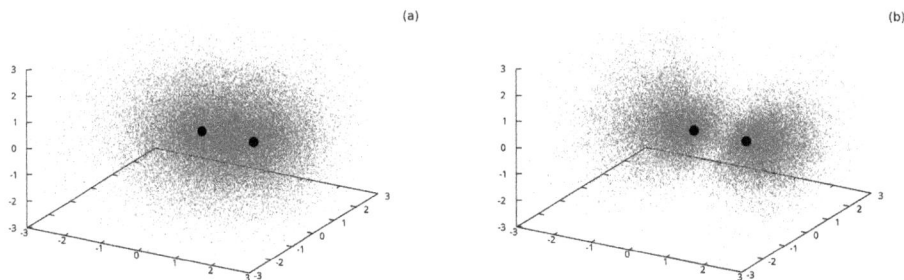

Figure 4.18. The HF bonding σ_g (a) and antibonding σ_u (b) MOs of H_2. The protons (black points) are 1.4 a.u. apart, and the positions of electrons 1 and 2 are red and blue, respectively. The reason why blue electrons appear to cluster near protons is an artifact of plotting first the red ones and then the blue ones. See the Python code H2_bonding_antibonding.py.

The wave function is an eigenstate of $\hat{S}^2$ and $\hat{S}_z$, and the spatial part is

$$\sigma_g(1)\sigma_g(2) = (\phi_{1s;A}(1) + \phi_{1s;B}(1))(\phi_{1s;A}(2) + \phi_{1s;B}(2)) \tag{4.345}$$

$$= \phi_{1s;A}(1)\phi_{1s;A}(2) \qquad \text{both electrons in } A \tag{4.346}$$

$$+ \phi_{1s;B}(1)\phi_{1s;B}(2) \qquad \text{both electrons in } B \tag{4.347}$$

$$+ \phi_{1s;A}(1)\phi_{1s;B}(2) + \phi_{1s;B}(1)\phi_{1s;A}(2) \qquad \text{one electron per atom}, \tag{4.348}$$

which shows that there is a nonzero probability amplitude of having both electrons in one atom. This predicts an unphysical charge distribution in dissociation, because the probability

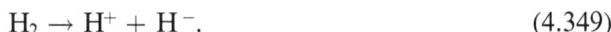

$$H_2 \rightarrow H^+ + H^-. \tag{4.349}$$

should be zero, not two out of four cases. The ground-state properties are predicted reasonably well, though. From the QMC point of view, the symmetric spatial part $\sigma_g(1)\sigma_g(2)$ is just another boson computation resembling the He atom ground-state calculation.

The spin-singlet state is a maximally entangled state, meaning that the measurement of one electron spin immediately determines the other spin.[11] Spin entanglement prevails in dissociation, so two H atoms far apart with entangled spins do not have quite the same energy as two H atoms without entangled electron spins. The energy difference due to the exchange interaction is minute: for H atoms $5a_0$ apart, it is of the order of a few micro-electronvolts, about 1×10^{-8} Ha. As mentioned earlier, the chemical accuracy is

$$1 \text{ kcal mol}^{-1} \approx 43 \text{ meV} \approx 0.001 \text{ Ha}, \tag{4.350}$$

---

[11] The two electrons in a singlet state are in a superposition state, with no specific spin before we *set* the spin to a value by the act of measurement. In that sense, entangled spins are nothing like two socks, one red and one blue. Both socks have a definite color whether we look or not, and the act of measurement does not set their color. The fact that entangled spins behave in such a strange manner has been verified experimentally, where it shows up as a violation of Bell's inequalities.

so compared to this error goal, we can ignore entanglement. Aligned spins prevent H_2 molecule formation, and gaseous atomic H was observed in the famous experiment by Silvera and Walraven under a magnetic field of 7 T at a temperature of 270 mK [44].

H_2 molecule at equilibrium

If equilibrium values are all you need, then you can take the RHF wave function (4.344), keep the spins in a singlet state, and use whatever symmetric spatial part you find appropriate. A Jastrow factor to improve e–e correlations comes to mind first. Another improvement would be to relax the Born–Oppenheimer approximation and let proton wave functions spread in space. In 1991, Traynor, Anderson, and Boghosian [45] used the following spatial part (electron labels are 1 and 2, protons are 3 and 4):

$$\varphi_T(\mathbf{x}) = (e^{-\zeta r_{13}} + e^{-\zeta r_{14}})(e^{-\zeta r_{23}} + e^{-\zeta r_{24}}) \qquad \text{e-p correlations (RHF)}$$

$$\times\, e^{\frac{br_{12}}{1+br_{12}}} \qquad \text{e-e correlations (Jastrow)} \tag{4.351}$$

$$\times\, e^{-d(r_{34}-c)^2} \qquad \text{p-p correlations (harmonic)}$$

The e–p correlation factor is the familiar RHF binding MO, $\sigma_g(1)\sigma_g(2)$.

The code `H2_DMC.jl` has a slightly different Jastrow factor, $e^{\frac{r_{12}}{2(1+br_{12})}}$. Optimization gives (the code has a line search optimization and a linear optimization),

$$\text{VMC:}\, \zeta = 1.286\, \text{a. u.} \tag{4.352}$$

$$b = 0.398 \tag{4.353}$$

$$c = 1.399\, \text{a. u.} = 0.740\, \text{Å} \tag{4.354}$$

$$d = 10.5 \tag{4.355}$$

$$\langle V \rangle = -2.2947 \pm 0.0006\, \text{Ha} \tag{4.356}$$

$$\langle E \rangle = -1.145\,59 \pm 0.000\,07\, \text{Ha.} \tag{4.357}$$

With these parameters, VMC gives $R_{eq} = 1.424$ a. u. $= 0.7536$ Å; the small deviation from the parameter c is caused by e–p correlations.

In [45], the ground-state energy (dissociation energy) was found to be $E_0 = -1.163\,97 \pm 0.000\,05$ Ha using DMC and $E_0 = -1.164\,024 \pm 0.000\,09$ Ha using GFMC.[12] After extrapolation, the code `H2_DMC.jl` gives $E_0 = -1.163\,96 \pm 0.000\,07$ Ha.

Electron positions do not commute with the Hamiltonian, so no position-dependent quantity can be evaluated directly from the DMC walker positions: the

---

[12] GFMC, Green's function Monte Carlo, uses Green's functions in imaginary-time evolution rather than diffusion. GFMC is computationally more demanding than DMC, and it struggles with the fermion sign problem because a simple fixed-node approach is not available in GFMC.

walkers are in the 'wrong positions.' This affects the DMC evaluation of quantities such as R_{eq}, the electron density $\langle n(\mathbf{r}) \rangle$, and the potential energy $\langle V \rangle$—the kinetic energy is $\langle T \rangle = \langle E \rangle - \langle V \rangle$. One way to evaluate such expectation values is the extrapolated estimate, introduced in section 4.4.6. For example, the electron density can be approximated using the linear extrapolation

$$\langle n(\mathbf{r}) \rangle_{\text{Extrapolated}} = 2\langle n(\mathbf{r}) \rangle_{\text{DMC}} - \langle n(\mathbf{r}) \rangle_{\text{VMC}}, \tag{4.358}$$

where the VMC positions give the baseline and the DMC walker positions give a mixed estimate. If the stochastic error in the DMC data is σ_{DMC} and that in VMC is σ_{VMC}, then the stochastic error in the extrapolation is given by error propagation rules,[13]

$$\sigma_{\text{Extrapolated}} = \sqrt{(2\sigma_{\text{DMC}})^2 + (\sigma_{\text{VMC}})^2}, \tag{4.359}$$

which does not take into account possible systematic error in the extrapolation itself. If the trial wave function is as good as in the H_2 molecule case, the extrapolated estimate is reliable.[14]

The electron density profile calculated using H2_DMC.jl is shown in figure 4.19. The protons are moved, so to fix the scale, their distance was normalized to one. The electron density $\langle n(x) \rangle$ was collected from a cylinder of radius 0.1 whose center passes through the protons. The VMC result is essentially the HF density profile, now collected as an average.

Using the parameters given earlier, the extrapolated estimate of the equilibrium distance is

$$\text{VMC} \langle R \rangle_{eq} = 1.424\ 36 \pm 0.000\ 04 \text{ a. u.} \tag{4.360}$$

$$\text{DMC} \langle R \rangle_{eq} = 1.4364 \pm 0.0001 \text{ a. u. }, \quad \tau = 0.02 \text{ Ha}^{-1} \tag{4.361}$$

$$\text{DMC} \langle R \rangle_{eq} = 1.4390 \pm 0.0001 \text{ a. u. }, \quad \tau = 0.01 \text{ Ha}^{-1} \tag{4.362}$$

$$\text{DMC} \langle R \rangle_{eq} = 1.4345 \pm 0.0002 \text{ a. u. }, \quad \tau = 0.005 \text{ Ha}^{-1} \tag{4.363}$$

$$\text{DMC} \langle R \rangle_{eq} = 1.4329 \pm 0.0004 \text{ a. u. } \quad \tau = 0.001 \text{ Ha}^{-1} \tag{4.364}$$

$$\text{DMC} \langle R \rangle_{eq} = 1.4269 \pm 0.0004 \text{ a. u. }, \quad \tau = 0.0001 \text{ Ha}^{-1} \tag{4.365}$$

$$\text{DMC} \langle R \rangle_{eq} = 1.4350 \pm 0.0002 \text{ a. u. extrapolation to } \tau = 0 \tag{4.366}$$

$$\text{Extrap.} \langle R \rangle_{eq} = 1.4456 \pm 0.0003 \text{ a. u. } = 0.765 \pm 0.0002 \text{ Å.} \tag{4.367}$$

---

[13] If $z = f(x, y)$, where x and y are independent stochastic variables with standard deviations σ_x and σ_y, respectively, then $\sigma_z^2 = \sqrt{(\frac{\partial f}{\partial x}\sigma_x)^2 + (\frac{\partial f}{\partial y}\sigma_y)^2}$.

[14] One could use a pure estimate, which weights the walker's data with the number of descendants the walker has. The process becomes unstable if continued over too many generations, and just when to stop has to be analyzed case by case.

H$_2$ molecule electron density profile

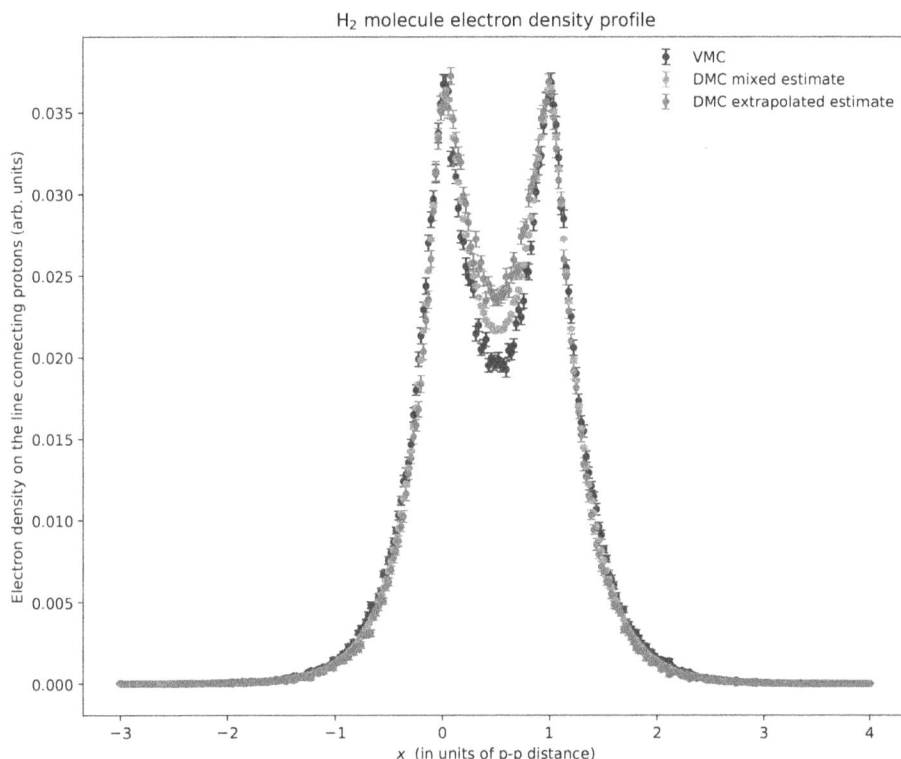

Figure 4.19. The electron density profile along the line of the two protons. Proton positions have been normalized to $x = 0$ and $x = 1$, and the electron density is a histogram of binned values in a cylinder of radius 0.1.

Similarly, the potential energy can be extrapolated to give $\langle V \rangle = -2.3368 \pm 0.0009$ Ha, so $\langle T \rangle = \langle E \rangle - \langle V \rangle = 1.1728 \pm 0.0009$ Ha, close to the virial theorem prediction of $\langle T \rangle = -\langle V \rangle/2 = 1.1684 \pm 0.0013$.

In [46], the extrapolated DMC bond length was found to be $R_{eq} = 1.401$ a.u. ≈ 0.7414 Å, which is also the experimental value cited by the Computational Chemistry Comparison and Benchmark DataBase 2022 [47]. Our calculated bond length is about 3% higher, which can be attributed to inaccuracies in the extrapolation. Notice that position-dependent averages need two linear extrapolations: first, the finite-τ DMC data is extrapolated to $\tau = 0$, and then the result is used to extrapolate from the VMC value to the final result. For a better trial wave function, the extrapolation error can be radically reduced.

Dissociation of H$_2$

In molecular quantum mechanics, one often assumes that the electronic and nuclear motions are fully decoupled,

$$\Psi(\mathbf{R}, \mathbf{r}) \approx \Psi_{\text{electronic}}(\mathbf{r}; \mathbf{R}) \Psi_{\text{nuclear}}(\mathbf{R}) \qquad \text{Born–Oppenheimer}, \qquad (4.368)$$

where the electronic wave function contains ion positions $\mathbf{R}$ as parameters.[15] As the proton–proton distance R changes in H_2, the electron cloud deforms, and the VMC calculations are repeated for each R. In our trial wave function, the deformation is the single adjustable parameter ζ, which at large p–p separations should go to the value of two ground-state H atoms, i.e. 1.0. The fact that the model fails at large R is obvious in figure 4.20; however, the culprit is not ζ but the wrong ionized H terms in the binding orbital RHF wave function $\sigma_g(1)\sigma_g(2)$.

In the code H2_DMC.jl, one can also see how the optimized ζ depends on R, but finding the optimal ζ becomes increasingly difficult with increasing R. This is expected because the overlap of the electron clouds is reduced, and the energy becomes insensitive to ζ.

A better trial wave function for H_2 was suggested in 1933 by James and Coolidge [48], who extended the Hylleraas wave function to molecules. The wave function was further advanced in 1960 by Kolos and Roothaan [49, 50]. The H_2 symmetry is most naturally captured by elliptic coordinates (more precisely, their 3D extension to

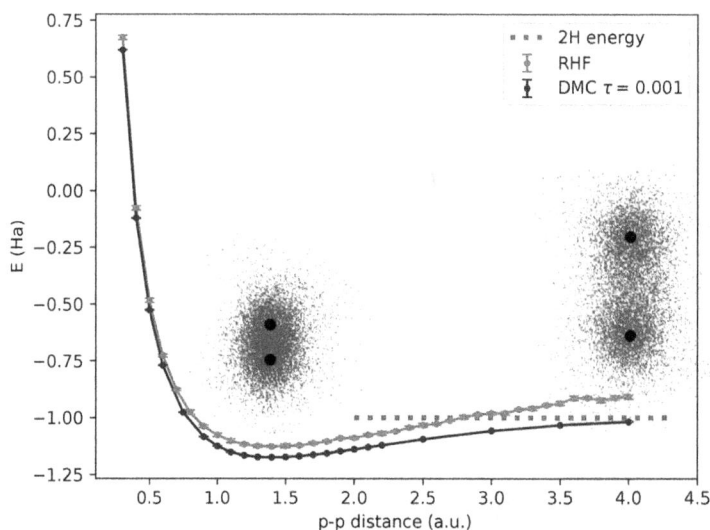

Figure 4.20. Upper curve (gray) is the H_2 dissociation curve predicted by RHF binding MO σ_g in equation (4.340), now calculated using VMC. The AO parameter ζ is re-optimized at each p–p separation. The horizontal dashed line shows the expected limit of two hydrogen atoms at large p–p separation, which RHF fails to meet. The lower curve (maroon) is the DMC dissociation curve computed using $\tau = 0.001$ Ha^{-1}; the curve is very close to the exact result, and the 2H limit is correctly reproduced. This is due to the flexibility of the Kalos–Roothaan wave function: the insets show the VMC electron positions of a H_2 molecule at $R = 1.4$ a. u. and at $R = 4.0$ a. u., i.e. near the dissociation limit.

---

[15] The *adiabatic approximation* allows some coupling, such as vibrational energy transfer, between electrons and nuclei.

prolate spherical coordinates); if electrons are denoted by 1 and 2, protons are denoted by 3 and 4, and the bond length is R: $=r_{34}$, then[16]

$$\xi_1 = (r_{13} + r_{14})/R \text{ electron 1} \qquad (4.369)$$

$$\xi_2 = (r_{23} + r_{24})/R \text{ electron 2} \qquad (4.370)$$

$$\eta_1 = (r_{13} - r_{14})/R \text{ electron 1} \qquad (4.371)$$

$$\eta_2 = (r_{23} - r_{24})/R \text{ electron 2}. \qquad (4.372)$$

The spin-singlet state is multiplied by the spatial part (equations (3) and (4) in reference [49]):

$$\varphi_{\text{Kolos–Roothaan}}(\mathbf{x}) = [\phi(\xi_1, \eta_1)\Psi(\xi_2, \eta_2) + \Psi(\xi_1, \eta_1)\phi(\xi_2, \eta_2)]\chi(r_{12}) \qquad (4.373)$$

$$\phi(\xi_j, \eta_j): = \sum_{i=0}^{m} a_i u_i(\xi_j, \eta_j) \text{ for electron } j = 1, 2 \qquad (4.374)$$

$$\psi(\xi_j, \eta_j): = \sum_{i=0}^{m} b_i u_i(\xi_j, \eta_j) \text{ for electron } j = 1, 2, \qquad (4.375)$$

and the e–e correlations are given by

$$\chi(r_{12}) = \sum_{\mu=0}^{n} c_\mu v_\mu(r_{12}). \qquad (4.376)$$

The functions u and v are (equation (5) in reference [49])

$$u_i(\xi_j, \eta_j) = \xi_j^{p_i} \eta_j^{q_i} e^{-\alpha\xi_j} \text{ for electron } j = 1, 2 \qquad (4.377)$$

$$v_\mu(r_{12}) = r_{12}^{\mu}. \qquad (4.378)$$

The code H2_Kolos_Roothaan.jl uses vectors of five integers,

$$p_i = [0, 1, 2, 0, 1] \qquad (4.379)$$

$$q_i = [0, 0, 0, 2, 2] \qquad (4.380)$$

$$\mu_i = [0, 1, 2, 3, 4]. \qquad (4.381)$$

The cylindrical symmetry of H_2 dictates that the powers of η, i.e. the integers q_i, must be even. The variational parameters are α in the exponent, and the expansion coefficients a_i, b_i, and c_i now have five elements each. The flexibility of the Kolos–Roothaan form is far beyond that of the RHF. Keeping R fixed (the Born–Oppenheimer approximation) gives a very accurate H_2 dissociation curve, shown

---

[16] Hylleraas used them for the He atom.

in figure 4.20. As depicted in the inset, the optimized $\varphi_{\text{Kolos–Roothaan}}$ gives qualitatively reasonable electron distributions both in the molecular state and near the dissociation limit. Adding a further DMC energy reduction gives essentially exact results.

Calculating the ground state of a two-electron system is particularly fast, as a nodeless wave function can be applied. Furthermore, a molecule that has a cylindrical symmetry that begs for cylindrical coordinates is just what it sounds like: a lucky strike that does not promise an easy extension to other molecules.

Before leaving the Kolos–Roothaan wave function, I would like to point out how simple QMC is compared to the computation of all integrals by other means, as done in [49]. The workload in the latter is mostly human effort, with some numerics, while in QMC almost all the work is done by a computer. In QMC, all you need is a function that evaluates $\varphi_{\text{Kolos–Roothaan}}$. If you crave speed, you can find the analytical gradients and code them as well, but if you are lazy, AD can do that without any loss of accuracy.

The history of finding the H_2 dissociation energy is a competition between ever-improving experiments and calculations.[17] One of the culmination points was in 1968 when Kolos and Wolniewicz [51] calculated the adiabatic ground-state energy to be $E_0 = -1.174\ 474\ 983\ 017\ 30$ Ha. The DMC result obtained by Hongo *et al* agreed quite well; they found $E_0 = -1.174\ 47(4)$ Ha [46]. Even the tiny energy difference between the energy of ortho-H_2, which has parallel *proton* spins, and that of para-H_2 (antiparallel proton spins) can be measured nowadays; that of ortho-H_2 is 0.000 539 865 769 a. u. greater than that of para-H_2 [52]. For recent theoretical, non-QMC results, see [53].

The H_2 dissociation curve in figure 4.21 includes the proton–proton Coulomb energy $1/R$, but since, in the Born–Oppenheimer approximation, the nuclei are held fixed, it makes sense to consider the energy as a function of R but without the $1/R$ contribution. Figure 4.21 shows that in the limit $R \to 0$, we recover the He atom ground-state energy of -2.9037 Ha in the Born–Oppenheimer approximation. Indeed, a nuclear charge of $Z = 2$ and two electrons *are* a He atom, because ignoring the two neutrons in a real He atom makes no difference in the Born–Oppenheimer approximation. Ignoring the strong nuclear force has no ill effects, either. It is not every day you observe nuclear fusion and fission ($H+H \rightleftharpoons He$) in action! And yes, a fissioned Born–Oppenheimer Li atom ends up as He + H or as H + H + H, a Be atom fissions to He + He or to Li + H, and so on. Molecules have less symmetry and less degeneracy than atoms, which makes molecular wave functions more complicated than those of atoms. In principle, if you have the Born–Oppenheimer QMC code for a charge-neutral molecule, you can fuse the nuclei and obtain the atomic ground-state energy as a free byproduct.

There is more to H_2 than the isolated molecule discussed here. A much tougher problem for DMC is the dissociation of H_2 on metal surfaces, an important process in heterogeneous catalysis. Recent DMC results for dissociation on Cu(111) and Pt (111) have achieved chemical accuracy and agree well with other methods. To give some perspective on the required accuracy, the dissociation barrier for hydrogen on Pt(111) was found to be 5.4 kcal mol^{-1} in [54], which is about 0.234 eV mol^{-1} or

---

[17] Two separate H atoms have $E = -1.0$ Ha, so the dissociation energy is $D(H_2) = E_0 - (-1.0\ \text{Ha})$.

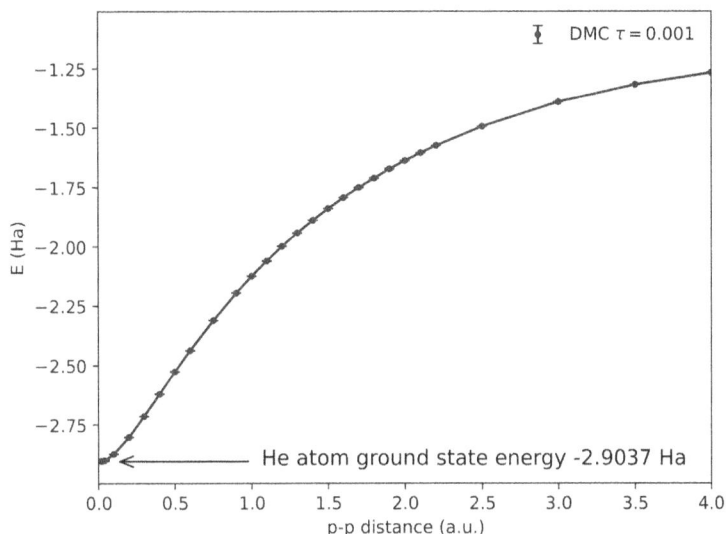

Figure 4.21. The energy of a H_2 molecule without the proton–proton Coulomb energy as a function of proton–proton distance. In the Born–Oppenheimer approximation, an H_2 molecule is a split He atom.

0.008 59 Ha mol^{-1}. According to the DMC calculations in [55], based on the CASINO code and a Slater–Jastrow wave function, the dissociation barrier on a Mg (0001) surface is 1.18 ± 0.03 eV ≈ 0.0434 Ha, a close match for the DFT-RBPE result; however, the reverse barrier of 1.28 ± 0.03 eV is as much as 0.3 eV higher than any of the comparable DFT results with various functionals. Such an energy discrepancy has a large impact on reaction rates and underlines the necessity for computationally more expensive DMC calculations.

As another example, the high-pressure phase diagram of hydrogen was computed using DMC by Mazzola *et al* [56]. DMC provides a better description of dissociation than DFT, which struggles with similar issues to those we encountered earlier with HF. Neither LDA nor GGA exchange-correlation functionals capture bond stretching correctly. Mazzola *et al* also list other problems with DFT, such as underestimated electron gaps, the resulting easier destabilization of molecular phases, and the fact that DFT solid-phase predictions depend on the choice of density functional. Considering the lack of experimental reference data under high-pressure conditions, DMC offers the most reliable results for the phase transition between a molecular fluid and an atomic fluid, although it has not provided a definitive answer to the long-standing question about the metallicity of hydrogen gas.

4.8.3 Antisymmetrized geminal power wave functions

The workhorse of multielectron wave functions is a sum of Slater determinants $\det(\phi_i(\mathbf{r}_j))$ of single-electron orbitals, where $\mathbf{r}_j$ is an electron position w.r.t. a nucleus. As shown in the Introduction, any many-body wave function can be obtained from single-particle orbitals. The e–e interaction, which is the ultimate source of e–e correlations, is present in the Hamiltonian, so energy optimization chooses the single-electron orbital parameters and the sum of Slater determinants that correlate electrons with each other. This, however, relies on the completeness of the Slater determinant basis; in practice, one has to sum up quite a few Slater determinants $\det(\phi_i(\mathbf{r}_j))$ to get accurate e–e correlations.

It therefore seems we could add e–e correlations to the wave function in a more effective manner. We have already used one shortcut by writing the e–e correlations as

$$u(r_{12}) = \frac{r_{12}}{s(1+br_{12})}, \text{ where } \begin{cases} s = 2 \text{ antiparallel spins} \\ s = 4 \quad \text{parallel spins} \end{cases}, \tag{4.382}$$

or as

$$u(r_{12}) = \frac{F}{2}(1 - e^{-r_{12}/F}), \tag{4.383}$$

both with one variational parameter. From $u(r_{12})$, we can build a Jastrow factor, which covers some of the e–e correlations but is not yet antisymmetric. Earlier, in the Slater–Jastrow wave function, antisymmetry was enforced with the Slater determinant of single-electron orbitals.

One can also take another route and, starting from something like 'paired electrons,' group the single-electron orbitals together pairwise. Wave functions made of such pairwise e–e correlations are called *geminals*, from the word *gemini* meaning 'twins.' Antisymmetrization of the geminal orbitals ensures the Pauli principle is satisfied, and the result goes under the name **antisymmetric geminal power (AGP)**.

The AGP wave function, suggested by A J Coleman [57], is expected to be a good description of systems with strong electron–electron correlations. The idea is closely related to the resonating valence bond (RVB) theory in condensed-matter physics, invented by Linus Pauling in the 1930s to describe metals, in a series of ground-breaking papers titled 'The nature of the chemical bond' and later extended by Philip Anderson to a theory of superconductivity [58].[18] RVB emphasizes the importance of correlations of electrons in different molecules or across a lattice. In a geminal, the pair of electrons is in a singlet state, and antisymmetrization is what makes the pairs 'resonate.'

An AGP wave function is an antisymmetrized product of spin orbitals [59, 60],

$$\Psi_{\text{AGP}}(x) = \hat{\mathcal{A}}[\phi_{\text{AGP}}(x_1, x_2)\phi_{\text{AGP}}(x_3, x_4). ..\phi_{\text{AGP}}(x_{N-1}, x_N)], \tag{4.384}$$

where $\hat{\mathcal{A}}$ is the antisymmetrization operator and N is the number of electrons. As mentioned earlier, electron pairs are in a singlet state, and the spatial part can be expressed as a Slater determinant,

$$\Psi_{\text{AGP}}(\mathbf{x}) = \det\big(\phi_{\text{AGP}}(\mathbf{r}_i, \mathbf{r}_j)\big), \tag{4.385}$$

---

[18] AGPs are also related to many other approaches for strongly correlated electrons, such as the super-conductivity theory devised by Schafroth in the 1950s and the Bardeen–Cooper–Schrieffer (BCS) theory developed in 1957.

where the ith electron is spin up and the jth is spin down. The geminal orbitals can be written as

$$\phi_{\text{AGP}}(\mathbf{r}_i, \mathbf{r}_j) = \sum_{b,m}\sum_{a,l} c_{(a,\,l),(b,\,m)}\phi_{a,l}(\mathbf{r}_i)\phi_{b,m}(\mathbf{r}_j): \; = \sum_{\mu}\sum_{\nu} c_{\mu\nu}\,\phi_{\mu}(\mathbf{r}_i)\phi_{\nu}(\mathbf{r}_j), \qquad (4.386)$$

where the atomic orbitals (indexed by l and m, respectively) are those of nuclei a and b, respectively. The last form introduces combined indices $\mu = (a,\,l)$ and $\nu = (b,\,m)$.[19] The matrix c can be diagonalized using an orthogonal transformation $\chi_{\mu}(\mathbf{r}) = \sum_{\nu} U_{\mu\nu}\phi_{\nu}(\mathbf{r})$ (see [60] for details), and the geminal can be written in a compact form,

$$\phi_{\text{AGP}}(\mathbf{r}_i, \mathbf{r}_j) = \sum_{\mu}\lambda_{\mu}\chi_{\mu}(\mathbf{r}_i)\chi_{\mu}(\mathbf{r}_j). \qquad (4.387)$$

Even with a single determinant, AGP corresponds to a sum of multiple single-particle Slater determinants; hence, it can be called an *implicitly multideterminant ansatz*. To demonstrate the connection between AGP and multideterminant wave functions, Zen *et al* [61] showed that the AGP state can be written as (a simplified notation)

$$
\begin{aligned}
|\Psi_{\text{AGP}}\rangle = c_0|\Psi_0\rangle &+ \sum_{\text{excited pairs}} c_{\text{excited pair}}\,|\Psi_{\text{excited pair}}\rangle \\
&+ \sum_{\text{excited two-pairs}} c_{\text{excite two-pairs}}\,|\Psi_{\text{excited two-pairs}}\rangle + \cdots,
\end{aligned}
\qquad (4.388)
$$

where the terms are built similarly to a CI sum of Slater determinant states, except that now electrons are not excited one by one but in pairs. The multideterminant expansion coefficients c_i can be readily computed from the AGP coefficients λ_i, for example, $c_0 = \prod_{i\in pairs} \lambda_i$ and $c_{\text{excited pair}} = c_0\lambda_a/\lambda_i$, where a pair from the geminal orbital i is excited to the geminal orbital a, etc. The AGP, as an implicitly multideterminant ansatz, takes into account all pair excitations, a subset of all excitations in the CI expansion.

The AGP wave function is commonly multiplied by a Jastrow factor, hence the acronym JAGP. Neuscamman [62] has made an interesting remark, that while one is used to adding factors to a wave function in order to give it new properties, the JAGP wave function does something he calls 'subtractive manufacturing': the AGP has unwanted features subtracted by the Jastrow factor. One unwanted feature of AGP is that it is not size consistent, but a Jastrow factor can penalize the size-inconsistent parts in dissociation [63, 64]. As an example, we saw how the RHF wave function of H_2, $\sigma_g(1)\sigma_g(2)$ in equation (4.348), contains unwanted ionic terms, and those can be 'subtracted' using a suitable factor.

The JAGP wave function has been benchmarked for H_4 molecule distortion, H_2O stretching, N_2 stretching, the C_2 triplet state, and unassisted gas phase insertion

---

[19] I admit that using AOs in geminal orbitals seems like betraying the cause. After all, we were supposed to get rid of nucleus-based orbitals.

of H_2 into ethene's double bond and has been shown to perform really well [65]. To keep the number of parameters in the AGP sum in equation (4.387) feasible, the sum can be truncated at some n without too much damage. Recently, such a modified JAGP, named JAGPn, was successfully applied to planar and twisted ethylenes, the hydrocarbons CH_4, C_2H_4, C_2H_6, C_6H_6, $C_{10}H_8$, $C_{14}H_{10}$, $C_{18}H_{12}$, and $C_{20}H_{10}$, the C_{60} fullerene, and a water–methane dimer—the largest molecule had 240 electrons. Again, the Jastrow factor was used to keep the wave function size consistent.

References

[1] Lemons D S and Gythiel A 1908 Paul Langevin's 1908 paper 'On the theory of Brownian motion' ['Sur la théorie du mouvement Brownien,' C. R. Acad. Sci. (Paris) 146, 530–3 (1908)] *Am. J. Phys.* **65** 1079–81

[2] Caffarel M and Claverie P 1988 Development of a pure diffusion quantum Monte Carlo method using a full generalized Feynman-Kac formula. II. Applications to simple systems *J. Chem. Phys.* **88** 1100–9

[3] Bosá I and Rothstein S M 2004 Unbiased expectation values from diffusion quantum Monte Carlo simulations with a fixed number of walkers *J. Chem. Phys.* **121** 4486–93

[4] Assaraf R, Caffarel M and Khelif A 2000 Diffusion Monte Carlo methods with a fixed number of walkers *Phys. Rev.* E **61** 4566–75

[5] Reynolds P J, Ceperley D M, Alder B J and Lester W A 1982 Fixed-node quantum Monte Carlo for molecules *J. Chem. Phys.* **77** 5593–603

[6] Trotter H F 1959 On the product of semi-groups of operators *Proc. Amer. Math. Soc.* **10** 545–51

[7] Suzuki M 1976 Generalized Trotter's formula and systematic approximants of exponential operators and inner derivations with applications to many-body problems *Commun. Math. Phys.* **51** 183–90

[8] Vrbik J and Rothstein S M 1986 Quadratic accuracy diffusion Monte Carlo *J. Comput. Phys.* **63** 130–9

[9] Hastings W K 1970 Monte Carlo sampling methods using Markov chains and their applications *Biometrika* **57** 97–109

[10] Ceperley D, Kalos M H and Lebowitz J L 1981 Computer simulation of the static and dynamic properties of a polymer chain *Macromolecules* **14** 1472–9

[11] Chin S A 1990 Quadratic diffusion Monte Carlo algorithms for solving atomic many-body problems *Phys. Rev.* A **42** 6991–7005

[12] Ceperley D M and Kalos M H 1986 Quantum Many-Body Problems *Monte Carlo Methods in Statistical Physics* (Berlin: Springer) pp 145–94

[13] Liu K S, Kalos M H and Chester G V 1974 Quantum hard spheres in a channel *Phys. Rev.* A **10** 303–8

[14] Casulleras J and Boronat J 1995 Unbiased estimators in quantum Monte Carlo methods: application to liquid ^{4}He *Phys. Rev.* B **52** 3654–61

[15] Forest E and Ruth R D 1990 Fourth-order symplectic integration *Physica* D **43** 105–17

[16] Sheng Q 1989 Solving linear partial differential equations by exponential splitting *IMA J. Numer. Anal.* **9** 199–212

[17] Suzuki M 1991 General theory of fractal path integrals with applications to many-body theories and statistical physics *J. Math. Phys.* **32** 400

[18] Suzuki M 1995 Hybrid exponential product formulas for unbounded operators with possible applications to Monte Carlo simulations *Phys. Lett.* A **201** 425–8

[19] Takahashi M and Imada M 1984 Monte Carlo calculation of quantum systems. II. Higher order correction *J. Phys. Soc. Jpn.* **53** 3765–9

[20] Li X-P and Broughton J Q 1987 High-order correction to the Trotter expansion for use in computer simulation *J. Chem. Phys.* **86** 5094–100

[21] Chin S A 2004 Quantum statistical calculations and symplectic corrector algorithms *Phys. Rev.* E **69** 046118

[22] Chin S A 1997 Symplectic integrators from composite operator factorizations *Phys. Lett.* A **226** 344–8

[23] Forbert H A and Chin S A 2001 Fourth-order diffusion Monte Carlo algorithms for solving quantum many-body problems *Phys. Rev.* B **63** 144518

[24] Umrigar C J, Nightingale M P and Runge K J 1993 A diffusion Monte Carlo algorithm with very small time-step errors *J. Chem. Phys.* **99** 2865–90

[25] Zen A, Sorella S, Gillan M J, Michaelides A and Alfè D 2016 Boosting the accuracy and speed of quantum Monte Carlo: size consistency and time step *Phys. Rev.* B **93** 241118

[26] Anderson J B 1976 Quantum chemistry by random walk *J. Chem. Phys.* **65** 4121–7

[27] Ceperley D M and Alder B J 1984 Quantum Monte Carlo for molecules: Green's function and nodal release *J. Chem. Phys.* **81** 5833–44

[28] Ceperley D M 1991 Fermion nodes *J. Stat. Phys.* **63** 1237–67

[29] Plante I, Devroye L and Cucinotta F A 2014 Calculations of distance distributions and probabilities of binding by ligands between parallel plane membranes comprising receptors *Comput. Phys. Commun.* **185** 697–707

[30] Mishchenko Y 2006 Remedy for the fermion sign problem in the diffusion Monte Carlo method for few fermions with antisymmetric diffusion process *Phys. Rev.* E **73** 026706

[31] Foulkes W M C, Mitas L, Needs R J and Rajagopal G 2001 Quantum Monte Carlo simulations of solids *Rev. Mod. Phys.* **73** 33–83

[32] Bogojeski M, Vogt-Maranto L, Tuckerman M E, Müller K-R and Burke K 2020 Quantum chemical accuracy from density functional approximations via machine learning *Nat. Commun.* **11** 5223

[33] Poole D, Galvez Vallejo J L and Gordon M S 2020 A new kid on the block: application of Julia to Hartree-Fock calculations *J. Chem. Theory Comput.* **16** 5006–13

[34] Aroeira G J R, Davis M M, Turney J M and Schaefer H F 2022 Fermi.jl: a modern design for quantum chemistry *J. Chem. Theory Comput.* **18** 677–86

[35] Bressanini D, Ceperley D M and Reynolds P J 2002 What do we know about wave function nodes *Recent Advances in Quantum Monte Carlo Methods* Part II (Singapore: World Scientific) pp 3–11

[36] Mitas L 2006 Structure of Fermion nodes and nodal cells *Phys. Rev. Lett.* **96** 240402

[37] Bressanini D 2012 Implications of the two nodal domains conjecture for ground state fermionic wave functions *Phys. Rev.* B **86** 115120

[38] Needs R J, Towler M D, Drummond N D, López Ríos P and Trail J R 2020 Variational and diffusion quantum Monte Carlo calculations with the CASINO code *J. Chem. Phys.* **152** 154106

[39] Wagner L K, Bajdich M and Mitas L 2009 QWalk: a quantum Monte Carlo program for electronic structure *J. Comput. Phys.* **228** 3390–404

[40] Kim J *et al* 2018 QMCPACK: an open source *ab initio* quantum Monte Carlo package for the electronic structure of atoms, molecules and solids *J. Phys.: Condens. Matter* **30** 195901

[41] Roos B O, Taylor P R and Sigbahn P E M 1980 A complete active space SCF method (CASSCF) using a density matrix formulated super-CI approach *Chem. Phys.* **48** 157–73

[42] Jensen F 2007 *Introduction to Computational Chemistry* (New York: Wiley)

[43] Townsend J, Kirkland J K and Vogiatzis K D 2019 Post-Hartree–Fock methods: configuration interaction, many-body perturbation theory, coupled-cluster theory *Mathematical Physics in Theoretical Chemistry, Developments in Physical and Theoretical Chemistry* ed S M Blinder and J E House (Amsterdam: Elsevier) ch 3 pp 63–117

[44] Silvera I F and Walraven J T M 1980 Stabilization of atomic hydrogen at low temperature *Phys. Rev. Lett.* **44** 164–8

[45] Traynor C A, Anderson J B and Boghosian B M 1991 A quantum Monte Carlo calculation of the ground state energy of the hydrogen molecule *J. Chem. Phys.* **94** 3657–64

[46] Hongo K, Kawazoe Y and Yasuhara H 2007 Diffusion Monte Carlo study of correlation in the hydrogen molecule *Int. J. Quant. Chem.* **107** 1459–67

[47] NIST Computational Chemistry Comparison and Benchmark Database, NIST Standard Reference Database Number 101, Release 22, 2022

[48] James H M and Coolidge. A S 1933 The ground state of the hydrogen molecule *J. Chem. Phys.* **1** 825–35

[49] Kolos W and Roothaan C C J 1960 Correlated orbitals for the ground state of the hydrogen molecule *Rev. Mod. Phys.* **32** 205–10

[50] Kolos W and Roothaan C C J 1960 Accurate electronic wave functions for the H_2 molecule *Rev. Mod. Phys.* **32** 219–32

[51] Kolos W and Wolniewicz L 1968 Improved theoretical ground-state energy of the hydrogen molecule *J. Chem. Phys.* **49** 404–10

[52] Cheng C-F *et al* 2018 Dissociation energy of the hydrogen molecule at 10^{-9} accuracy *Phys. Rev. Lett.* **121** 013001

[53] Puchalski M, Komasa J, Czachorowski P and Pachucki K 2019 Nonadiabatic QED correction to the dissociation energy of the hydrogen molecule *Phys. Rev. Lett.* **122** 103003

[54] Hoggan P E 2018 Quantum Monte Carlo calculations for industrial catalysts: accurately evaluating the H_2 dissociation reaction barrier on Pt(111) *Novel Electronic Structure Theory: General Innovations and Strongly Correlated Systems* (Advances in Quantum Chemistry vol 76) ed P E Hoggan (New York: Academic) pp 271–8

[55] Pozzo M and Alfè D 2008 Hydrogen dissociation on Mg(0001) studied via quantum Monte Carlo calculations *Phys. Rev.* B **78** 245313

[56] Mazzola G, Yunoki S and Sorella S 2014 Unexpectedly high pressure for molecular dissociation in liquid hydrogen by electronic simulation *Nat. Commun.* **5** 3487

[57] Coleman A J 1963 Structure of Fermion density matrices *Rev. Mod. Phys.* **35** 668–86

[58] Anderson P W, Baskaran G, Zou Z and Hsu T 1987 Resonating-valence-bond theory of phase transitions and superconductivity in La_2CuO_4 -based compounds *Phys. Rev. Lett.* **58** 2790–3

[59] Anderson. P W 1987 The resonating valence bond state in La_2CuO_4 and superconductivity *Science* **235** 1196–8

[60] Casula M, Attaccalite C and Sorella S 2004 Correlated geminal wave function for molecules: an efficient resonating valence bond approach *J. Chem. Phys.* **121** 7110–26

[61] Sorella S, Casula M and Rocca D 2007 Weak binding between two aromatic rings: feeling the van der Waals attraction by quantum Monte Carlo methods *J. Chem. Phys.* **127** 014105

[62] Nakano K, Sorella S, Alfè D and Zen. A 2024 Beyond single-reference fixed-node approximation in *ab initio* diffusion Monte Carlo using antisymmetrized geminal power applied to systems with hundreds of electrons *J. Chem. Theory Comput.* **20** 4591–604

[63] Zen A, Coccia E, Luo Y, Sorella S and Guidoni L 2014 Static and dynamical correlation in diradical molecules by quantum Monte Carlo using the jastrow antisymmetrized geminal power ansatz *J. Chem. Theory Comput.* **10** 1048–61

[64] Neuscamman E 2016 Subtractive manufacturing with geminal powers:making good use of a bad wave function *Mol. Phys.* **114** 577–83

[65] Neuscamman E 2012 Size consistency error in the antisymmetric geminal power wave function can be completely removed *Phys. Rev. Lett.* **109** 203001

[66] Sørensen L K 2022 On the size consistency problem for anti-symmetrised geminal power wave function ansatz *Molecular Physics* **120** e2049385

[67] Neuscamman. E 2013 The Jastrow antisymmetric geminal power in Hilbert space: theory, benchmarking, and application to a novel transition state *J. Chem. Phys.* **139** 194105

Chapter 5

Path integral Monte Carlo

In path integral Monte Carlo (PIMC), it is customary to write the Hamiltonian as

$$\hat{\mathcal{H}} = \hat{\mathcal{T}} + \hat{\mathcal{V}} = -\lambda \sum_{i=1}^{N} \nabla_i^2 + \hat{\mathcal{V}}, \tag{5.1}$$

where the constant λ: $= \hbar^2/(2m)$. In diffusion Monte Carlo (DMC), we used the notation D for the same quantity to reflect the fact that there it takes the role of a diffusion constant. In the statistical physics of equilibrium quantum systems, the two frequently encountered conditions are the **canonical ensemble**, where energy and particle number are conserved, and the **grand canonical ensemble**, where the particle number fluctuates. Let us begin with the canonical ensemble.

In the canonical ensemble, thermal averages can be computed from the canonical partition function

$$Z(\beta) = \sum_i e^{-\beta E_i} = \sum_i \langle i | e^{-\beta \hat{\mathcal{H}}} | i \rangle \tag{5.2}$$

where $\hat{\mathcal{H}} | i \rangle = E_i | i \rangle$, $\beta = 1/(k_B T)$, k_B is the Boltzmann constant, and T is the temperature. The partition function holds information about the statistical behavior of the system at a finite temperature. The **density operator** and its representation in the position basis or in the state basis, the **density matrix**, are defined as

$$\hat{\rho}: = e^{-\beta \hat{\mathcal{H}}} \tag{5.3}$$

$$\rho(\mathbf{x}', \mathbf{x}; \beta): = \langle \mathbf{x}' | e^{-\beta \hat{\mathcal{H}}} | \mathbf{x} \rangle \tag{5.4}$$

$$\rho(i, j; \beta): = \langle i | e^{-\beta \hat{\mathcal{H}}} | j \rangle. \tag{5.5}$$

The partition function is the trace of the density matrix; in the position basis, a trace means an integral over positions,

doi:10.1088/978-0-7503-6310-5ch5

$$Z(\beta) = \int d\mathbf{x} \sum_i \langle i|e^{-\beta\hat{\mathcal{H}}}|\mathbf{x}\rangle\langle\mathbf{x}|i\rangle = \int d\mathbf{x} \sum_i \langle\mathbf{x}|i\rangle\langle i|e^{-\beta\hat{\mathcal{H}}}|\mathbf{x}\rangle$$
$$= \int d\mathbf{x}\langle\mathbf{x}|e^{-\beta\hat{\mathcal{H}}}|\mathbf{x}\rangle = \int d\mathbf{x}\rho(\mathbf{x}, \mathbf{x}; \beta). \tag{5.6}$$

While the partition function determines most equilibrium properties, the density matrix gives a more detailed view of quantum statistics, such as the populations of individual states (diagonal elements) and coherences between states (off-diagonal elements).

In most cases, the density matrices $\rho(\mathbf{x}', \mathbf{x}; \beta)$ are unknown, except for a reasonably accurate small-β approximation. A small β (high T) in PIMC is equivalent to a small time step in DMC. Since $\hat{\mathcal{H}}$ commutes with itself, we can do an exact operator splitting,

$$\rho(\mathbf{x}', \mathbf{x}; \beta) = \langle\mathbf{x}'|e^{-\beta\hat{\mathcal{H}}}|\mathbf{x}\rangle = \langle\mathbf{x}'|e^{-\frac{\beta}{2}\hat{\mathcal{H}}}e^{-\frac{\beta}{2}\hat{\mathcal{H}}}|\mathbf{x}\rangle. \tag{5.7}$$

We repeat this and insert complete position states between operators,

---

Convolution property of the density matrix

$$\rho(\mathbf{x}', \mathbf{x}; \beta) = \underbrace{\langle\mathbf{x}'|e^{-\beta\hat{\mathcal{H}}}|\mathbf{x}\rangle}_{\text{density matrix at } T} \tag{5.8}$$

$$= \int d\mathbf{x}_1...d\mathbf{x}_{M-1}\langle\mathbf{x}'|e^{-\frac{\beta}{M}\hat{\mathcal{H}}}|\mathbf{x}_1\rangle\underbrace{\langle\mathbf{x}_1|e^{-\frac{\beta}{M}\hat{\mathcal{H}}}|\mathbf{x}_2\rangle}_{\text{density matrix at } MT}...\langle\mathbf{x}_{M-1}|e^{-\frac{\beta}{M}\hat{\mathcal{H}}}|\mathbf{x}\rangle. \tag{5.9}$$

The partition function is

$$Z(\beta) = \int d\mathbf{x}\langle\mathbf{x}|e^{-\beta\hat{\mathcal{H}}}|\mathbf{x}\rangle$$
$$= \int d\mathbf{x}d\mathbf{x}_1...d\mathbf{x}_{M-1}\langle\mathbf{x}|e^{-\frac{\beta}{M}\hat{\mathcal{H}}}|\mathbf{x}_1\rangle\langle\mathbf{x}_1|e^{-\frac{\beta}{M}\hat{\mathcal{H}}}|\mathbf{x}_2\rangle...\langle\mathbf{x}_{M-1}|e^{-\frac{\beta}{M}\hat{\mathcal{H}}}|\mathbf{x}\rangle. \tag{5.10}$$

The last equation is interpreted as a **path integral**: the partition function is a sum over all possible paths that start from positions $\mathbf{x} = (\mathbf{r}_1, ..., \mathbf{r}_N)$, travel through M intermediate points, and return to the same positions $\mathbf{x}$. **Particles in the system are represented by closed loops in imaginary time.** Paths are also called **world lines**.

The whole point of this exercise, writing the integral over $\mathbf{x}$ as a seemingly more complicated multidimensional integral, is that the unknown density matrix at temperature T (inverse temperature β) is replaced with a product of density matrices at a higher temperature MT (inverse temperature β/M), which may be approximated. For shortness, we define

$$\boxed{\tau := \beta/M} \tag{5.11}$$

to signify the imaginary-time step in a single density matrix factor. The full imaginary-time path of length β is split into M small segments of length τ.

Identical particles, bosons or fermions, are special because we have no physical reason to say that a particle that started the loop at a point r_1 and ended the loop at r_1 is 'the same particle,' because identical particles are all 'the same.' This opens up the possibility of **exchange loops** or **loop fusion**, where two single-particle loops of length β combine to form a single two-particle loop of length 2β, and also longer loops with more particles. We expect to see observable physical effects—**superfluidity**—when the largest fused loops have macroscopic numbers of particles. Loop fusion takes place at low temperatures, so we need to push the inverse temperature β to a high value to see any quantum effects.

The task of computing approximate $Z(\beta)$ has some challenges. First, infinitely many paths cannot be numerically integrated, so we need to rely on Monte Carlo (MC) sampling of paths—we need *path sampling* and PIMC. The second problem is that if the interesting temperature regime is around $T \sim 1K$ and the density matrix is approximately known at $T \approx 1000K$, one needs $M \sim 1000$ divisions to compute $Z(\beta)$. Such paths have very complicated structures, and the efficiency of path sampling is reduced. The general goals are now clear:

- Find an efficient way of sampling closed paths in such a way that loop fusion is also possible.
- Find a way to approximate the density matrix at a temperature T, so that the target temperature T/M can be achieved with as small an M as possible.

The first point is about ergodicity: the algorithm should sample possible paths without systematically overlooking certain kinds of paths. The second is about numerical efficiency: the more complicated the paths are, the longer the calculations take.

5.0.1 Boson and fermion path integrals

The density matrix at inverse temperature τ has the eigenfunction expansion

$$\rho(\mathbf{x}', \mathbf{x}; \tau) = \langle \mathbf{x}'|e^{-\tau\hat{\mathcal{H}}}|\mathbf{x}\rangle = \sum_i e^{-\tau E_i}\phi_i^*(\mathbf{x}')\phi_i(\mathbf{x}) \quad \text{(distinguishable particles).} \tag{5.12}$$

The Hamiltonian we have been using does not account for particle statistics, so we need to require, at this late point, that since we cannot tell identical particles apart, their labeling cannot have any physical meaning. For that purpose, we could use Slater determinants (fermions) or permanents (bosons) of the 'particle-labeled' eigenfunctions $\phi_i(\mathbf{x})$, but in PIMC, it is often simpler to think in terms of permutations. The (anti)symmetric eigenfunctions can be constructed as a sum over permutations,

$$\frac{1}{N!}\sum_{perm}(\pm 1)^P\phi_i(P\mathbf{x})\begin{cases} + \text{ bosons} \\ - \text{ fermions} \end{cases}, \tag{5.13}$$

where P permutes particle labels and counts pair permutations.

For $N = 3$, there are $N! = 6$ permutations. Starting from the order 123 (any particle numbering will do), we count how many pair swaps are needed to get the new order. The even permutations are:

$P(123) = 123$	$P=0$	identity
$P(123) = 231$	$P=2$	two swaps, (1,2) and (1,3)
$P(123) = 312$	$P=2$	two swaps, (1,2) and (2,3)

and the odd permutations are:

$P(123) = 213$	$P=1$	one swap, (1,2)
$P(123) = 132$	$P=1$	one swap, (2,3)
$P(123) = 321$	$P=1$	one swap, (1,3).

Including the permutation sum, the density matrix of identical particles is[1]

$$\rho_{B/F}(\mathbf{x}', \mathbf{x}; \tau) = \sum_i e^{-\beta E_i} \phi_i^*(\mathbf{x}') \left[\frac{1}{N!} \sum_{perm} (\pm 1)^P \phi_i(P\mathbf{x}) \right].$$ (5.14)

Putting the permutation sum in front, one finds that boson and fermion density matrices can be written as sums over $\rho(\mathbf{x}', P\mathbf{x}; \tau)$, the permuted density matrix of distinguishable particles, but fermion terms have the factor $(-1)^P$,

$$\rho_B(\mathbf{x}', \mathbf{x}; \tau) = \frac{1}{N!} \sum_{perm} \rho(\mathbf{x}', P\mathbf{x}; \tau) \qquad \text{bosons}$$ (5.15)

$$\rho_F(\mathbf{x}', \mathbf{x}; \tau) = \frac{1}{N!} \sum_{perm} (-1)^P \rho(\mathbf{x}', P\mathbf{x}; \tau) \qquad \text{fermions.}$$ (5.16)

The latter formula is an algorithm killer. In $Z(\beta)$, one counts all path and all particle permutations, but now those with even P are positive and those with odd P are negative, so the sum has terms with comparable absolute values but different signs, and it is not even an alternating sum. The signal-to-noise ratio will become poor with an increasing number of fermions, and this happens really fast because there are $N!$ terms. This is a manifestation of the fermion sign problem, and circumventing it is now much more involved than the effective, albeit mundane, DMC solution that relies on the nodes of a trial wave function. The boson path integrals are not trivial either, but all terms in the sum are positive, and there are effective algorithms to sample boson paths.

---

[1] It does not matter whether we permute $\mathbf{x}$ or $\mathbf{x}'$, or both.

5.1 From real-time path integrals to imaginary-time path integrals

If you are familiar with real-time path integrals from, for example, the book *Quantum Mechanics and Path Integrals* by R Feynman and A Hibbs [1], there is a connection to finite-temperature path integrals. If real-time path integrals are not your cup of tea, I suggest you skip this section. The imaginary-time path integrals were already derived without any reference to the real-time path integral formulation of quantum mechanics. Quantum statistics at the basic level given in the introduction is totally adequate; however, take note that some commonly used terms have their origin in the real-time path integral formulation.

The Wick rotation, which replaces a real time t with the imaginary time $i\tau$, changes the time evolution operator $\exp(it\hat{\mathcal{H}}/\hbar)$ to the Boltzmann factor $exp(-\tau\hat{\mathcal{H}})$ with inverse temperature τ. The Wick rotation is a mathematically sound process. It can be viewed as a rotation in the complex plane and as a special case of analytical continuation widely used in complex analysis. The fact that it now leads from a branch of well-established physics to another, seemingly unrelated, branch of well-established physics may not be a coincidence but rather a sign of some kind of duality between the real-time dynamics and the finite-temperature, imaginary-time dynamics.[2]

While the canonical quantization is a conversion of the classical Hamilton's function to a Hamiltonian operator, the path integral quantization is based on the classical Lagrangian. Integrating the Lagrangian, one finds the *action S*,

$$S = \int dt \left(\frac{1}{2}m\left(\frac{d\mathbf{x}}{dt}\right)^2 - V(\mathbf{x}) \right) \qquad \text{classical action.} \qquad (5.17)$$

Quantum mechanics boils down to stating that the probability amplitude of a particle getting from $(\mathbf{x}, t)$ to $(\mathbf{x}', t')$ is

$$\langle \mathbf{x}', t'|\mathbf{x}, t \rangle = \int \mathcal{D}[\text{paths}] e^{iS([\text{path}])/\hbar}, \qquad (5.18)$$

where the integral covers all possible paths from start to finish. The measure $\mathcal{D}$ is a shorthand way of saying that we take all paths and normalize them properly. Something like this is to be expected on the basis of the double-slit experiment, where quantum mechanics says it is impossible to tell through which slit the electron went and observes that a superposition of the two possibilities leads to an interference pattern on the screen. Nothing prevents us from making a third slit and repeating the 'all paths' argument. Adding more and more slits so that there is no matter left between the electron source and the screen means we need to count all the paths the electron can take. Feynman's intuition was that the paths must have weights related to their classical action, and Dirac noticed that if one goes over to classical physics by, say, letting $\hbar$ diminish, the oscillatory weights make all but the

---

[2] The relationship between motion and time is well known from special relativity; however, a relation between acceleration and temperature has also been suggested by S Hawking, W G Unruh, and others.

classical least-action path very improbable, and finally all that is left is the classical particle trajectories given by the **least-action principle**.

The Wick rotation changes the action to

$$S = \int_0^\beta d\tau \left(\frac{1}{2} m \left(\frac{d\mathbf{x}}{d\tau} \right)^2 + V(\mathbf{x}) \right) \qquad \text{imaginary time action} \qquad (5.19)$$

and the path weights are now the decaying factor $\exp(-S)$ with a positive sign on the potential term, all of which is crucial for the convergence of imaginary-time path integrals. Each path $\mathbf{x}(\tau)$ is counted with the weight given by the action on that path, and the closing of loops is given by the condition $\mathbf{x}(\beta) = \mathbf{x}(0)$.

5.2 High-temperature density matrix

We would not be any the wiser if we could not evaluate the individual segments of the paths. It turns out that high-temperature density matrices can be approximated. We could use the approximate operator splitting used in DMC, namely the *primitive approximation* (PA)

$$e^{-\tau\hat{\mathcal{H}}} = e^{-\tau(\hat{T}+\hat{V})} \approx e^{-\tau\hat{T}} e^{-\tau\hat{V}} + \mathcal{O}(\tau^2), \qquad (5.20)$$

but, for symmetry reasons that soon become clear, it is better to use the second-order-accurate (1/2–1–1/2) splitting, the so-called *symmetrized PA*

$$e^{-\tau\hat{\mathcal{H}}} \approx e^{-\frac{\tau}{2}\hat{V}} e^{-\tau\hat{T}} e^{-\frac{\tau}{2}\hat{V}} + \mathcal{O}(\tau^3). \qquad (5.21)$$

In coordinate space,

$$\rho(\mathbf{x}', \mathbf{x}; \tau) \approx \langle \mathbf{x}' | e^{-\frac{\tau}{2}\hat{V}} e^{-\tau\hat{T}} e^{-\frac{\tau}{2}\hat{V}} | \mathbf{x} \rangle \qquad (5.22)$$

$$= \int d\mathbf{x}_1 d\mathbf{x}_2 \langle \mathbf{x}' | e^{-\frac{\tau}{2}\hat{V}} | \mathbf{x}_2 \rangle \langle \mathbf{x}_2 | e^{-\tau\hat{T}} | \mathbf{x}_1 \rangle \langle \mathbf{x}_1 | e^{-\frac{\tau}{2}\hat{V}} | \mathbf{x} \rangle. \qquad (5.23)$$

We have already evaluated such matrix elements in DMC (see equations (4.39) and (4.52)). The potential matrix elements are

$$\langle \mathbf{x}' | e^{-\tau\hat{V}} | \mathbf{x} \rangle = e^{-\tau V(\mathbf{x})} \delta(\mathbf{x}' - \mathbf{x}). \qquad (5.24)$$

The kinetic matrix elements in free space are

$$\langle \mathbf{x}' | e^{-\tau\hat{T}} | \mathbf{x} \rangle = (4\pi\lambda\tau)^{-3N/2} e^{-\frac{(\mathbf{x}'-\mathbf{x})^2}{4\lambda\tau}}. \qquad (5.25)$$

The variance σ^2 of these Gaussians is

$$\sigma^2 = 2\lambda\tau \Leftrightarrow \sigma = \sqrt{2\lambda\tau} = \sqrt{\frac{\hbar^2}{m}\tau} = \sqrt{\frac{\hbar^2}{mk_B T}}, \qquad (5.26)$$

at the inverse temperature $\tau = 1/(k_B T)$. We notice that the width of the Gaussians σ is proportional to the **thermal wavelength** of particles of mass m at temperature T, defined as

$$\lambda_T := \sqrt{\frac{2\pi\hbar^2}{mk_B T}}. \qquad (5.27)$$

If you take a look at the derivation of the kinetic matrix elements in appendix B, you notice that something could go wrong because of the finite size of the simulated system. The sum in equation (B.6) was approximated by an integral, and it turns out to rely on the assumption that

$$\sigma \ll L \Leftrightarrow \lambda_T \ll L, \qquad (5.28)$$

in a simulation box with sides L. If this condition is not met, the Gaussian (5.25) will reach across the simulation box and cause spurious effects. From a DMC perspective, the kinetic matrix elements are free diffusion, and this cannot be right if a diffusing particle meets the box edges. In DMC, we dealt with the problem using branching, but in PIMC we cannot cut particle paths. However, the problem is not that severe in PIMC because τ has to be small, and we can often apply periodic boundary conditions (PBCs) to keep the imaginary-time paths continuous.

The high-temperature density matrix can now be written in the form

---

Density matrix in the PA

$$\rho(\mathbf{x}', \mathbf{x}; \tau) \approx \langle \mathbf{x}' | e^{-\tau\hat{T}} | \mathbf{x} \rangle e^{-\frac{\tau}{2}(V(\mathbf{x}') + V(\mathbf{x}))} = (4\pi\lambda\tau)^{-3N/2} e^{-\frac{(\mathbf{x}'-\mathbf{x})^2}{4\lambda\tau} - \frac{\tau}{2}(V(\mathbf{x}') + V(\mathbf{x}))}. \qquad (5.29)$$

Based on the eigenfunction expansion in equation (5.12), the exact density matrix is symmetric in the exchange $\mathbf{x} \leftrightarrow \mathbf{x}'$, and now the approximate density matrix also comes out symmetric thanks to the second-order accurate operator splitting.

Stacking M high-temperature density matrices at the inverse temperature $\beta = M\tau$ gives

---

Density matrix path integral in the PA

$$\rho(\mathbf{x}', \mathbf{x}; \beta) = \int d\mathbf{x}_1 d\mathbf{x}_2, \ldots, d\mathbf{x}_{M-1} (4\pi\lambda\tau)^{-3NM/2}$$

$$\times e^{-\frac{(\mathbf{x}'-\mathbf{x}_1)^2}{4\lambda\tau}} e^{-\frac{(\mathbf{x}_1-\mathbf{x}_2)^2}{4\lambda\tau}} \cdot \ldots e^{-\frac{(\mathbf{x}_{M-1}-\mathbf{x})^2}{4\lambda\tau}}$$

$$\times e^{-\tau[\frac{1}{2}V(\mathbf{x}') + V(\mathbf{x}_{M-1}) + \cdots + V(\mathbf{x}_2) + V(\mathbf{x}_1) + \frac{1}{2}V(\mathbf{x})]}. \qquad (5.30)$$

The partition function is

$$Z(\beta) = \int d\mathbf{x}_1 d\mathbf{x}_2, \ldots, d\mathbf{x}_M (4\pi\lambda\tau)^{-3NM/2}$$
$$\times e^{-\frac{(\mathbf{x}_1 - \mathbf{x}_2)^2}{4\lambda\tau}} e^{-\frac{(\mathbf{x}_2 - \mathbf{x}_3)^2}{4\lambda\tau}} \ldots e^{-\frac{(\mathbf{x}_{M-1} - \mathbf{x}_M)^2}{4\lambda\tau}} e^{-\frac{(\mathbf{x}_M - \mathbf{x}_1)^2}{4\lambda\tau}} \qquad (5.31)$$
$$\times e^{-\tau[V(\mathbf{x}_1) + V(\mathbf{x}_2) + \cdots + V(\mathbf{x}_M)]}.$$

which can be written more compactly as

Partition function path integral in the PA

$\mathbf{x}_{M+1} = \mathbf{x}_1$ (for identical particles, add permutations)

$$Z(\beta) = \int \prod_{m=1}^{M} d\mathbf{x}_m \times \prod_{m=1}^{M} (4\pi\lambda\tau)^{-3N/2} e^{-\frac{(\mathbf{x}_m - \mathbf{x}_{m+1})^2}{4\lambda\tau}} \times \prod_{m=1}^{M} e^{-\tau V(\mathbf{x}_m)}. \qquad (5.32)$$

We will add the identical particle permutations later. Path integrals are made of diffusion factors that act like harmonic strings, multiplied by potential factors that add weights to paths.

5.2.1 Isomorphism of PIMC paths and classical polymers

The exact density matrix can be written in terms of a **link action**,

$$S_m := S(\mathbf{x}_m, \mathbf{x}_{m-1}; \tau) := -\ln \rho(\mathbf{x}_m, \mathbf{x}_{m-1}; \tau) \qquad \text{link action}. \qquad (5.33)$$

The notion of action is borrowed from the path integral formulation of real-time quantum mechanics. The exact low-temperature density matrix, split into M time slices, can now be written compactly as

$$\rho(\mathbf{x}_0, \mathbf{x}_M; \beta) = \int d\mathbf{x}_1 \ldots d\mathbf{x}_{M-1} e^{-\sum_{m=1}^{M} S_m}. \qquad (5.34)$$

The link action S_m is the physics in a density matrix. In the PA, the link action comprises a **kinetic action** K_m and an **interaction** U_m,

$$S_m^{\text{prim. approx.}} = \underbrace{\frac{3M}{2} \ln(4\pi\lambda\tau) + \frac{(\mathbf{x}_m - \mathbf{x}_{m-1})^2}{4\lambda\tau}}_{\text{kinetic action } K_m} + \underbrace{\frac{\tau}{2}(V(\mathbf{x}_m) + (\mathbf{x}_{m-1}))}_{\text{interaction } U_m} \qquad (5.35)$$

$$= \text{'spring term'} + \text{'potential term.'} \qquad (5.36)$$

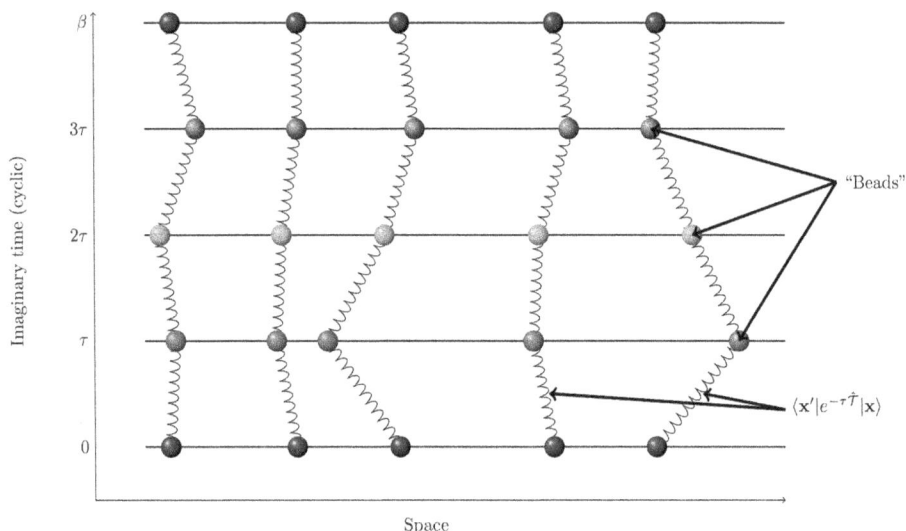

Figure 5.1. The partition function is like a ring polymer, where the springs represent free diffusion. The bead positions at imaginary time β, marked in blue, are the same as those at time zero because, in $Z(\beta)$, the imaginary time is periodic; each spring-connected path is a closed loop. The cyclic nature of imaginary time is more apparent in figure 5.2.

The path integral expression for the partition function is made of closed loops that behave like **classical ring polymers**: the spring term holds the polymer together and the potential term gives the interpolymer interaction.

Figure 5.1 shows an example of possible paths in the partition function $Z(\beta) = \int d\mathbf{x}\rho(\mathbf{x}, \mathbf{x}; \beta)$ for five particles and $M = 4$ imaginary-time slices; the partition function is an average over all such paths. Particle positions in space and imaginary time are called **beads**, and each particle path forms a ring polymer in imaginary time. Free diffusion acts like a spring that connects positions of the same particle at different imaginary-time slices. Figure 5.2 shows the same scenario from the spatial point of view.

5.3 Path generation

5.3.1 Collective slice coordinates x

If particle paths are split into M imaginary-time slices—slices for short—then we would need the particle index j and the slice index m, something like $\mathbf{r}_{j,m}$ for a single bead coordinate of the jth particle on slice m. This notation becomes cumbersome; therefore, it is common to use collective coordinates for whole slices $\mathbf{x}_m$,

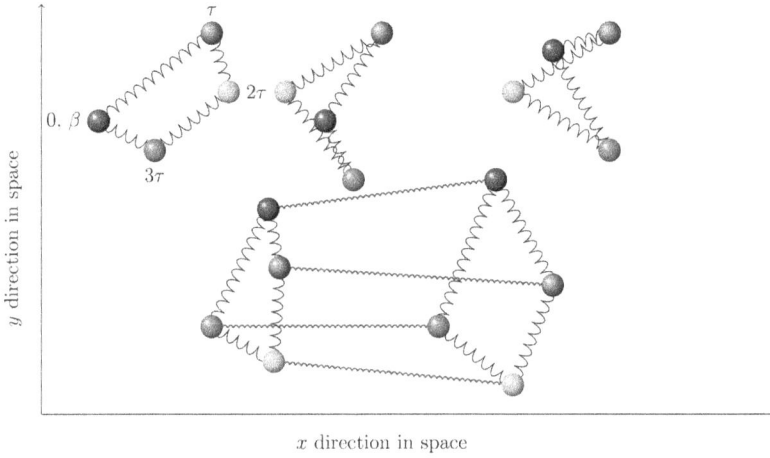

Figure 5.2. The same ring polymer as that of figure 5.1 projected onto space. The closed loops of the imaginary-time paths are more evident. Only beads with similar colors are present in the same time slice and feel each other's presence through the potential $V(\mathbf{x})$; the interaction between just two of the polymers is shown as wiggly dark red lines. The figure shows one possible term that contributes to the partition function.

$$\boxed{\mathbf{x}_m \equiv (\mathbf{r}_{1,m}, \mathbf{r}_{2,m}, \ldots, \mathbf{r}_{N,m}) \quad \text{coordinates of } N \text{ particles on slice } m.} \tag{5.37}$$

In the literature, slice coordinates are sometimes marked $\mathbf{R}_m$. Unless noted otherwise, the system is assumed to be three-dimensional; hence, $\mathbf{x}$ has $3N$ coordinates.

5.3.2 Path generation for free particles

To become familiar with path integrals, let us take the case of free particles in free space. Using the collective slice coordinates $\mathbf{x}$, the density matrix of free particles is (here, slices are evenly spaced)

$$
\begin{aligned}
\rho_0(\mathbf{x}', \mathbf{x}; K\tau) &= \langle \mathbf{x}' | e^{-K\tau \hat{T}} | \mathbf{x} \rangle \\
&= \int d\mathbf{x}_1 \ldots d\mathbf{x}_{K-1} \langle \mathbf{x}' | e^{-\tau \hat{T}} | \mathbf{x}_1 \rangle \langle \mathbf{x}_1 | e^{-\tau \hat{T}} | \mathbf{x}_2 \rangle \ldots \langle \mathbf{x}_{M-1} | e^{-\tau \hat{T}} | \mathbf{x} \rangle.
\end{aligned}
\tag{5.38}
$$

Assuming further that the particles are in free space, we can pick the free-space matrix elements from DMC. For a single particle,

$$\rho_0(\mathbf{r}', \mathbf{r}; \tau) \overset{\text{free space}}{=} (4\pi\lambda\tau)^{-3N/2} e^{-\frac{(\mathbf{r}'-\mathbf{r})^2}{4\lambda\tau}}. \tag{5.39}$$

The density matrix of N particles is the product of these single-particle density matrices,

$$\rho_0(\mathbf{x}', \mathbf{x}; K\tau) = (4\pi\lambda K\tau)^{-3N/2} e^{-\frac{(\mathbf{x}'-\mathbf{x})^2}{4\lambda K\tau}} = \prod_{j=1}^{N} (4\pi\lambda\tau)^{-3/2} e^{-\frac{(\mathbf{r}'-\mathbf{r}_j)^2}{4\lambda K\tau}}. \tag{5.40}$$

In DMC, this merely says that free particles diffuse independently, and the only difference is that in PIMC, we rather speak of paths of independent particles. If we now split all paths into M segments, we need to integrate over all slices:

Free-particle density matrix (in free space)

$$\rho_0(\mathbf{x}', \mathbf{x}; K\tau) = (4\pi\lambda K\tau)^{-3N/2} e^{-\frac{(\mathbf{x}'-\mathbf{x})^2}{4\lambda K\tau}}$$

$$= (4\pi\lambda\tau)^{-3NK/2} \int d\mathbf{x}_1 ... d\mathbf{x}_{K-1} [e^{-\frac{(\mathbf{x}'-\mathbf{x}_1)^2}{4\lambda\tau}}][e^{-\frac{(\mathbf{x}_1-\mathbf{x}_2)^2}{4\lambda\tau}}] ... [e^{-\frac{(\mathbf{x}_{M-1}-\mathbf{x})^2}{4\lambda\tau}}] \quad (5.41)$$

$$= \int \prod_{m=1}^{K} d\mathbf{x}_m \prod_{m=1}^{K} (4\pi\lambda\tau)^{3N/2} e^{-\frac{(\mathbf{x}_m-\mathbf{x}_{m+1})^2}{4\lambda\tau}}, \text{ where } \mathbf{x}_1 := \mathbf{x}' \text{ and } \mathbf{x}_K := \mathbf{x}. \quad (5.42)$$

The equality is exact, as the convolution of Gaussians is a Gaussian. In DMC, we interpreted Gaussians as diffusion, so the equation states that diffusion from a fixed point $\mathbf{x}$ to another fixed point $\mathbf{x}'$ in the duration β is the same as the sum of all possible diffusive paths of length $\tau = \beta/M$ that span from $\mathbf{x}$ to $\mathbf{x}'$. Each such path is known as a **Brownian bridge**.[3]

Density matrices are symmetric in spatial coordinates,

$$\rho_0(\mathbf{x}', \mathbf{x}; \beta) = \rho_0(\mathbf{x}, \mathbf{x}'; \beta), \quad (5.43)$$

and, naturally, we can also relabel the integration variables in equation (5.42) as we please.

The path integral in equation (5.42) is computed using MC, and we next find out how path segments of arbitrary length

$$(\text{fixed } \mathbf{x}') \rightarrow \mathbf{x}_1 \rightarrow \mathbf{x}_2 \rightarrow ... \rightarrow \mathbf{x}_{K-1} \rightarrow (\text{fixed } \mathbf{x}), \quad (5.44)$$

can be generated so that the integral in ρ_0 is sampled exactly. For future use, we will also relax the assumption that slices are evenly spaced.

5.3.3 Staging algorithm

We need an algorithm that, given two fixed endpoints, generates diffusive paths between them. Paths in PIMC are often generated for a single particle, so here we use $\mathbf{r}_m$ for coordinates in slice m without marking the particle index.

---

[3] The path generation process is a special case of a Gaussian Lévy process [2].

Let us look at a path segment of K evenly spaced slices. The free-particle density matrix in equation (5.42) for one particle is

$$\rho_0(\mathbf{r}_1, \mathbf{r}_{K+1}; K\tau) = \int \prod_{m=1}^{K} d\mathbf{x}_m \prod_{m=1}^{K} (4\pi\lambda\tau)^{-3N/2} e^{-\frac{(\mathbf{r}_m - \mathbf{r}_{m+1})^2}{4\lambda\tau}}, \qquad (5.45)$$

where $\mathbf{r}_1$ and $\mathbf{r}_{K+1}$ are known fixed points. The constant factors $(4\pi\lambda\tau)^{-3N/2}$ have no effect on path generation.

The staging algorithm can be viewed as a coordinate transformation, known as the staging transformation [3]. The derivation presented below is based on a simple reorganization of Gaussian factors.

First, we would like to sample $\mathbf{r}_2$ from equation (5.45), but it appears in two factors, one attached to the known $\mathbf{r}_1$ and the other attached to the as-yet unknown $\mathbf{r}_3$. This approach is problematic; however, we can write the integrand in a different form. Take, for example, a path with points $\mathbf{r}_1$, $\mathbf{r}_2$, $\mathbf{r}_3$, $\mathbf{r}_4$, and $\mathbf{r}_5$, so that $K = 4$ in equation (5.45) and the segments span 4τ in imaginary time. Multiply and divide by the new factors, marked in matching colors [4],

$$e^{-\frac{(\mathbf{r}_1 - \mathbf{r}_2)^2}{4\lambda\tau}} e^{-\frac{(\mathbf{r}_2 - \mathbf{r}_3)^2}{4\lambda\tau}} e^{-\frac{(\mathbf{r}_3 - \mathbf{r}_4)^2}{4\lambda\tau}} e^{-\frac{(\mathbf{r}_4 - \mathbf{r}_5)^2}{4\lambda\tau}} \qquad (5.46)$$

$$= e^{-\frac{(r_1 - r_5)^2}{4\lambda(4\tau)}} \left[\frac{e^{-\frac{(\mathbf{r}_1 - \mathbf{r}_2)^2}{4\lambda\tau}} e^{-\frac{(\mathbf{r}_2 - \mathbf{r}_5)^2}{4\lambda(3\tau)}}}{e^{-\frac{(r_1 - r_5)^2}{4\lambda(4\tau)}}} \right] \left[\frac{e^{-\frac{(\mathbf{r}_2 - \mathbf{r}_3)^2}{4\lambda\tau}} e^{-\frac{(r_3 - r_5)^2}{4\lambda(2\tau)}}}{e^{-\frac{(\mathbf{r}_2 - \mathbf{r}_5)^2}{4\lambda(3\tau)}}} \right] \left[\frac{e^{-\frac{(\mathbf{r}_3 - \mathbf{r}_4)^2}{4\lambda\tau}} e^{-\frac{(\mathbf{r}_4 - \mathbf{r}_5)^2}{4\lambda\tau}}}{e^{-\frac{(r_3 - r_5)^2}{4\lambda(2\tau)}}} \right]. \qquad (5.47)$$

Start here

The leftmost bracket contains only $\mathbf{r}_2$ and the known endpoints $\mathbf{r}_1$ and $\mathbf{r}_5$; furthermore,

$$\left[\frac{e^{-\frac{(\mathbf{r}_1 - \mathbf{r}_2)^2}{4\lambda\tau}} e^{-\frac{(\mathbf{r}_2 - \mathbf{r}_5)^2}{4\lambda(3\tau)}}}{e^{-\frac{(r_1 - r_5)^2}{4\lambda(4\tau)}}} \right] = e^{-\frac{1}{4\lambda}\left(\frac{1}{\tau} + \frac{1}{3\tau}\right)\left(\mathbf{r}_1 - \left(\frac{3\mathbf{r}_1 + \mathbf{r}_5}{4}\right)\right)^2} \qquad (5.48)$$

$$:= e^{-\frac{1}{4\lambda\tau_2^*}(\mathbf{r}_2 - \mathbf{r}_2^*)^2}, \qquad (5.49)$$

where

$$\frac{1}{\tau_2^*} := \frac{1}{\tau} + \frac{1}{3\tau} \Leftrightarrow \tau_2^* := \frac{3}{4}\tau, \quad \mathbf{r}_2^* := \frac{3\mathbf{r}_1 + \mathbf{r}_5}{4}. \qquad (5.50)$$

Now the point $\mathbf{r}_2$ is easy to sample: compute $\mathbf{r}_2^*$ deterministically, and let

$$\mathbf{r}_2 = \mathbf{r}_2^* + \sqrt{2\lambda\tau_2^*}\,\eta \qquad (5.51)$$

with a normally distributed random vector η. Next, sample $\mathbf{r}_3$ from the previous bracket using the previously sampled $\mathbf{r}_2$, etc.

In the staging algorithm, the intermediate points are found in stages: first $\mathbf{r}_2$, then $\mathbf{r}_3$, and so on. Figure 5.3 shows how staging builds the random walk path piece by piece.

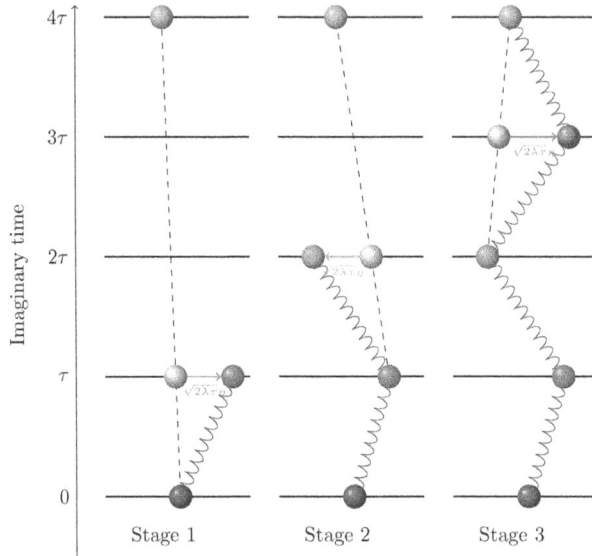

Figure 5.3. Building a Brownian bridge between fixed endpoints at imaginary times zero and $m\tau$ (blue and cyan). The result is a single-particle path segment, one possible term in $\rho_0(\mathbf{r}_1, \mathbf{r}_5; 4\tau)$.

Generalization:

Staging algorithm for evenly spaced imaginary-time slices
Path segment $(\mathbf{r}_1, \mathbf{r}_2, \ldots, \mathbf{r}_K, \mathbf{r}_{K+1})$, where endpoints $\mathbf{r}_1$ and $\mathbf{r}_{K+1}$ are known.
Sample points in the order $m = 2, 3, \ldots, K$.

$$\mathbf{r}_m = \mathbf{r}_m^* + \sqrt{2\lambda\tau_m^*}\,\boldsymbol{\eta}$$
$$\tau_m^* := \frac{K - m + 1}{K - m + 2}\tau$$
$$\mathbf{r}_m^* := \frac{(K - m + 1)\mathbf{r}_{m-1} + \mathbf{r}_{K+1}}{K - m + 2}.$$

(5.52)

The imaginary-time slices do not have to be evenly spaced. We may split the density operator unevenly,

---

Arbitrary time slices:

$$e^{-\beta\hat{\mathcal{H}}} = e^{-\tau_1\hat{\mathcal{H}}}e^{-\tau_2\hat{\mathcal{H}}}\ldots e^{-\tau_M\hat{\mathcal{H}}}, \quad \sum_{i=1}^{M}\tau_i \equiv \beta.$$

(5.53)

To take variable-width time slices into account, define imaginary-time distances

$$\tau_{i,j} := |\tau_i - \tau_j| (\text{mod}\,\beta). \tag{5.54}$$

Each new bead m is computed using the previous bead $m - 1$ at the distance $\tau_{n-1,n}$, and the end bead at the distance $\tau_{m,K+1}$. The staging time step is

$$\tau_m = \frac{1}{\dfrac{1}{\tau_{m-1,m}} + \dfrac{1}{\tau_{m,K+1}}} = \frac{\tau_{m-1,m}\tau_{m,K+1}}{\tau_{m-1,K+1}}, \tag{5.55}$$

and the midpoint position is a linear interpolation,

$$\mathbf{r}_m^* = \frac{\tau_{m,K+1}\mathbf{r}_{m-1} + \tau_{m-1,m}\mathbf{r}'}{\tau_{m-1,m} + \tau_{m,K+1}} = \frac{\tau_{m,K+1}\mathbf{r}_{m-1} + \tau_{m-1,m}\mathbf{r}'}{\tau_{m-1,K+1}}. \tag{5.56}$$

Staging algorithm for arbitrary imaginary-time slices
Path segment $(\mathbf{r}_1(\tau_1))$, $\mathbf{r}_2(\tau_2)$, ..., $\mathbf{r}_K(\tau_K)$, $(\mathbf{r}_{K+1}(\tau_{K+1}))$, where endpoints $\mathbf{r}_1$ and $\mathbf{r}_{K+1}$ are known.
Sample points in the order $\mathbf{r}_2$, $\mathbf{r}_3$, ..., $\mathbf{r}_K$

$$\mathbf{r}_m = \mathbf{r}_m^* + \sqrt{2\lambda\tau_m^*}\,\eta$$
$$\tau_m^* := \frac{\tau_{m-1,m}\,\tau_{m,K+1}}{\tau_{m-1,K+1}}$$
$$\mathbf{r}_m^* := \frac{\tau_{m,K+1}\mathbf{r}_{m-1} + \tau_{m-1,m}\mathbf{r}_{K+1}}{\tau_{m-1,K+1}}$$
$$\tau_{i,j} := |\tau_i - \tau_j|(\text{mod}\,\beta). \tag{5.57}$$

- The staging algorithm can be applied for any imaginary-time span. We will use it to sample a random walk path between any two points in (position, imaginary-time) space.
- The staging algorithm produces positions $\mathbf{r}_2$, ..., $\mathbf{r}_K$ that sample exactly the *integral* in the free-particle density matrix. For evenly spaced slices,

$$\rho_0(\mathbf{r}_1, \mathbf{r}_{k+1}; K\tau) \propto e^{-\frac{(\mathbf{r}_1 - \mathbf{r}_{K+1})^2}{4\lambda K\tau}} \underbrace{\int d\mathbf{r}_2 \cdots d\mathbf{r}_K e^{-\frac{(\mathbf{r}_2 - \mathbf{r}_2^*)^2}{4\lambda\tau_2^*}} \cdots e^{-\frac{(\mathbf{r}_K - \mathbf{r}_{K+1}^*)^2}{4\lambda\tau_K^*}}}_{\text{staging paths compute this integral}}. \tag{5.58}$$

Notice the 'long-diffusion' factor $e^{-\frac{(\mathbf{r}_1 - \mathbf{r}_{K+1})^2}{4\lambda K\tau}}$. While this factor has no effect on the paths between the endpoints, it is there because we first have to set the endpoints $\mathbf{r}_1$ and $\mathbf{r}_{K+1}$, and *their* probability is given by the long-diffusion factor: $\mathbf{r}_1$ and $\mathbf{r}_{K+1}$ prefer to be near each other, at least within the time frame

$K\tau$. Later, the same factor will haunt us in the worm algorithm probabilities as

$$\rho_0(\mathbf{r}_H, \mathbf{r}_T; \tau_{HT}): = (4\pi\lambda\tau_{HT})^{3/2}e^{-\frac{(\mathbf{r}_H - \mathbf{r}_T)^2}{4\lambda\tau_{HT}}}, \tag{5.59}$$

where $\mathbf{r}_H$ and $\mathbf{r}_T$ are the coordinates of the worm's 'head' H and 'tail' T.

Figure 5.4 shows five possible paths found using the staging algorithm. The lower the temperature, the more the imaginary-time paths spread in space.

Figure 5.5 examines how the closed paths in the partition function $Z(\beta)$ behave as a function of temperature. The high-temperature paths are tightly clamped around the initial positions, and the loop structure begins to appear with decreasing temperature.

The path loops in $Z(\beta)$ get larger in space as the temperature drops.

At some low temperature, loops start to overlap in space, and identical particles may experience loop fusion.

The idea that lower temperatures expand loops can be put to real use. One can take the 'center of mass' of each loop as a reference point and scale each loop's coordinates w.r.t. this reference point in order to get a larger loop that corresponds to a lower temperature.

Notice that figure 5.5 shows only one possible path for each particle at a fixed endpoint. To sample $Z(\beta) = \int d\mathbf{x}\langle\mathbf{x}|e^{-\beta\hat{T}}|\mathbf{x}\rangle$, we would have to sample many paths *and* many endpoints $\mathbf{x}$. However, in free space, the endpoints $\mathbf{x}$ can be anywhere, so this calculation would give $Z(\beta)$ for six particles in infinite space, that is, zero density. A more physically relevant case would be one with constant density, which can be achieved using a box, or even better, a box with PBCs.

5.3.4 Bisection algorithm

The bisection algorithm first samples the point halfway between the fixed endpoints and progresses to find the next two midpoints until all is solved; the number of sampled points follows the progression 1, 3, 7, 15, ..., that is,

$$\text{number of sampled points} = 2^i - 1, \text{ for } i = 1, 2, ... \tag{5.60}$$

For example, the free-particle density matrix for the path segment $(\mathbf{r}_1, \mathbf{r}_2, \mathbf{r}_3, \mathbf{r}_4, \mathbf{r}_5)$ with fixed endpoints $\mathbf{r}_1$ and $\mathbf{r}_5$ is, as in staging,

$$\rho_0(\mathbf{r}_1, \mathbf{r}_5; 4\tau) = (4\pi\lambda\tau)^{-3N4/2}\int d\mathbf{r}_2...d\mathbf{r}_4 e^{-\frac{(\mathbf{r}_1 - \mathbf{r}_2)^2}{4\lambda\tau}} e^{-\frac{(\mathbf{r}_2 - \mathbf{r}_3)^2}{4\lambda\tau}} e^{-\frac{(\mathbf{r}_3 - \mathbf{r}_4)^2}{4\lambda\tau}} e^{-\frac{(\mathbf{r}_4 - \mathbf{r}_5)^2}{4\lambda\tau}}. \tag{5.61}$$

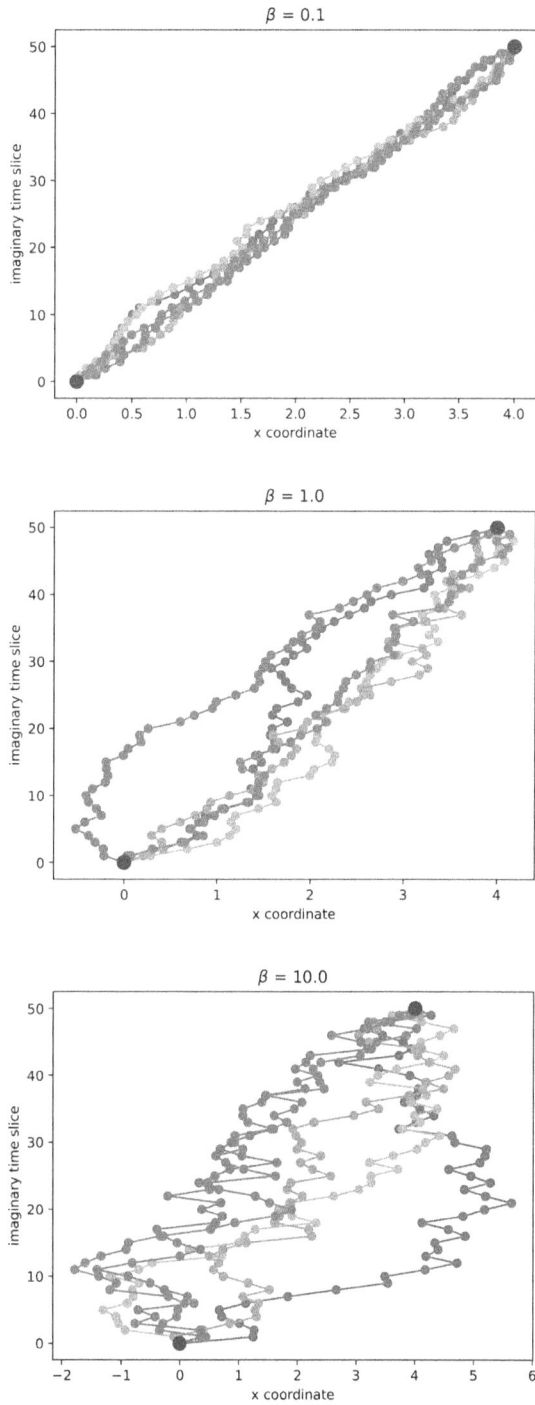

Figure 5.4. Five possible paths between fixed endpoints at $x = 0$ and $x = 4$ (large blue circles) at three inverse temperatures β. See the Python code `staging.py`.

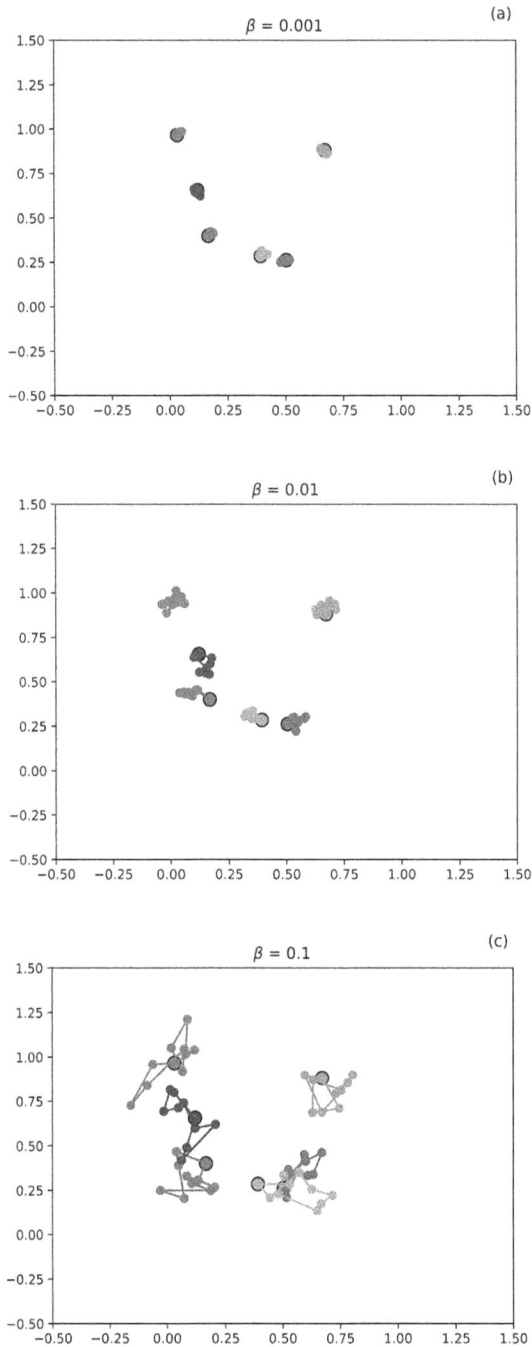

Figure 5.5. Free-particle paths of six particles in two dimensions; just one possible path is shown for each particle. The large circles indicate the positions of the particles in the first and last time slices; these are the same in all three inverse temperatures shown in figures (a), (b), and (c). See the Python code `paths_-free_diffusion.py`.

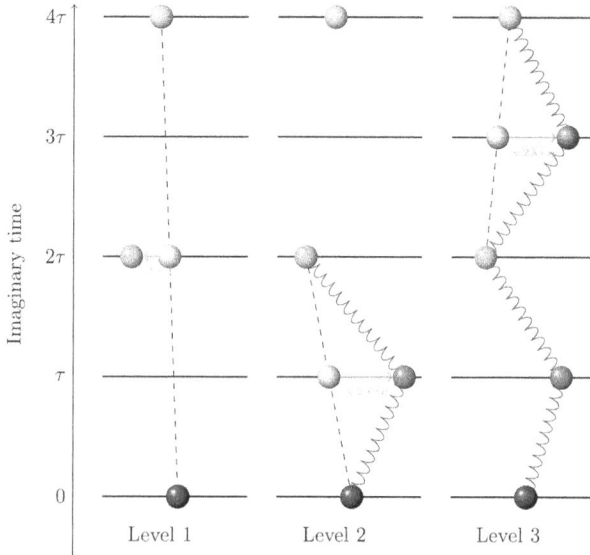

Figure 5.6. Building a Brownian bridge between fixed endpoints at imaginary times zero and 4τ, shown in blue and cyan, respectively. The result is a single-particle path sampled from $\rho_0(\mathbf{r}', \mathbf{r}; 4\tau)$.

Multiply and divide by suitable factors so that $\mathbf{r}_3$ can be sampled first:

$$
\rho_0(\mathbf{r}_1, \mathbf{r}_5; 4\tau) = (4\pi\lambda\tau)^{-3N4/2} e^{-\frac{(\mathbf{r}_1-\mathbf{r}_5)^2}{4\lambda(4\tau)}} \int d\mathbf{r}_2 \ldots d\mathbf{r}_4 \left[\frac{e^{-\frac{(\mathbf{r}_1-\mathbf{r}_3)^2}{4\lambda(2\tau)}} e^{-\frac{(\mathbf{r}_3-\mathbf{r}_5)^2}{4\lambda(2\tau)}}}{e^{-\frac{(\mathbf{r}_1-\mathbf{r}_5)^2}{4\lambda(4\tau)}}} \right]
$$

Start here (5.62)

$$
\times \left[\frac{e^{-\frac{(\mathbf{r}_1-\mathbf{r}_2)^2}{4\lambda\tau}} e^{-\frac{(\mathbf{r}_2-\mathbf{r}_3)^2}{4\lambda\tau}}}{e^{-\frac{(\mathbf{r}_1-\mathbf{r}_3)^2}{4\lambda(2\tau)}}} \right] \left[\frac{e^{-\frac{(\mathbf{r}_3-\mathbf{r}_4)^2}{4\lambda\tau}} e^{-\frac{(\mathbf{r}_4-\mathbf{r}_5)^2}{4\lambda\tau}}}{e^{-\frac{(\mathbf{r}_3-\mathbf{r}_5)^2}{4\lambda(2\tau)}}} \right].
$$

The factors in the leftmost bracket can be written in terms of $\mathbf{r}_3^*$ and τ_3^*, and then $\mathbf{r}_3$ can be sampled just as we did in staging. Proceed to sample the next bisection points $\mathbf{r}_2$ and $\mathbf{r}_4$. An example of bisection path generation is shown in figure 5.6.

Bisection algorithm for evenly spaced imaginary-time slices

Constant τ; path segment $(\mathbf{r}_1, \mathbf{r}_2, \ldots, \mathbf{r}_K, \mathbf{r}_{K+1})$, $K = 2^{N_{levels}}$, where endpoints $\mathbf{r}_1$ and $\mathbf{r}_{K+1}$ are known and fixed.

Sample points $\mathbf{r}_m$ halfway between two known points at that level, with $beg = 1$ and $end = K + 1$,

$$
m = \underbrace{\frac{beg + end}{2}: = a}_{\text{level 1}}; \quad \underbrace{\frac{beg + a}{2}: = b, \frac{a + end}{2}: = c}_{\text{level 2}}; \quad \underbrace{\frac{beg + b}{2}, \frac{b + a}{2}, \frac{a + c}{2}, \frac{c + end}{2}}_{\text{level 3}}; \ldots
$$

$$\mathbf{r}_m = \mathbf{r}_m^* + \sqrt{2\lambda\tau_\ell^*}\eta \qquad (5.63)$$

$$\tau_\ell^*: = \frac{K}{2^{\ell+1}}\tau \text{ for level } \ell = 1, \ldots, N_{levels} \qquad (5.64)$$

$$\mathbf{r}_m^*: = \frac{\mathbf{r}_{prev} + \mathbf{r}_{next}}{2}, \text{ for } m \text{ sampled between } prev \text{ and } next \text{ known points.} \qquad (5.65)$$

The bisection algorithm also works for unevenly spaced imaginary-time slices.

Bisection algorithm for arbitrary imaginary-time slices

Path segment $(\mathbf{r}_1(\tau_1), \mathbf{r}_2(\tau_2), \ldots, \mathbf{r}_K(\tau_k), \mathbf{r}_{K+1}(\tau_{K+1}))$, where endpoints $\mathbf{r}_1$ and $\mathbf{r}_{K+1}$ are known and fixed.

Sample points $\mathbf{r}_m$ halfway between two known points at that bisection level, $beg = 1$ and $end = K + 1$,

$$m = \underbrace{\frac{beg + end}{2}: =a}_{\text{level 1}}; \underbrace{\frac{beg + a}{2}: =b, \frac{a + end}{2}: =c}_{\text{level 2}}; \underbrace{\frac{beg + b}{2}, \frac{b + a}{2}, \frac{a + c}{2}, \frac{c + end}{2}}_{\text{level 3}}; \ldots$$

The halfway point is found by counting slices while ignoring their widths.

$\mathbf{r}_m$ is sampled between known $\mathbf{r}_{prev}$ and $\mathbf{r}_{next}$.

$$\mathbf{r}_m = \mathbf{r}_m^* + \sqrt{2\lambda\tau_m^*}\eta \qquad (5.66)$$

$$\tau_m^*: = \frac{\tau_{prev, m} \; \tau_{m,next}}{\tau_{prev, next}} \qquad (5.67)$$

$$\mathbf{r}_m^*: = \frac{\tau_{m,next}\mathbf{r}_{prev} + \tau_{prev, m}\mathbf{r}_{next}}{\tau_{prev, next}} \qquad (5.68)$$

$$\tau_{i,j}: = |\tau_i - \tau_j|(\mathrm{mod}\,\beta). \qquad (5.69)$$

The code `bisection.py` samples seven points using the bisection algorithm.

We are not done yet. We need to add the potential factors (the interaction U) and identical particle permutations. The staging and bisection algorithms stay in our toolbox and will be invoked whenever we need to generate a path from scratch:

The staging algorithm and the bisection algorithm sample **exact** paths from $\rho_0(\mathbf{r}', \mathbf{r}; \beta)$, i.e. the spring factors are sampled exactly.

5.3.5 Bisection with interaction in the primitive approximation

Bisection is reasonably effective in interacting systems because a newly sampled bead position can be discarded early if it leads to a large interaction (large potential energy). To show an example of how this **early rejection** can be done, start from the three sampled points in equation (5.62). Generalization to longer path segments is straightforward. In the PA, the interaction affecting the path segment in question is (see equation (5.35))

$$U = \sum_{m=1}^{4} U_m = \sum_{m=1}^{4} \frac{\tau}{2}(V(\mathbf{x}_m) + V(\mathbf{x}_{m-1}))$$
$$= \frac{\tau}{2}(V(\mathbf{x}_4) + V(\mathbf{x}_0)) + \tau(V(\mathbf{x}_1) + V(\mathbf{x}_2) + V(\mathbf{x}_3)). \tag{5.70}$$

Instead of using the interaction factor e^{-U} only after the whole path segment has been sampled, we can include it in the bisection levels given in brackets:

$$\rho(\mathbf{x}_4, \mathbf{x}_0; m\tau) = (4\pi\lambda\tau)^{-3N4/2} e^{\frac{(\mathbf{x}_4 - \mathbf{x}_0)^2}{4\lambda\tau}} e^{-\frac{\tau}{2}(V(\mathbf{x}_4)+V(\mathbf{x}_0))} \int d\mathbf{x}_1 ...d\mathbf{x}_3 \left[e^{-\frac{(\mathbf{x}_2 - \mathbf{x}_2^*)^2}{4\lambda\tau_2^*}} e^{-\tau V(\mathbf{x}_2)} \right]$$
$$\times \left[e^{-\frac{(\mathbf{x}_3 - \mathbf{x}_3^*)^2}{4\lambda\tau_3^*}} e^{-\tau V(\mathbf{x}_3)} \right] \left[e^{-\frac{(\mathbf{x}_1 - \mathbf{x}_1^*)^2}{4\lambda\tau_1^*}} e^{-\tau V(\mathbf{x}_1)} \right]. \tag{5.71}$$

In the collective notation, the slice $\mathbf{x}_m$ contains $3N$ coordinates (in 3D). To make the formulas more explicit, consider sampling only bead positions ($\mathbf{r}_1$, $\mathbf{r}_2$, and $\mathbf{r}_3$). If the other beads in the three slices are at ($\tilde{\mathbf{x}}_1$, $\tilde{\mathbf{x}}_2$, and $\tilde{\mathbf{x}}_3$), we get

$$\rho(\mathbf{x}_m, \mathbf{x}_0; m\tau) = (4\pi\lambda\tau)^{-3N4/2} e^{\frac{(\mathbf{x}_4 - \mathbf{x}_0)^2}{4\lambda\tau}} e^{-\frac{\tau}{2}(V(\mathbf{x}_4)+V(\mathbf{x}_0))}$$
$$\times \int d\tilde{\mathbf{x}}_1 d\tilde{\mathbf{x}}_2 d\tilde{\mathbf{x}}_3 \left\{ \left[e^{-\frac{(\tilde{\mathbf{x}}_2 - \tilde{\mathbf{x}}_2^*)^2}{4\lambda\tau_2^*}} \right] \left[e^{-\frac{(\tilde{\mathbf{x}}_1 - \tilde{\mathbf{x}}_1^*)^2}{4\lambda\tau_1^*}} \right] \left[e^{-\frac{(\tilde{\mathbf{x}}_3 - \tilde{\mathbf{x}}_3^*)^2}{4\lambda\tau_3^*}} \right] \right.$$
$$\left. \times \int d\mathbf{r}_1 d\mathbf{r}_2 d\mathbf{r}_3 \left[e^{-\frac{(\mathbf{r}_2 - \mathbf{r}_2^*)^2}{4\lambda\tau_2^*}} e^{-\tau V(\mathbf{x}_2)} \right] \left[e^{-\frac{(\mathbf{r}_3 - \mathbf{r}_3^*)^2}{4\lambda\tau_3^*}} e^{-\tau V(\mathbf{x}_3)} \right] \left[e^{-\frac{(\mathbf{r}_1 - \mathbf{r}_1^*)^2}{4\lambda\tau_1^*}} e^{-\tau V(\mathbf{x}_1)} \right] \right\}. \tag{5.72}$$

Bisection level 1   Bisection level 2a   Bisection level 2b

In evaluating the last integrals, the other bead positions ($\tilde{\mathbf{x}}_1$, $\tilde{\mathbf{x}}_2$, and $\tilde{\mathbf{x}}_3$) have already been sampled and fixed, and we can concentrate on sampling ($\mathbf{r}_1$, $\mathbf{r}_2$, and $\mathbf{r}_3$). Notice that the potential factors still contain all coordinates in each time slice. Bisection level one can be sampled using the ordinary bisection update,

$$\mathbf{r}_2 = \mathbf{r}_2^* + \sqrt{2\pi\tau_2^*}\,\eta, \tag{5.73}$$

followed by the computation of the weight ratio[4]

---

[4] For a pair potential, the weight ratio can be evaluated quickly, since only the distances from sampled beads to other beads change in the update.

$$\frac{W(new)}{W(old)} = \frac{e^{-\tau V(x_2(new))}}{e^{-\tau V(x_2(old))}}. \tag{5.74}$$

The bisection algorithm with early rejection now asks the Metropolis question based on this weight ratio and accepts the suggested bead position r_2 with the probability

$$P = \max\left[1, \frac{W(new)}{W(old)}\right]. \tag{5.75}$$

To illustrate how weights propagate, consider a two-level bisection algorithm with a Metropolis question at each level. Mark the original bead positions 1, 2, 3, 4, and 5, where 1 and 5 are fixed. After the bisection algorithm has been run, the bead positions are 1, $2'$, $3'$, $4'$, and 5, where $'$ (prime) indicates a new position. At the first bisection level, the Metropolis question uses the ratio

$$\frac{W(new)}{W(old)}_{\text{level 1}} = \frac{e^{-S(2, 3', 4)}}{e^{-S(2, 3, 4)}}, \tag{5.76}$$

and separating the action S into kinetic and potential actions yields

$$\frac{W(new)}{W(old)}_{\text{level 1}} = \frac{e^{-K(2, 3', 4)}e^{-U(2, 3', 4)}}{e^{-K(2, 3, 4)}e^{-U(2, 3, 4)}}. \tag{5.77}$$

Coming to level two, the 'old' positions are 2, $3'$, and 4, the suggested new ones are $2'$, $3'$, and $4'$, and the Metropolis question weight ratio is

$$\frac{W(new)}{W(old)}_{\text{level 2}} = \frac{e^{-K(2', 3', 4')}e^{-U(2', 3', 4')}}{e^{-K(2, 3', 4)}e^{-U(2, 3', 4)}}. \tag{5.78}$$

Combined, the decision to accept a level-two update is based on the weight

$$\frac{W(new)}{W(old)}_{\text{level 1}} \times \frac{W(new)}{W(old)}_{\text{level 2}} = \frac{e^{-K(2,3',4)}e^{-U(2,3',4)}}{e^{-K(2, 3, 4)}e^{-U(2, 3, 4)}} \times \frac{e^{-K(2', 3', 4')}e^{-U(2', 3', 4')}}{e^{-K(2,3',4)}e^{-U(2,3',4)}}, \tag{5.79}$$

so the decision to accept the whole suggested bisection update is

$$\frac{W(new)}{W(old)} = \frac{e^{-K(2', 3', 4')}}{e^{-K(2, 3, 4)}}\frac{e^{-U(2', 3', 4')}}{e^{-U(2, 3, 4)}} \equiv \frac{e^{-S(new)}}{e^{-S(old)}} = e^{-\Delta S}, \tag{5.80}$$

which is the correct answer. The algorithm samples the free-particle kinetic actions exactly, so the K ratios can be dropped, and the bisection levels are accepted or rejected based on the interactions U.

We cheated a bit in the previous calculation, as the actual acceptance probabilities are (if kinetic actions are sampled exactly):

$$P_{\text{level 1}} = \min\left[1, \frac{e^{-U(2, 3', 4)}}{e^{-U(2, 3, 4)}}\right] \tag{5.81}$$

$$P_{\text{level 2}} = \min\left[1, \frac{e^{-U(2', 3', 4')}}{e^{-U(2, 3', 4)}}\right], \tag{5.82}$$

and the total acceptance probability is

$$P = P_{\text{level 1}} \times P_{\text{level 2}}. \tag{5.83}$$

Obviously, it is possible that a path segment rejected at an early stage might have been accepted at the final level. However, the probability of such events is usually so low that the error is negligible and well below the level of statistical error.

Using fake acceptance conditions in the intermediate levels

The PA leads to simple weights, but other operator splittings may give weights whose evaluation may be slow. To save time, it is possible to use a simple, approximate early rejection condition; then the acceptance in the final step with a full path segment cancels the approximate weights and uses the correct weight. This ensures that the path segment is accepted with the correct weight.

DMC-style importance sampling?

So far, we have mostly discussed the sampling of a new bead position $\mathbf{x}_i$ in bisection based on the free-particle factor $e^{-\frac{(\mathbf{x}_i - \mathbf{x}_i^*)^2}{4\lambda\tau_i^*}}$, followed by a Metropolis question for weight $e^{-\tau V(\mathbf{x}_i)}$. The first is a diffusion step, while the latter reminds us of the branching step we used in importance sampling DMC. To elaborate further, let us introduce a many-body trial wave function $\varphi_T(\mathbf{x})$. I emphasize that this is *not* what one usually does in PIMC: instead of a trial wave function, PIMC algorithms sometimes use a *trial density matrix*. A trial wave function is easier to invent, so it is worth a try.

In DMC, the trial wave function served two purposes. First, it kept the walkers away from unfavorable potential energy regions; second, it forced the wave function to have a certain symmetry (that is why we never needed to worry about permutations in DMC). Trial wave functions are generally constructed to approximate the ground state, while the finite-temperature state is a combination of the ground state and excited states. Nevertheless, a reasonably good ground-state trial wave function could improve the acceptance in bisection by keeping the sampled beads in favorable potential energy regions.

In bisection, the diffusion step starts from $\mathbf{r}_i^*$,

$$\mathbf{r}_i = \mathbf{r}_i^* + \sqrt{2\lambda\tau_i^*}\,\boldsymbol{\eta} + \lambda\tau_i^*\mathbf{F}(\mathbf{x}(old)), \tag{5.84}$$

where the components of the drift are, as usual,

$$\mathbf{F}_k(\mathbf{x}):\ = \frac{2\,\boldsymbol{\nabla}_k\,\varphi_T(\mathbf{x})}{\varphi_T(\mathbf{x})}. \tag{5.85}$$

The expression for diffusion+drift is correct to the first order in τ_i^*, which in bisection varies from level to level.

5.3.6 Harmonic oscillator

The eigenvalues and eigenvectors of a harmonic oscillator (HO) are known, so the exact density matrix is also known [1]. The density matrix can be given in several forms using the identity

$$\tanh(y/2) = \frac{\cosh(y) - 1}{\sinh(h)} = \frac{\sinh(y)}{\cosh(y) + 1}. \tag{5.86}$$

Exact density matrix of N HOs in 3D, mass m, $\hat{\mathcal{H}} = \hat{\mathcal{T}} + \frac{1}{2}m\omega^2\hat{\mathbf{x}}^2$

$$\rho_{\text{h.o.}}(\mathbf{x}', \mathbf{x}; \beta)$$

$$= \left(\frac{m\omega}{2\pi\hbar\,\sinh(\hbar\omega\beta)}\right)^{3N/2} e^{-\frac{m\omega}{2\hbar\,\sinh(\hbar\omega\tau)}[(\mathbf{x}'^2 + \mathbf{x}^2)\cosh(\hbar\omega\tau) - 2\mathbf{x}'\cdot\mathbf{x}]} \tag{5.87}$$

$$= \left(\frac{m\omega}{2\pi\hbar\,\sinh(\hbar\omega\beta)}\right)^{3N/2} e^{-\frac{m\omega}{2\hbar}\tanh(\frac{1}{2}\hbar\omega\tau)(\mathbf{x}'^2 + \mathbf{x}^2) - \frac{m\omega}{2\hbar\,\sinh(\hbar\omega\tau)}(\mathbf{x}' - \mathbf{x})^2} \tag{5.88}$$

$$= \left(\frac{m\omega}{2\pi\hbar\,\sinh(\hbar\omega\beta)}\right)^{3N/2} e^{-\frac{1}{\frac{2\hbar}{m\omega}\tanh(\hbar\omega\tau)}\left(\mathbf{x}' - \frac{\mathbf{x}}{\cosh(\hbar\omega\tau)}\right)^2} \times e^{-\mathbf{x}^2\frac{m\omega}{2\hbar}\tanh(\hbar\omega\tau)}. \tag{5.89}$$

Equalities are checked in the Python code `harm_osc_density_matrix_t-ests.py`. In the last form, the density matrix contains a Gaussian factor, and as a consequence, the HO paths can be sampled directly without any need for a Metropolis accept/reject step.[5]

To see how HO paths can be sampled using staging, we start with multiplications and divisions, as we did for equation (5.47),

$$\prod_{m=1}^{K} \rho_{\text{h.o.}}(\mathbf{x}_m, \mathbf{x}_{m+1}; \tau)$$

$$= \rho_{\text{h.o.}}(\mathbf{x}_1, \mathbf{x}_{K+1}; K\tau) \prod_{m=2}^{K} \frac{\rho_{\text{h.o.}}(\mathbf{x}_{m-1}, \mathbf{x}_m; \tau)\rho_{\text{h.o.}}(\mathbf{x}_m, \mathbf{x}_{K+1}; (K-m)\tau)}{\rho_{\text{h.o.}}(\mathbf{x}_{m-1}, \mathbf{x}_{K+1}; (K-m+1)\tau)} \tag{5.90}$$

$$\overset{\text{Gaussian}}{=} \rho_{\text{h.o.}}(\mathbf{x}_1, \mathbf{x}_{K+1}; K\tau) \prod_{m=2}^{K} (2\pi\sigma_m^2)^{-3N/2} e^{-\frac{(\mathbf{x}_m - \mathbf{x}_m^*)^2}{2\sigma_m^2}}.$$

The first equality is actually valid for any density matrix, while the second is valid for Gaussians such as free particles or HOs. For HOs, the midpoint is again the

---

[5] Propagating from $\mathbf{x}$ to $\mathbf{x}'$, the exponent factor with only $\mathbf{x}$ is just a weight.

weighted sum of the known points, but, unlike the free-particle case, the variance is slice dependent,

$$\mathbf{x}_m^* = \frac{\sinh(\hbar\omega(K-m)\tau)\mathbf{x}_{m-1} + \sinh(\hbar\omega\tau)\mathbf{x}_{K+1}}{\sinh(\hbar\omega(K-m+1)\tau)} \tag{5.91}$$

$$\sigma_m^2 = \frac{\hbar}{m\omega}\frac{\sinh(\hbar\omega\tau)\sinh(\hbar\omega(K-m)\tau)}{\sinh(\hbar\omega(K-m+1)\tau)}, \tag{5.92}$$

and the points $\mathbf{x}_2, \ldots, \mathbf{x}_K$ can be sampled exactly using

$$\mathbf{x}_m = \mathbf{x}_m^* + \sigma_m\boldsymbol{\eta}. \tag{5.93}$$

To summarize, the paths in the HO partition function

$$Z_{\text{h.o}}(\beta = M\tau) = \int \prod_{m=1}^{M} d\mathbf{x}_m \prod_{m=2}^{M} (2\pi\sigma_m^2)^{-3N/2} e^{-\frac{(\mathbf{x}_m - \mathbf{x}_m^*)^2}{2\sigma_m^2}} \times e^{-\frac{m\omega}{\hbar}\tanh(\frac{1}{2}\hbar\omega\tau)\mathbf{x}_1^2}, \tag{5.94}$$

can be sampled exactly.

We have two cases where paths can be sampled directly: free particles $\hat{\mathcal{H}}_0$ and HOs $\hat{\mathcal{H}}_{\text{HO}}$. These serve as a base to build upon if the system Hamiltonian is $\hat{\mathcal{H}} = \hat{\mathcal{H}}_0 + \text{additional terms}$, or if $\hat{\mathcal{H}} = \hat{\mathcal{H}}_{\text{HO}} + \text{additional terms}$. The HO density matrix is a perfect reference system in the quantum Drude model, which describes the polarizability of atoms and molecules [5]. The quantum Drude model captures the long-range part of non-additive dispersion interactions arising from many-body effects beyond the dipole approximation, although the short-range part is incorrect.

5.4 Worm algorithm and permutation sampling

For bosons or fermions, we need to hide the particle identity, so we sum over all permutations of bosons or fermions,

$$Z(\beta) = \sum_P (\pm 1)^P \int d\mathbf{x}\rho(\mathbf{x}, P\mathbf{x}; \beta). \tag{5.95}$$

It is tedious to keep repeating the permutation sum in all formulas, so let us save space:

Convention: Permutation $P\mathbf{x}$ implies summation over permutations with the appropriate boson or fermion sign.

Now we can write

$$Z(\beta) = \int d\mathbf{x}\rho(\mathbf{x}, P\mathbf{x}; \beta). \tag{5.96}$$

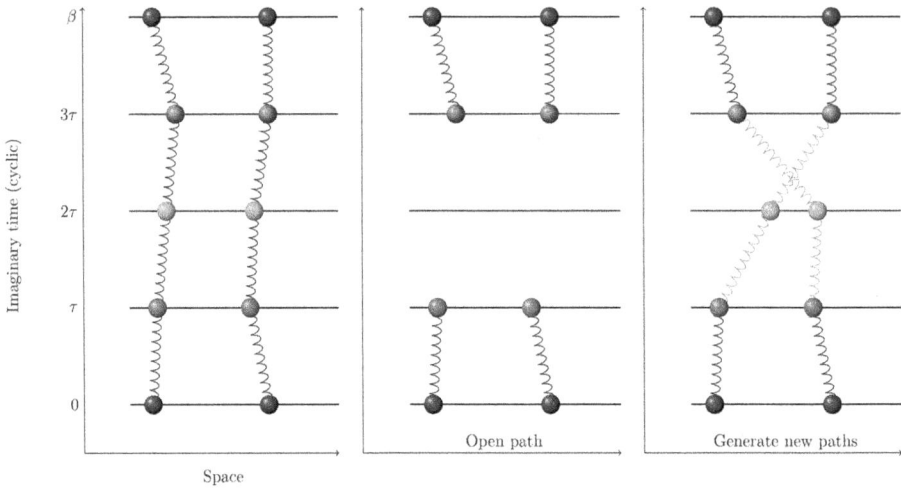

Figure 5.7. Fusion of two loops in imaginary time. The path generation in the last step can be done using the bisection algorithm.

The physical terms that contribute to the partition function are loops. As always, MC is effective only if one can sample a wide variety of different states. It is simple to fuse two loops using trial and error, and such brute-force permutation sampling can be implemented in the bisection algorithm. An example is shown in figure 5.7, where two nearby loops are picked as candidates for fusion.

Whether the change is accepted depends on the values of $\rho(\mathbf{x}, P\mathbf{x}; \beta)$ before and after the update. As stated at the end of the previous section, the bisection algorithm samples exactly the free-particle spring terms. In an interacting system, the interaction can still cause the update to be rejected. I will not give detailed formulas for the acceptance of permutation sampling using bisection, because we will not be using it. The reason for abandoning brute-force permutation sampling is that it has trouble generating *long exchange loops*. Once the number of particles exceeds $N \sim 100$, a loop that contains a large fraction of the particles is extremely hard to come by.

To sample Z effectively, we have to generate loops whose length varies a lot, and working with just loops is too restrictive. Prokof'ev, Svistunov, and Tupitsin suggested a **continuous-space worm algorithm** [6], which was later refined by Boninsegni, Prokof'ev, and Svistunov [7].[6] The principle is that

Local worm updates lead to large global changes.

[6] Lattice versions of worm algorithms were invented in the 1980s by Swendsen and Wang and by Wolff, to efficiently sample states in the MC simulations of the Ising model near criticality.

The intermediate states, where the worm is present, are not valid terms in Z. Let us introduce two kinds of configurations:

- The Z **sector** or **diagonal sector** has closed paths; the configuration is a valid term in Z.
- The G **sector** or an **off-diagonal sector** has an unclosed path that is called a worm.

Relaxing the 'stay in valid Z' constraint gives us an enormous advantage in sampling because we can work with open paths as long as we please and close them later to get a completely different term in Z. An open path has a particle that vanishes in one bead and reappears in another bead, and these points are here called **the Tail and the Head of the worm**, marked T and H, respectively. Technically, the Tail is a creation operator $\hat{\Psi}^{\dagger}(\mathbf{r}_T; \tau_T)$, and the Head is an annihilation operator $\hat{\Psi}(\mathbf{r}_H; \tau_H)$.

The rules that define how a worm can move come from two sources. One is the requirement that closing loops must be **valid terms in the partition function** with correct weights, not just any set of closed paths. The other is the decision to work in either the **canonical ensemble** and sample Z_N for exactly N particles, or in the **grand canonical ensemble** and sample the partition function $\mathcal{Z}$, where N fluctuates around an average whose value is governed by the chemical potential μ. The relation between the two partition functions is

$$\mathcal{Z} = \sum_{N=0}^{\infty} e^{\mu N} Z_N := \sum_{N=0}^{\infty} z^N Z_N, \tag{5.97}$$

where z is the fugacity. In PIMC, μ can be iteratively adjusted to keep a targeted average particle number $\langle N \rangle$. An interesting possibility is to perform a grand canonical PIMC and record, as a histogram, the probability $P(N)$ of having N particles at a fixed temperature. The quantity to measure is the canonical chemical potential, defined as

$$\mu(N, T) := F(N + 1, T) - F(N, T) = \mu - k_B T \ln \frac{P(N + 1)}{P(N)}, \tag{5.98}$$

where F is the free energy and μ is the grand canonical chemical potential. The free energy cost of particle insertion and removal shows signals of first-order phase transitions. The histogram estimator was used for liquid He^4 [8] and recently also for a warm dense uniform electron gas [9].

A sufficient set of worm updates consists of the operations: open, close, move Head or Tail, and (the most important) the swap update.

5.4.1 Worm open update

Start from a configuration in the Z sector and move to the G sector by deleting a path segment. First, decide where to put Head and Tail, and record how the decision was made. The open update is written for the grand canonical ensemble; to perform the canonical update, set $\mu = 0$. The adjustable parameters are marked in blue. From now on, *above* means up, and *below* means down in imaginary time; remember, imaginary time is periodic with period β.

Open
- Set Head to a random bead (there are N_{beads} to choose from).
- Pick $k \in U[1, K]$ and set Tail k steps above Head; $\tau_{HT} \equiv \tau_H - \tau_T$.
- Remove beads between Head and Tail.
- Accept with probability

$$P_{\text{open}} = \min\left\{1, \frac{CKN_{\text{beads}}e^{-\Delta U - \mu\tau_{HT}}}{\rho_0(\mathbf{r}_H, \mathbf{r}_T; \tau_{HT})}\right\}. \tag{5.99}$$

where the interaction difference is $\Delta U :\, = U(new) - U(old)$.

The open update is shown in figure 5.8.

The acceptance is calculated using the **detailed balance condition**, based on weights given by $\mathcal{Z}$ and remembering how we ended up suggesting this update. In Metropolis–Hastings, the acceptance of update $S \to S'$ is (see equation (4.134))

$$A(old \to new) = \frac{W(new)T(new \to old)}{W(old)T(old \to new)}. \tag{5.100}$$

For every way to try an update $T(old \to new)$, we need to figure out how many ways there are to try the reverse update, $T(new \to old)$. The G sector function we are sampling is the partition function with a worm, an 'extended partition function' that can be written as

$$\mathcal{Z}_{\text{worm}} = \sum_{N=0}^{\infty}\left[Z_N e^{\beta\mu N} + C\sum_{i_H, i_T}\int d\mathbf{r}_H d\mathbf{r}_T(Z_N e^{\beta\mu N} - \text{path H-T})\right].$$

The parameter C controls how much simulation time we want to spend in the G sector. A small C means we are not seeing many worms and will not get their full benefit. A large C means we hardly ever get a term that counts as one in the partition function. Somewhere in between, there is an optimal C.

The Head was picked from N_{beads} beads, which is the current value of NM (prepare for the case where N fluctuates); and the reverse, the unpicking, can be done in one way. Tail was picked from K beads, unpicking can be done in one way. So far, we have

$$\frac{T(new \to old)}{T(old \to new)} = \frac{1 \times 1}{1/K \times 1/N_{\text{beads}}} = KN_{\text{beads}} = KNM. \tag{5.101}$$

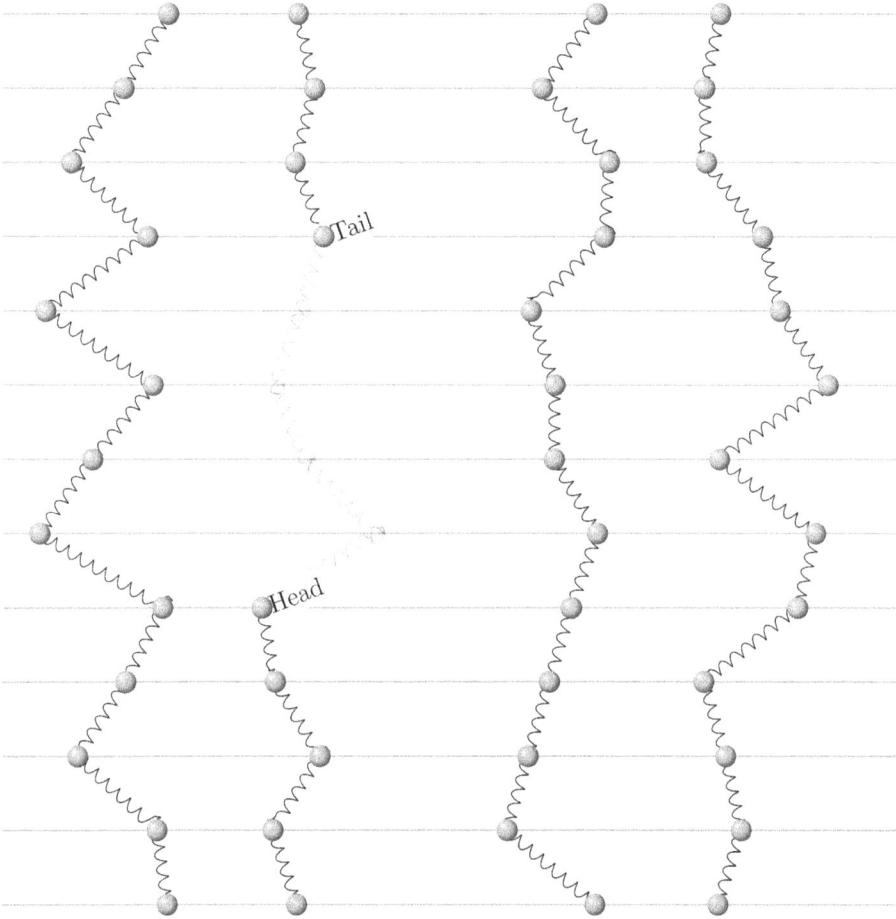

Figure 5.8. Worm open update.

From the chemical potential, the old weight is $e^{\beta\mu N}$. The new weight has a factor C and there are $N-1$ particles for the duration τ_{HT} and N particles for the duration $\beta - \tau_{HT}$,

$$\frac{W(new)}{W(old)} = \frac{Ce^{\mu N(\beta - \tau_{HT}) + \mu(N-1)\tau_{HT}}}{e^{\beta\mu N}} = Ce^{-\mu\tau_{HT}}. \tag{5.102}$$

Next, we make an assumption that simplifies the formulas:

A new path segment between Tail and Head is generated using free-particle random walk, using, for example, staging or bisection. This ensures that the spring factors are exactly sampled.

The new path segment between Tail and Head has no spring factors (they were used up in sampling), while the old has $\rho_0(\mathbf{r}_H, \mathbf{r}_T; \tau_{HT})$. The acceptance from spring factors is

$$\frac{W(new)}{W(old)} = \frac{1}{\rho_0(\mathbf{r}_H, \mathbf{r}_T; \tau_{HT})}. \tag{5.103}$$

The change in the interaction instigated by a new path segment gives yet another acceptance factor,

$$\frac{W(new)}{W(old)} = \frac{e^{-U(new)}}{e^{-U(old)}} = e^{-(U(new)-U(old))}: = e^{-\Delta U}. \tag{5.104}$$

The weights were written assuming the imaginary time is evenly split into steps τ. Collecting all factors together, we find the acceptance of the open worm update.

5.4.2 Worm close update

Close

1. Reject the update if Tail is zero or more than K steps above Head.
2. Generate a path from Head to Tail (free-particle spring factors).
3. Accept with probability

$$P_{\text{close}} = \min\left\{1, \frac{\rho_0(\mathbf{r}_H, \mathbf{r}_T; \tau_{HT})}{CKN_{\text{beads}}e^{\Delta U - \mu\tau_{HT}}}\right\}. \tag{5.105}$$

N_{beads} is the number of beads *after* closing, and $\Delta U = U(new) - U(old)$.

The probability of worm closing may become very small if the random walk distance between Head and Tail is large, i.e. $\rho_0(\mathbf{r}_H, \mathbf{r}_T; \tau_{HT})$ is very small. Such worms are rarely created, but they are also hard to close. In order to avoid such sticky worms, Boninsegni et al used a hard limit that keeps $\rho_0(\mathbf{r}_H, \mathbf{r}_T; \tau_{HT})$ within reasonable bounds: if

$$\frac{(\mathbf{r}_H - \mathbf{r}_T)^2}{4\lambda\tau_{HT}} > X, \tag{5.106}$$

opening and closing are rejected.[7] The limit is deterministic and used symmetrically in opening and closing, so $T(new \rightarrow old) = T(old \rightarrow new)$, and no harm is done to detailed balance.

---

[7] Boninsegni et al used $X = 4$ in liquid He calculations.

Worm closing is the reverse of worm opening. Be careful in testing: the configurations before and after the open update must exactly match those of the close update, but in reverse order.

5.4.3 Worm insert and remove updates

In the grand canonical ensemble, seed a new world-line strand of length k:

Insert
1. Set Tail to a random $\mathbf{r}$ and a random imaginary-time slice; we assume the simulation volume Ω is finite.
2. Pick $k \in U[1, K]$ and generate a k-step path above Tail; put Head there.
3. Accept with probability

$$P_{\text{insert}} = \min\{1, CKM\Omega e^{\Delta U - \mu \tau_{HT}}\}. \tag{5.107}$$

The acceptance P_{insert} assumes that the new path segment is generated from exact sampling of ρ_0; either the staging or the bisection algorithm does the job.

Remove
1. If the worm is longer than K, reject removal.
2. Accept with probability

$$P_{\text{remove}} = \min\left\{1, \frac{e^{\Delta U - \mu \tau_{HT}}}{CKM\Omega}\right\}, \tag{5.108}$$

Figure 5.9 shows how a new path segment is inserted among the existing ones; removal is the reverse update.

5.4.4 Worm advance and recede updates

The acceptances of advance and recede moves given below assume that you sample the kinetic action exactly. The new Head is sampled as

$$\mathbf{r}_{\text{new Head}} = \mathbf{r}_{\text{Head}} + \sqrt{2\lambda \Delta \tau_{\text{Head, new Head}}}\, \eta, \tag{5.109}$$

which samples exactly $\rho_0(\mathbf{r}_{\text{Head}}, \mathbf{r}_{\text{new Head}}, \Delta\tau_{\text{Head, new Head}})$. After that, the path from $\mathbf{r}_{\text{Head}}$ to $\mathbf{r}_{\text{new Head}}$ is sampled exactly from the free-particle kinetic action using staging or bisection. This leaves only the interaction difference ΔU to be evaluated.

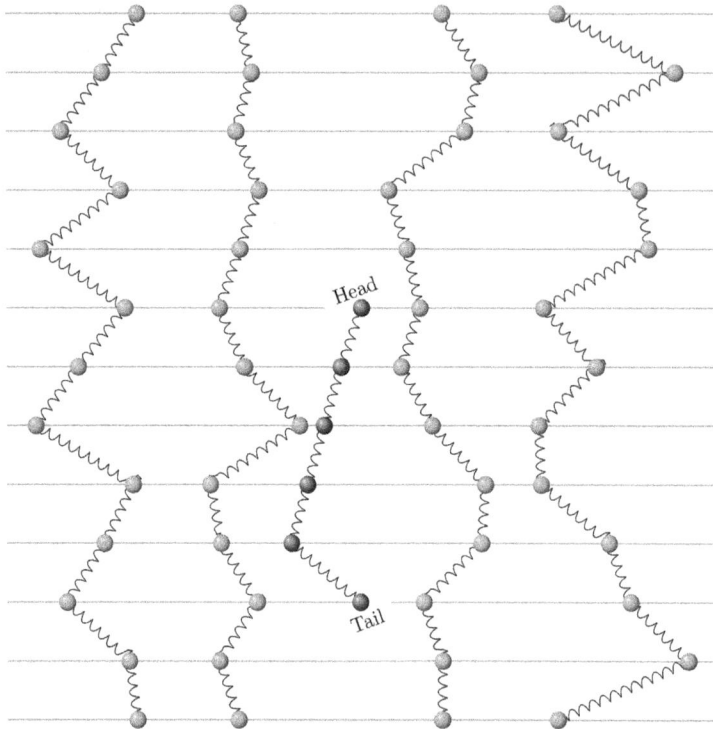

Figure 5.9. Worm insert/remove update: if the inserted worm closes into a loop, the partition function is evaluated at one more particle than before, in the spirit of the grand canonical ensemble.

Try moving Head up in imaginary time as shown in figure 5.10:

Advance
1. Pick $k \in U[1, K]$
2. Generate a k-step path from old Head up in imaginary time and set the new Head there.
3. Accept advance with probability

$$P_{\text{advance}} = \min\{1, e^{\Delta U + \mu \tau_{HT}}\}. \tag{5.110}$$

Try moving Head down in imaginary time:

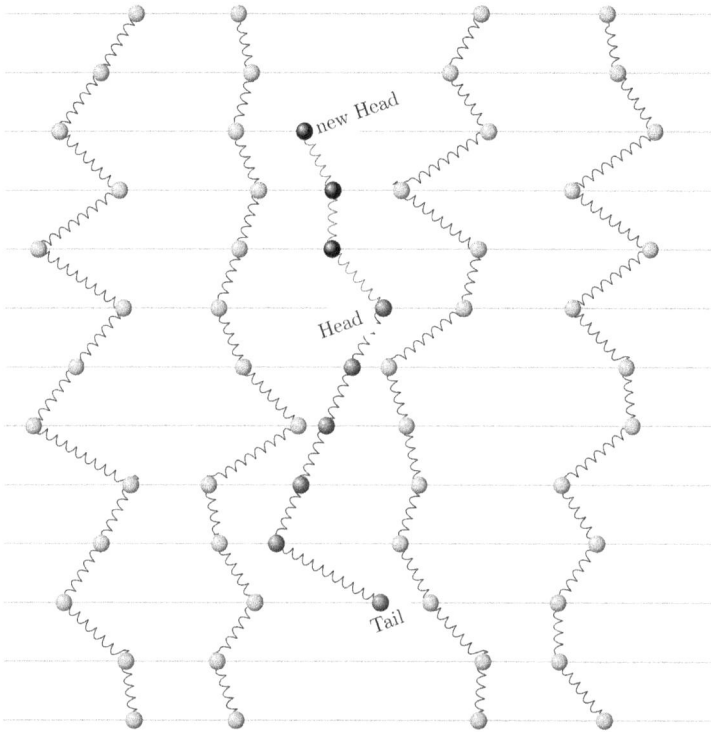

Figure 5.10. Worm advance update; the recede update is the reverse. In the grand canonical ensemble, Head can advance past Tail.

Recede
1. Pick $k \in U[1, K]$. If the worm has less than k beads, reject the update.
2. Move Head k steps down in imaginary time, deleting beads on the way.
3. Accept with probability

$$P_{\text{recede}} = \min\{1, e^{\Delta U - \mu \tau_{HT}}\}. \tag{5.111}$$

In both advance and recede acceptances, $\Delta U = U_{\text{new}} - U_{\text{old}}$, with new and old referring to the configurations after the move and before the move, respectively. If you advance, say, $k = 3$ slices, and immediately recede $k = 3$ slices, then $\Delta U(\text{advance}) = -\Delta U(\text{recede})$; this is useful in debugging.

5.4.5 Worm wiggle tail update

In the canonical ensemble, one can move Tail (or Head) and part of the path with it to new positions. The wiggle tail update, shown in figure 5.11, is often combined

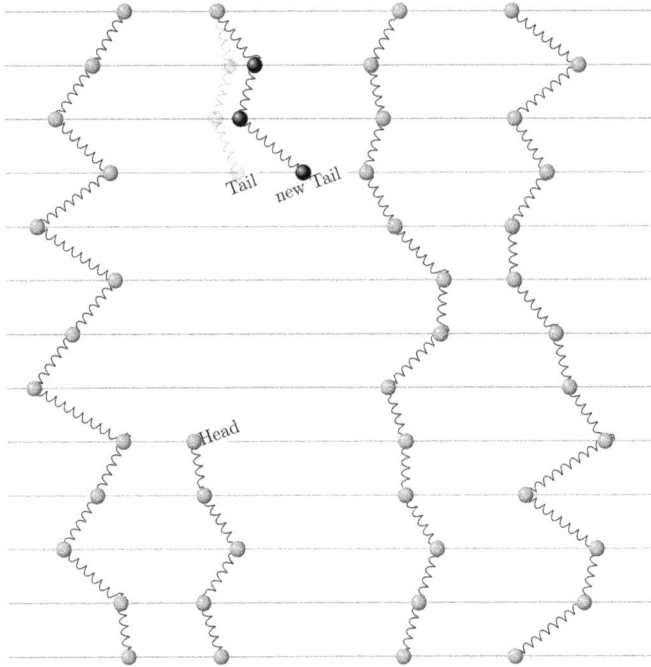

Figure 5.11. Worm wiggle tail update increases the chance that the worm closes.

with worm advance/recede and close attempts. To keep the worm close probability reasonable, it is common practice to set a hard limit similar to equation (5.106) to prevent Head and Tail from getting too far from each other.

Let us assume we want a new path from a base bead to the new Tail. The new path segment can be sampled exactly from the free-particle kinetic action, that is, from the spring factors. In this case, the new Tail is sampled from the factor

$$\rho_0(\text{base,new Tail}; \Delta\tau_{\text{base,new Tail}}), \quad (5.112)$$

meaning

$$\mathbf{r}_{\text{new Tail}} = \mathbf{r}_{\text{base}} + \sqrt{2\lambda\Delta\tau_{\text{base,new Tail}}}\ \boldsymbol{\eta}, \quad (5.113)$$

where $\boldsymbol{\eta}$ is a normally distributed random vector. The path from base to Tail is a Brownian bridge that can be generated using staging or bisection. Now, all that remains in the Metropolis acceptance probability is the ratio of the potential factors,

$$P_{\text{wiggle}} = \min\{1, e^{\Delta U}\}. \quad (5.114)$$

Another option is to sample the position of the new Tail near the old Tail, for example, as

$$\mathbf{r}_{\text{new Tail}} = \mathbf{r}_{\text{Tail}} + \sigma\ \boldsymbol{\eta}, \quad (5.115)$$

where σ is a free parameter. This is a symmetrical choice, so it cancels from the acceptance. If the Brownian bridge is again sampled using staging or bisection, there is one part that is not yet accounted for, namely the ρ_0 factor from the base to the new Tail. One has to adjust the acceptance using Metropolis–Hastings with

$$T(\rightarrow) = \rho_0(\text{base,new Tail}; \Delta\tau_{\text{base,new Tail}}) \tag{5.116}$$

$$T(\leftarrow) = \rho_0(\text{base,Tail}; \Delta\tau_{\text{base,Tail}}). \tag{5.117}$$

Here, $\Delta\tau_{\text{base, new Tail}} = \Delta\tau_{\text{base, Tail}}$, so

$$\frac{T(\leftarrow)}{T(\rightarrow)} = \frac{\exp\left(-\dfrac{\Delta r^2_{\text{base, Tail}}}{4\lambda\Delta\tau_{\text{base, Tail}}}\right)}{\exp\left(-\dfrac{\Delta r^2_{\text{base, new Tail}}}{4\lambda\Delta\tau_{\text{base, new Tail}}}\right)} = \exp\left(\frac{\Delta r^2_{\text{base, new Tail}} - \Delta r^2_{\text{base, Tail}}}{4\lambda\Delta\tau_{\text{base, Tail}}}\right), \tag{5.118}$$

and the Metropolis–Hastings acceptance of the update is

$$P_{\text{wiggle}} = \min\left\{1, \exp\left(\Delta U + \frac{\Delta r^2_{\text{base, new Tail}} - \Delta r^2_{\text{base, Tail}}}{4\lambda\Delta\tau_{\text{base, Tail}}}\right)\right\}. \tag{5.119}$$

5.4.6 Worm swap update

For identical particles, the swap update is the real reason to use the worm algorithm because it does permutations in a very natural way. No longer do we need to suggest permutations pair by pair.

Swap
1. Pick $k \in U[1, K]$. Build a list $\{S_i\}$ of beads that are k time steps above Head (if needed, go back to the bottom using time periodicity).
2. Build a weight list $\{W_{H \rightarrow S_i}\} = \{\rho_0(\mathbf{r}_H, \mathbf{r}_{S_i}, \tau_{HS_i})\}$.
3. Pick a bead S from the weight list $\{W_{H \rightarrow S_i}\}$.
4. Find bead T (here T does not refer to the Tail) on the timeline k steps below the swap bead S. Reject the update if Tail is met along the way. Bead T will become the new Head if the swap update is accepted.
5. Build another weight list $\{W_{T \rightarrow S_i}\} = \{\rho_0(\mathbf{r}_T, \mathbf{r}_{S_i}, \tau_{TS_i})\}$. Here, τ_{TS_i} is the same as τ_{HS_i} because T and Head are on the same time slice.
6. Generate a free-particle random path (Brownian bridge) from Head to S.
7. Accept the update with probability

$$P_{\text{swap}} = \min\left\{1, e^{\Delta U}\frac{\sum_i W_{H \rightarrow S_i}}{\sum_i W_{T \rightarrow S_i}}\right\}. \tag{5.120}$$

Bead S in step 3 can be picked using the following algorithm, which has a higher probability of selecting higher-weighted S_is; it favors beads that are fairly close to Head.[8]

Step 3: Picking bead S
1. For weights in the list $\{W_{H \to S_i}\}$, compute $\sum_i W_{H \to S_i}$ (also needed in step 7).
2. Pick a random number $r \in U[0, 1]$ and compute $R = r^* \sum_i W_{H \to S_i}$.
3. Compute the cumulative sum of values $W_{H \to S_i}$, and when the sum becomes greater than or equal to R, choose the current S_i to be S.

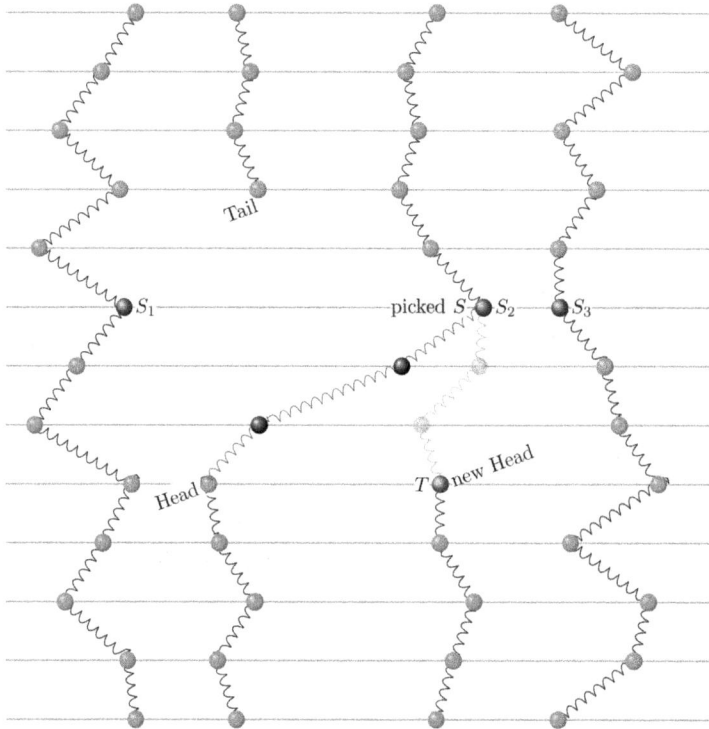

Figure 5.12. Worm swap update. A **swap bead** S is sampled from the list of beads $\{S_1, S_2, \dots\}$, where each bead has a weight $W_{H \to S_i}$. The new path from Head to S is marked in dark purple. Part of the old path is erased, and Head moves to the new Head T. For detailed balance, one must also keep in mind the reverse swap update, where the swap bead is sampled from weights $W_{T \to S_i}$.

Figure 5.12 shows an example of a swap update. The swap update is its own reverse, and therefore the acceptance needs more calculations. The list of possible

---

[8] Depending on how you picture this, it is a wheel of fortune or tower sampling.

swap beads $\{S_i\}$ is needed in both the swap update and the reverse swap update. Suppose the swap bead S were chosen based on its free-particle distance from Head, that is, from the list $\{\rho_0(H, S_i, \Delta\tau)\}$. The reverse update would then pick the swap bead from the distance list $\{\rho_0(T, S_i, \Delta\tau)\}$, and the two distance lists would not be the same. These asymmetric tries must be compensated in the acceptance probability.

If the swap is accepted, Head moves to a new position, and the new worm is longer. In figure 5.12, note that when moving up from Tail to the new Head, the worm covers the positions of *two* particles. If the worm closes after the swap, the resulting world line will be an exchange loop (a permutation loop). There can be many accepted swaps before the worm closes, resulting in multiple particles in exchange loops. Swap updates are local changes that accumulate to form **global** changes.

Swap updates can change the whole topology of the world lines.

5.4.7 Periodic boundary conditions

PBCs in a 3D box L^3 are the coordinate transformations

$$
\begin{aligned}
x + L &= x \\
y + L &= y \\
z + L &= z.
\end{aligned}
$$
(5.121)

With PBCs, the path integral representation of the partition function is periodic in *every* direction, both in space and in imaginary time. Every dimension is wrapped. Figure 5.13 shows how PBCs cause paths that leave from one side of the simulation box to enter from the opposite side.

One purpose of PBCs is to dilute the effects of the box edges. The simulation volume is always finite. However, PBCs mimic an infinite system better than a box with hard walls because every bead has available space around it in all directions. A side effect of PBCs is that bead–bead correlations beyond the distance $L/2$ are unphysical. Beads near the distance $L/2$ are equally influenced by both periodic images. For example, in a simulation of a molecule on an infinite surface, the direction normal to the surface is not really periodic, but PBCs can be used if there is enough empty space between the molecule and the periodic image of the surface lurking above the molecule.

5.5 Approximate action

So far, we have been discussing the primitive action (PA, based on the primitive approximation). In the DMC chapter, we showed a few ways to split the operator $e^{-\tau\hat{\mathcal{H}}}$ into kinetic and potential parts. Some of the fourth-order splittings were the

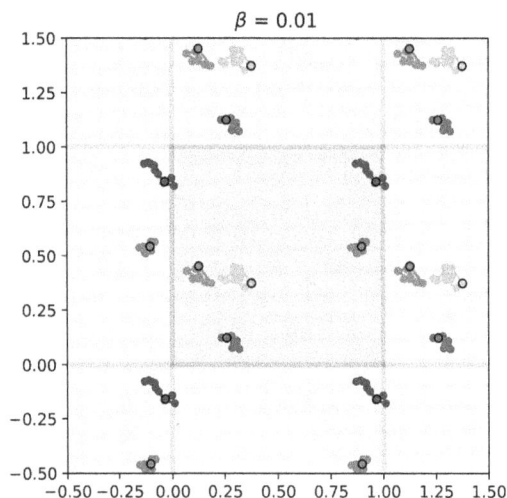

Figure 5.13. A few free-particle paths under PBCs. The center box is the actual simulation box, while the rest are periodic images.

Takahashi–Imada action (TIA, equation (4.212)) and the Suzuki–Chin action (SCA, equation (4.216)). The latter has been slightly improved to produce the Chin action (CA, [10]). These actions were compared in [11] for H_2O and HCN–HNC systems; CA was found to be typically twice as efficient as TIA and approximately ten times more efficient than PA, regardless of temperature.

Given that PA is pretty useless for real PIMC calculations, it is worth investing in a better action. However, PA has its value in introducing ideas without using long formulas and complicated notation.

Higher-order actions are not the only way to improve PIMC efficiency. The original He^4 simulations used the so-called **pair-product action**[12], which is based on the exact action of a pair of particles, $u(\mathbf{r}_{ij}, \mathbf{r}'_{ij}; \tau)$. The premise is that this is the most important quantity to get right, especially for particles with repulsive interactions. The pair-product action is quite efficient but system-specific, while the actions based on high-order operator splittings are more jacks of all trades.

5.5.1 Chin action

The CA is a state-of-the-art PIMC action, first applied to PIMC in 2009 by Sakkos, Casulleras, and Boronat [13]. The CA fourth-order operator splitting is [10]:

CA operator splitting

$$e^{-\tau\hat{\mathcal{H}}} \approx e^{-v_1\tau\hat{\mathcal{W}}_{a_1}}e^{-t_1\tau\hat{\mathcal{T}}}e^{-v_2\tau\hat{\mathcal{W}}_{1-2a_1}}e^{-t_1\tau\hat{\mathcal{T}}}e^{-v_1\tau\hat{\mathcal{W}}_{a_1}}e^{-2t_0\tau\hat{\mathcal{T}}}$$

$$\hat{\mathcal{W}}_{a_1} = \hat{\mathcal{V}} + \frac{u_0}{v_1}a_1\tau^2[\hat{\mathcal{V}}, [\hat{\mathcal{T}}, \hat{\mathcal{V}}]]$$

$$\hat{\mathcal{W}}_{1-2a_1} = \hat{\mathcal{V}} + \frac{u_0}{v_2}(1 - 2a_1)\tau^2[\hat{\mathcal{V}}, [\hat{\mathcal{T}}, \hat{\mathcal{V}}]].$$

(5.122)

The double commutator $[\hat{\mathcal{V}}, [\hat{\mathcal{T}}, \hat{\mathcal{V}}]]$ is the key to the high-order accuracy. It was computed in equation (4.206), where it was shown to be related to the classical force acting on a particle. CA has two free parameters, a_1 and t_0, which should be optimized for each system. These are restricted to the values

$$0 \leqslant a_1 \leqslant 1, \, 0 \leqslant t_0 \leqslant \frac{1}{2}\left(1 - \frac{1}{\sqrt{3}}\right). \tag{5.123}$$

The highest gain in optimization is sixth-order accuracy for a HO. The rest of the parameters are given by the relations

$$u_0 = \frac{1}{12}\left[1 - \frac{1}{1 - 2t_0} + \frac{1}{6(1 - 2t_0)^3}\right] \tag{5.124}$$

$$v_1 = \frac{1}{6(1 - 2t_0)^2} \tag{5.125}$$

$$v_2 = 1 - 2v_1 \tag{5.126}$$

$$t_1 = \frac{1}{2} - t_0. \tag{5.127}$$

The restrictions to a_1 and t_0 keep all parameters in (5.122) positive.

Every imaginary-time slice of length τ is further split into three 'subslices' of lengths $2t_0\tau$, $t_1\tau$, and $t_1\tau$, such that $(t_1 + t_2 + 2t_0)\tau = \tau$. As a first step, let us insert a complete position space $\mathbb{1} = \int |\mathbf{x}\rangle\langle\mathbf{x}|$ between every operator in (5.122). Since there are two subslices for each τ, one needs $M - 1$ integrations in the density matrix at temperature $\beta = M\tau/3$,[9]

$$\rho(\mathbf{x}', \mathbf{x}; \beta = M\tau/3) = \int d\mathbf{x}_1 .. d\mathbf{x}_{M-1} e^{-v_1\tau W_{a_1}(\mathbf{x}')}\langle\mathbf{x}'|e^{-t_1\tau\hat{T}}|\mathbf{x}_1\rangle e^{-v_2\tau W_{1-2a_1}(\mathbf{x}_1)}$$
$$\times \langle\mathbf{x}_1|e^{-t_1\tau\hat{T}}|\mathbf{x}_2\rangle e^{-v_1\tau W_{a_1}(\mathbf{x}_2)}\langle\mathbf{x}_2|e^{-2t_0\tau\hat{T}}|\mathbf{x}_3\rangle$$
$$... \times e^{-v_1\tau W_{a_1}(\mathbf{x}_{M-3})}\langle\mathbf{x}_{M-3}|e^{-t_1\tau\hat{T}}|\mathbf{x}_{M-2}\rangle e^{-v_2\tau W_{1-2a_1}(\mathbf{x}_{M-2})}$$
$$\times \langle\mathbf{x}_{M-2}|e^{-t_1\tau\hat{T}}|\mathbf{x}_{M-1}\rangle e^{-v_1\tau W_{a_1}(\mathbf{x}_{M-1})}\langle\mathbf{x}_{M-1}|e^{-2t_0\tau\hat{T}}|\mathbf{x}\rangle. \tag{5.128}$$

This formula is actually quite straightforward to implement: it alternates free propagation and potential factor evaluation. Figure 5.14 compares the PA and CA density matrix computations. The slightly more complicated CA pays off handsomely by dramatically reducing the number of slices M. Sometimes, one hears claims that, because of the three slices within each τ propagation, the fourth-order action is just 'Monte Carlo within Monte Carlo.' Technically, that is exactly what happens, but for every 'extra' slice, we get a lot more accuracy than by adding as many slices in PA. It is like solving a differential equation numerically using fourth-

---

[9] One could just as well define $\beta = M\tau$ and have $3M - 1$ integrations in the density matrix.

Primitive action:

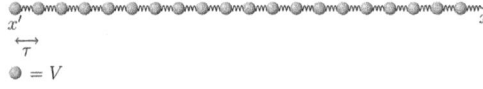

x' x

$\overset{\longleftrightarrow}{\tau}$

$\circ = V$

Chin action:

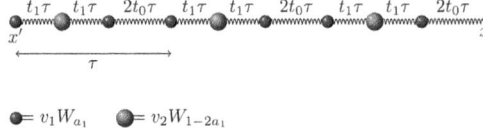

$t_1\tau$ $t_1\tau$ $2t_0\tau$ $t_1\tau$ $t_1\tau$ $2t_0\tau$ $t_1\tau$ $t_1\tau$ $2t_0\tau$

x' x

τ

$\bullet = v_1 W_{a_1}$ $\bullet = v_2 W_{1-2a_1}$

Figure 5.14. A schematic comparison of the density matrix calculations in the primitive action and the CA. The latter has uneven time slices and τ-dependent effective potentials W. The figure emphasizes the fact that PA needs a lot more beads than CA in order to get $\rho(\mathbf{x}', \mathbf{x}; \beta)$ with the same accuracy.

order Runge–Kutta versus the first-order Euler method. Admittedly, there is a price to pay in evaluating $|\frac{\partial}{\partial \mathbf{x}} V(\mathbf{x})|^2$ needed in effective potentials W.

Let us rearrange spring and potential factors:

$$\rho(\mathbf{x}', \mathbf{x}; \beta = M\tau/3) = \int d\mathbf{x}_1 ... d\mathbf{x}_{M-1} \langle \mathbf{x}' | e^{-t_1\tau\hat{T}} | \mathbf{x}_1 \rangle \langle \mathbf{x}_1 | e^{-t_1\tau\hat{T}} | \mathbf{x}_2 \rangle \langle \mathbf{x}_2 | e^{-2t_0\tau\hat{T}} | \mathbf{x}_3 \rangle$$

$$... \times \langle \mathbf{x}_{M-3} | e^{-t_1\tau\hat{T}} | \mathbf{x}_{M-2} \rangle \langle \mathbf{x}_{M-2} | e^{-t_1\tau\hat{T}} | \mathbf{x}_{M-1} \rangle \langle \mathbf{x}_{M-1} | e^{-2t_0\tau\hat{T}} | \mathbf{x} \rangle \quad (5.129)$$

$$\times e^{-v_1\tau W_{a_1}(\mathbf{x}')} e^{-v_2\tau W_{1-2a_1}(\mathbf{x}_1)} e^{-v_1\tau W_{a_1}(\mathbf{x}_2)}$$

$$... \times e^{-v_1\tau W_{a_1}(\mathbf{x}_{M-3})} e^{-v_2\tau W_{1-2a_1}(\mathbf{x}_{M-1})} e^{-v_1\tau W_{a_1}(\mathbf{x}_{M-1})}.$$

We would like to write this in a more algorithmic manner, so we define

$$t_m := \begin{cases} t_1 & \text{, if } m \bmod 3 \neq 0 \\ 2t_0 & \text{, if } m \bmod 3 = 0 \end{cases} \quad (5.130)$$

and replace $v_1 W_{a_1}(\mathbf{x}; \tau)$ and $v_2 W_{1-2a_1}(\mathbf{x}; \tau)$ with

$$\tilde{V}_m(\mathbf{x}; \tau) := \begin{cases} v_1 W_{a_1}(\mathbf{x}; \tau) & \text{, if } (m+1) \bmod 3 \neq 0 \\ v_2 W_{1-2a_1}(\mathbf{x}; \tau) & \text{, if } (m+1) \bmod 3 = 0 \end{cases}$$

$$= \begin{cases} v_1 V(\mathbf{x}) + \tau^2 u_0 a_1 \frac{\hbar^2}{m} |\frac{\partial}{\partial \mathbf{x}} V(\mathbf{x})|^2 & \text{, if } (m+1) \bmod 3 \neq 0 \\ v_2 V(\mathbf{x}) + \tau^2 u_0 (1 - 2a_1) \frac{\hbar^2}{m} |\frac{\partial}{\partial \mathbf{x}} V(\mathbf{x})|^2 & \text{, if } (m+1) \bmod 3 = 0 \end{cases} \quad (5.131)$$

For example, for slices $m = 1, 2, 3, 4, 5, 6$, the sequence of potentials is

$$v_1 W_{a_1}, \; v_2 W_{1-2a_1}, \; v_1 W_{a_1}, \; v_1 W_{a_1}, \; v_2 W_{1-2a_1}, \; v_1 W_{a_1}. \quad (5.132)$$

The potential 'correction' is proportional to $\tau^2 |\frac{\partial}{\partial \mathbf{x}} V(\mathbf{x})|^2$. If we think about how to keep beads away from high-potential-energy regions, the first thing to do would be to penalize bead positions $\mathbf{x}$ with high potential energy $V(\mathbf{x})$. The next precaution would be to avoid steeply rising potential energies, and for that we need a scalar contribution of $\frac{\partial}{\partial \mathbf{x}} V(\mathbf{x})$, which is either $\mathbf{x} \cdot \frac{\partial}{\partial \mathbf{x}} V(\mathbf{x})$ or $|\frac{\partial}{\partial \mathbf{x}} V(\mathbf{x})|^2$. The former is a virial term encountered in the virial energy estimator in section 5.6.3. Apparently, a virial term would not lead to a fourth-order approximation for $e^{-\tau \hat{\mathcal{H}}}$, so we make a correction based on the double commutator $[\hat{\mathcal{V}}, [\hat{\mathcal{T}}, \hat{\mathcal{V}}]] \propto |\frac{\partial}{\partial \mathbf{x}} V(\mathbf{x})|^2$. Paths with very small τ have plenty of (imaginary) time to react to rising $V(\mathbf{x})$, so there is less need for gradient corrections—hence $|\frac{\partial}{\partial \mathbf{x}} V(\mathbf{x})|^2$ should have a τ-dependent factor.

We can now write the density matrix and the partition function more compactly:

Density matrix, CA
$\mathbf{x}_0 \equiv \mathbf{x}', \mathbf{x}_M \equiv \mathbf{x}$

$$\rho(\mathbf{x}', \mathbf{x}; \beta = M\tau/3) = \int \prod_{m=1}^{M-1} d\mathbf{x}_m \times \prod_{m=0}^{M} \langle \mathbf{x}_m | e^{-t_m \tau \hat{\mathcal{T}}} | \mathbf{x}_{m+1} \rangle \times \prod_{m=0}^{M} e^{-\tau \tilde{V}_m(\mathbf{x}_m; \tau)} \quad (5.133)$$

Partition function, CA $\mathbf{x}_{M+1} \equiv \mathbf{x}_1$ or $P\mathbf{x}_1$

$$Z(\beta = M\tau/3) = \int \prod_{m=1}^{M} d\mathbf{x}_m \times \prod_{m=1}^{M} \langle \mathbf{x}_m | e^{-t_m \tau \hat{\mathcal{T}}} | \mathbf{x}_{m+1} \rangle \times \prod_{m=1}^{M} e^{-\tau \tilde{V}_m(\mathbf{x}_m; \tau)}. \quad (5.134)$$

An example of a path in the CA is shown in figure 5.15.

5.6 PIMC measurements

It might seem odd that we are not going to start from energy. PIMC, in the form we have introduced here, does not have a 'natural energy estimator.' PIMC is free of bias because it does not use a trial wave function; however, for the same reason, there is no local energy. Instead, there are action-dependent energy estimators that range from simple and poor to complicated and good.

Using bisection and worm algorithms, we can generate paths that are terms that appear in the partition function, so let us start with quantities that can be measured from paths and bead positions only. Structural data, such as density and the pair distribution function (a histogram of particle–particle distances), are readily available. These quantities are routinely measured and good for testing freshly made PIMC code, but they are not that different from molecular dynamics simulations.

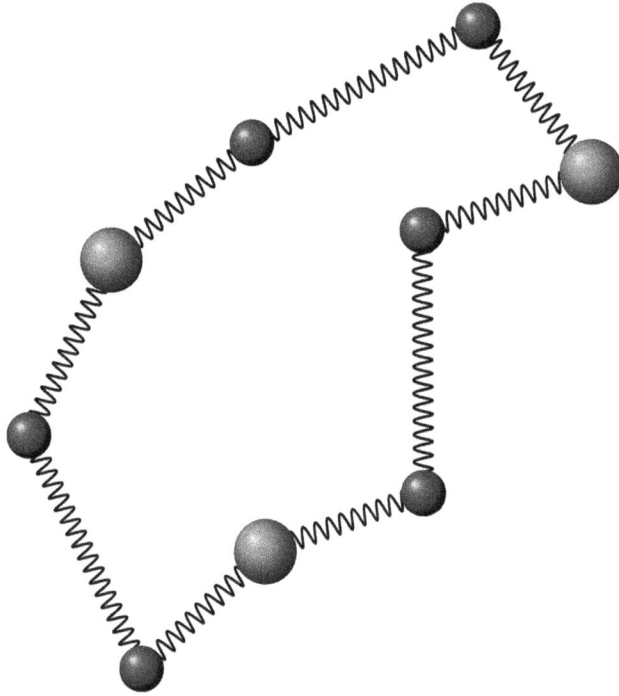

Figure 5.15. A loop of length 3τ in the CA partition function; $M = 3$ and there are $3M = 9$ beads. The next bead would be at $\mathbf{r}_{10}$, which cycles back to $\mathbf{r}_1$. The bead color and size show which effective potential W is applied in that time slice, and the color coding is the same as in figure 5.14. The springs between beads are the free-particle propagators $\langle \mathbf{r}_i | e^{-t_i \tau \hat{T}} | \mathbf{r}_{i+1} \rangle$, where t_i is given in equation (5.130).

Quantities naturally suited to PIMC include the superfluid density and Bose–Einstein condensate fraction, so let us start with them. The theory is lengthy, but once it is complete, measuring superfluid density takes about ten lines of code.

5.6.1 Superfluidity and winding paths

The theory of superfluidity dates back to the 1930s, and the story so far is covered in the review article by Leggett [14]. We will concentrate on a few key points relevant to PIMC.

Characteristic to superfluidity is nonviscous flow, while Bose–Einstein condensation (BEC) is the macroscopic occupation of the ground state by bosons. London [15, 16] suggested that superfluidity and BEC may be related.

Landau [17] described the resilience of superfluidity. Fluid in motion has excitations. Excitations carry momentum and mass; therefore, they can be called quasiparticles. Following Landau, the mass carried by quasiparticles is called normal fluid density ρ_n. The rest is superfluid density, ρ_s, so $\rho_n + \rho_s = \rho$; this model is called a **two-fluid model**. According to Landau, superfluidity is observed if the creation of quasiparticles is limited and stays below a certain threshold. If this happens, there is more mass in the system than the mass moving with quasiparticles; hence, part of the fluid is impervious to disturbances. At $T = 0$, all fluid is superfluid.

As a first thought, one might think that there has to be an energy gap in the spectrum $\epsilon(p) = \Delta + \cdots$, so that it takes a finite amount of energy to create quasiparticles. However, Landau showed that the actual criterion is that the quasiparticle spectrum $\epsilon(p)$ satisfies the condition

$$\epsilon(p) + \mathbf{p} \cdot \mathbf{v} < 0, \tag{5.135}$$

and since the quasiparticle momentum $\mathbf{p}$ acts in the opposite direction to the fluid velocity $\mathbf{v}$, there may be a nonzero critical velocity,

$$v < \frac{\epsilon(p)}{p}: =v_{\text{crit}} \qquad \text{Landau critical velocity.} \tag{5.136}$$

Ideal bosons have $\epsilon(p) = p^2/(2m)$, so $v_{\text{crit}} = 0$: an ideal Bose gas in motion is swarming with quasiparticles and cannot have superfluidity. Still, as statistical physics textbooks show, 100% of ideal bosons are in the condensate, so the relation between superfluidity and BEC is not straightforward.

The theory described so far is semimicroscopic. We want a formula that gives the superfluid density in a microscopic theory and in PIMC. In a series of articles in the 1950s, Feynman showed how superfluidity can be understood in the path integral formulation of quantum mechanics [18, 19]. The reasoning was compelling but qualitative, and it left a feeling that we might be missing something. Given the density matrix and an observable, one can compute the finite-temperature expectation value; there was no doubt about that. But what is the 'observable of superfluidity'?

The paper published in 1987 by the authors Pollock and Ceperley [20] sums up years of hard work. In the beginning, there was no clear-cut idea of what one was supposed to look for. Watching simulated PIMC paths, it dawned on them that superfluidity is related to the so-called **winding paths**, a connection that had been previously missed. There was no formula to use, so it had to be derived [21].

The PIMC results by Pollock and Ceperley were in excellent agreement with experimental data on the superfluid transition in liquid He^4. The He^4 superfluid transition is called the lambda transition, as the specific heat curve resembles the Greek letter λ, and that too was convincingly reproduced in PIMC simulations. PIMC results obtained using the worm algorithm (and faster computers) pinpoint the He^4 superfluid transition quite accurately [7].

It was a long and winding road, and many beautiful physical ideas had to be put together in order to come up with a mathematical formula that ties winding paths to superfluid density. The experimental observation of superfluidity and Landau's argumentation hint that we need *fluid in motion* and consequently *the Galilean transformation* applied to quantum mechanics. The final twist—and it is literally a twist—is measuring moving fluid properties using a PIMC simulation, where paths change but nothing moves. Such is the nature of equilibrium thermodynamics.

5.6.1.1 Superfluidity in experiments

How does one actually measure the superfluid density ρ_s? Not from a simulation but for real. Physicists define mass by finding the momentum p of an object, and if that is proportional to velocity v, the factor of proportionality is the mass m: $p = mv$.[10] The momentum–velocity relation gives the mass and density of anything in slow motion. The linear relation $p \propto v$ may not hold for large fluid velocities, because a fast-moving fluid does not resemble a slow-moving one; if it did, we would have solved the turbulence problem a long time ago. In addition, this superfluid scenario breaks down at some fluid velocity, following the formation of quantized vortices.

So put identical bosons, such as liquid He^4, in slow motion and find their momentum. In 1947, Andronikashvili performed experiments—per a suggestion by Landau—on how liquid He^4 moves between a stack of oscillating disks. The experiment is affectionately described as the rotating bucket experiment. Normal fluid experiences friction with walls, be they disks or a bucket, and the fluid soon moves at the same velocity as the walls. Momentum is transferred from the walls to the fluid; momentum transfer within the fluid is viscosity, and viscosity is what makes it normal. If you halt the bucket, normal fluid soon stops moving, too. However, the experiments on liquid He^4 show that below the superfluid critical temperature of $T_\lambda = 2.17$ K, part of the fluid keeps rotating. The part that still rotates experiences neither friction nor viscosity and is called superfluid. If, instead, you start with fluid below the critical temperature, part of the fluid refuses to follow after you start to rotate the bucket. We now have a definition for superfluid; the rest is normal fluid: $\rho = \rho_n + \rho_s$.

Nonclassical rotational inertia

Rotating fluid has rotational inertia, and a good way—perhaps the best way—to tell whether there is superfluid present is to compare the fluid's inertia with the classical inertia; the deviation is known as *nonclassical rotational inertia (NCRI)*. The observation of NCRI is, in principle, a sure-fire method of detecting super-fluidity. NCRI is the key quantity in superfluidity in **finite systems**. Winding paths, discussed in this chapter, are not *necessary* for superfluidity, and winding paths simply cannot happen in finite systems. Instead, superfluidity in droplets (clusters) is measured using an **area estimator** [22], which relates the superfluid fraction to the projected area of the path in the plane perpendicular to a rotational axis. The idea is that if some paths do not rotate with the rest, the moment of inertia is reduced, and the system is superfluid. As few as 60 He^4 atoms show superfluidity, and one can examine the rotations of molecules embedded in a superfluid droplet [22–24].

Supersolids

In recent years, many groups have searched for evidence of so-called *supersolids*, i.e. superfluidity in solid materials, as suggested by Leggett [25]. A supersolid is a fascinating contradiction, a zero-viscosity fluid that exhibits a solid's crystalline structure. In 2004, Kim and Chan [26] claimed to have observed superfluidity in pressurized solid He^4, but the anomalies in rotational inertia were later attributed to

---

[10] Quantum field theorists find the relationship between energy and momentum and read m from the relation $E^2 = p^2 c^2 + m^2 c^4$.

elastic effects in the lattice (by the same authors who performed the original measurements). *Real* supersolidity has been observed in ultracold atoms trapped in optical lattices [27], and dipolar atoms [28] are also supersolids, as confirmed by the observation of quantized vortices [29].

5.6.1.2 *Superfluidity in isotropic fluids*

Consider a finite-temperature many-body system—a fluid—in equilibrium. If the fluid is in slow motion, the expectation value of the total momentum should be proportional to the velocity, and we can determine the amount of moving mass. The expectation value of the total momentum is

$$\langle \hat{\mathcal{P}} \rangle \equiv \mathrm{Tr}\left[\hat{\mathcal{P}} \frac{1}{Z} e^{-\beta \hat{\mathcal{H}}} \right]. \tag{5.137}$$

Question: how does a lab-frame observer 'see' the fluid that moves with velocity **v**?

To model the experimental setup of a fluid rotating in a bucket or a torus, we can think of fluid between two flat plates or fluid between moving walls. To begin with, wave functions are frame dependent.

Fluid moves, but with respect to what?

There is a frequently cited physical reason for modeling a *rotating* system instead of just using two inertial frames, i.e. a lab frame and the fluid rest frame. The term 'lab frame' gives away the dilemma: fluid properties cannot depend on motion w.r.t. to an arbitrarily chosen lab frame. There must be some kind of matter in the specific lab frame interacting with the fluid. In an earlier narrative, we spoke of frictional forces between the fluid particles and the walls. If we do not want to add such complications (which has been done in [30]), then we can distinguish the frames by saying that one frame is inertial, while the other is in accelerated motion. The role of the bucket is reduced to turning the fluid velocity vector; thus, in a big bucket, **v** is almost a constant vector.

Let us return to fluid below the superfluid transition temperature. If the fluid has a nonzero momentum, we know it is carried by the normal fluid, and the superfluid is at rest. The superfluid rest frame is a well-defined lab frame. In this case, there is no need to assume rotational motion to distinguish frames, for the fluid itself carries two frames. If the fluid occupies a finite volume, we cannot have 'phase separation' between the normal fluid and the superfluid, so the confinement must interfere with the fluid quite a lot. If we do not want the box walls to affect the results of our thought experiment, we can consider a large bulk of fluid in a huge box. If we add a small amount of momentum to a small cube of fluid (e.g. by pushing it) in the middle of the box, the normal fluid carries the momentum to the next cube via viscous processes. To reach a steady state, we should add momentum to the cube at the same rate at which it dissipates. Momentum enters from one end of the cube and leaves from the other, which begins to resemble the application of PBCs.

The formulation in non-isotropic fluids needs some modification (see reference [31]).

> The superfluid calculations presented here apply only to isotropic fluids.[11]

There will be no mention of rotating fluid in the partition function; it is just **fluid moving with a constant velocity v**. Actually, in superfluidity, only part of the fluid is moving, and what prevents the two-fluid system from separating into two fluids—which would obviously be unphysical—is **PBCs**. We have now identified the main ingredients of our model, and it is time for some math.

5.6.1.3 Galilean covariance of the Schrödinger equation

Consider particles moving with velocity **v**, and call the lab frame $\mathcal{F}$ (unprimed quantities) and the rest frame of the particles $\mathcal{F'}$ (primed quantities). The Galilean transformation between the frames is

$$\mathbf{r}' = \mathbf{r} - \mathbf{v}t \tag{5.138}$$

$$t' = t. \tag{5.139}$$

The momentum operator of particle i in the lab frame is $\hat{p}_i = -i\hbar\,\boldsymbol{\nabla}_i$, and in the fluid rest frame, the operator is

$$\hat{p}'_i = \hat{p}_i - m\mathbf{v}. \tag{5.140}$$

What happens to the wave function if we change frames? For simplicity, consider one particle in one dimension. The wave function is not an observable, so there is no physical reason why the wave functions in the two frames should be equal. Instead, we must require that the probability of finding the particle at any (x, t) remains unchanged,

$$|\psi'(x, t)|^2 = |\psi(x, t)|^2. \tag{5.141}$$

In other words, observers in both frames must agree upon on the probability that there is a particle at a certain point (x, t). Notice that the coordinate (x, t) is given in the lab frame; a point in space-time is unique, and we can choose whatever frame we please (using (x, t) here and (x', t') there would be too confusing). Since the probabilities must be the same, the wave functions in the two frames can differ by at most a phase factor,

$$\psi'(x, t) = e^{i\phi(x, t)}\psi(x, t), \tag{5.142}$$

where $\phi(x, t)$ is a real function. Heuristically, the phase factor can be found by considering a free particle with momentum $p = \hbar k$ and energy $E = \hbar\omega = p^2/(2m)$. The Fourier components of the wave function $\psi'(x, t)$ transform as

---

[11] For non-isotropic fluids, see [31].

$$e^{i(k'x-\omega't)} = e^{\frac{i}{\hbar}(p'x-\frac{(p')^2}{2m}t)} = e^{\frac{i}{\hbar}((p-mv)x-\frac{(p-mv)^2}{2m}t)} \tag{5.143}$$

$$= e^{\frac{i}{\hbar}(px-mv\,x-(\frac{p^2}{2m}+pv-\frac{1}{2}mv^2)t)} \tag{5.144}$$

$$= e^{\frac{i}{\hbar}(-mv\,x-(pv-\frac{1}{2}mv^2)t)}e^{\frac{i}{\hbar}(px-\frac{p^2}{2m}t)} = e^{\frac{i}{\hbar}(\frac{1}{2}mv^2t)}e^{\frac{i}{\hbar}(-mv\,x-pvt)}e^{kx-\omega t}, \tag{5.145}$$

so the wave function transforms as

$$\psi'(x,\,t) = e^{\frac{i}{\hbar}(\frac{1}{2}mv^2t)}e^{\frac{i}{\hbar}(-mv\,x-pvt)}\psi(x,\,t). \tag{5.146}$$

The first factor is an uninteresting, constant phase shift. The second factor is the position- and time-dependent phase we were looking for.

Without a detailed derivation, the result in a many-body system at $t = 0$ is the following:

> **The Galilean covariance**
> **of the Schrödinger equation requires that**
> $$\Psi'(\mathbf{x}) = e^{-\frac{i}{\hbar}m\mathbf{v}\cdot\mathbf{x}}\Psi(\mathbf{x}), \quad \mathbf{v}\cdot\mathbf{x}: =(\mathbf{v}\cdot\mathbf{r}_1, \mathbf{v}\cdot\mathbf{r}_2, \ldots, \mathbf{v}\cdot\mathbf{r}_N) \tag{5.147}$$
> $\Psi'(\mathbf{x})$ is the wave function in the moving frame
> $\Psi(\mathbf{x})$ is the wave function in the lab frame.

The Galilean covariance of the Schrödinger equation also shows that *wave functions must be complex-valued functions!* One simply cannot stick to real-valued functions because moving observers would disagree with that choice and attach a complex phase factor to the wave function.

The principle that a momentum transformation $\mathbf{p}' = \mathbf{p} - m\mathbf{v}$ is accompanied by a wave function that picks the phase $e^{-\frac{i}{\hbar}m\mathbf{v}\cdot\mathbf{r}}$ is a valuable finding. Electrons in a magnetic field with a vector potential $\mathbf{A}$ have a momentum $\mathbf{p}' = \mathbf{p} - e\mathbf{A}$, and if they move in a full circle around a flux tube with flux Φ, their wave function picks up the so-called *Aharonov–Bohm phase*, $e^{-\frac{i}{\hbar}e\Phi}$.

From the phase factor, we can deduce that the Galilean transformation (at $t = 0$) is the unitary operator

$$\hat{\mathcal{U}} = e^{-\frac{i}{\hbar}m\mathbf{v}\cdot\hat{\mathbf{x}}} \quad \text{Galilean transformation to moving frame, at } t = 0 \tag{5.148}$$
$$\mathbf{v}\cdot\hat{\mathbf{x}} = (\mathbf{v}\cdot\hat{\mathbf{r}}_1, \mathbf{v}\cdot\hat{\mathbf{r}}_2, \ldots, \mathbf{v}\cdot\hat{\mathbf{r}}_N),$$

where $\hat{\mathbf{x}}$ is the many-body position operator. For example, the single-particle momentum operator in the moving frame is, as expected,

$$\hat{p}' = \hat{\mathcal{U}}^\dagger\hat{p}\hat{\mathcal{U}} = e^{\frac{i}{\hbar}m\mathbf{v}\cdot\hat{\mathbf{r}}}\hat{p}e^{-\frac{i}{\hbar}m\mathbf{v}\cdot\hat{\mathbf{r}}} = \hat{p} - m\mathbf{v}. \tag{5.149}$$

We now use Hadamard's lemma (some call it Hausdorff's lemma),

$$e^{\hat{A}}\hat{B}e^{-\hat{A}} = \hat{B} + [\hat{A},\,\hat{B}] + \frac{1}{2!}[\hat{A},\,[\hat{A},\,\hat{B}]] + \cdots, \tag{5.150}$$

which applies if the commutators exist and the series converges. Start from $f(t) = e^{t\hat{A}}\hat{B}e^{-t\hat{A}}$ and find the Taylor expansion around $t = 0$; finally, set $t = 1$. The formula is closely related to the Baker–Campbell–Hausdorff formula given in equation (4.57). Insert $\hat{A} = \frac{i}{\hbar}m\mathbf{v} \cdot \hat{\mathbf{r}}$ and $\hat{B} = \hat{\mathbf{p}}$, and use the canonical commutator $[\hat{\mathbf{r}}, \hat{\mathbf{p}}] = i\hbar \times$ unit operator.

Why are both $\hat{\mathcal{U}}$ and $\hat{\mathcal{U}}^{\dagger}$ always involved in the transformation of operators? There are several reasons. One is that the commutation relations remain valid. In the Schrödinger picture, only wave functions change, $|\psi'\rangle = \hat{\mathcal{U}}|\psi\rangle$, so the expectation values transform as

$$\langle\psi'|\hat{\mathcal{A}}|\psi'\rangle = \langle\psi|\hat{\mathcal{U}}^{\dagger}\hat{\mathcal{A}}\hat{\mathcal{U}}|\psi\rangle = \langle\psi|\hat{\mathcal{A}}'|\psi\rangle. \tag{5.151}$$

5.6.1.4 From moving fluid to boundary conditions

Let us forget time and motion for a while and think of $\mathbf{v}$ as a *parameter*. For one particle, the 'new wave function' is related to the 'old wave function' by the relation

$$\psi'(\mathbf{r}) = e^{-\frac{i}{\hbar}m\mathbf{v}\cdot\mathbf{r}}\psi(\mathbf{r}), \tag{5.152}$$

where $\mathbf{v}$ is just some vector. We can interpret that $\psi'(\mathbf{r})$ means '$\psi(\mathbf{r})$ with a boundary condition.' The so-called **twisted boundary conditions (TBCs)** fit the bill. Consider a wave function that obeys

$$\psi_{\text{TBC}}(\mathbf{r} + \mathbf{L}) = e^{i\boldsymbol{\theta}\cdot\mathbf{L}}\psi_{\text{TBC}}(\mathbf{r}), \tag{5.153}$$

where $\boldsymbol{\theta}$ is the twist-angle vector. A ψ_{TBC} that satisfies the TBC is

$$\psi_{\text{TBC}}(\mathbf{r}) = e^{i\boldsymbol{\theta}\cdot\mathbf{r}}\psi_{\text{PBC}}(\mathbf{r}), \tag{5.154}$$

with any $\psi_{\text{PBC}}(\mathbf{r})$ that satisfies the **PBCs**,

$$\psi_{\text{PBC}}(\mathbf{r} + \mathbf{L}) = \psi_{\text{PBC}}(\mathbf{r}). \tag{5.155}$$

Proof:

$$\psi_{\text{TBC}}(\mathbf{r} + \mathbf{L}) \stackrel{(5.153)}{=} e^{i\boldsymbol{\theta}\cdot(\mathbf{r}+\mathbf{L})}\psi_{\text{PBC}}(\mathbf{r} + \mathbf{L}) \stackrel{(5.155)}{=} e^{i\boldsymbol{\theta}\cdot\mathbf{r}}e^{i\boldsymbol{\theta}\cdot\mathbf{L}}\psi_{\text{PBC}}(\mathbf{r}) \stackrel{(5.153)}{=} e^{i\boldsymbol{\theta}\cdot\mathbf{L}}\psi_{\text{TBC}}(\mathbf{r}). \tag{5.156}$$

The choice of $\boldsymbol{\theta} = -\frac{i}{\hbar}m\mathbf{v}$ is equivalent to motion with velocity $\mathbf{v}$. Instead of actually setting the PIMC paths in motion—which would be awkward in a description of a system in thermodynamical equilibrium—we use TBCs. The twist gives the phase factor, which turns out to give the quantity to observe in a PIMC simulation, while the rest boils down to using PBCs in the simulations.

Partition function and free energy

Keeping track of how the fluid properties are observed in different frames can be confusing (at least to me), so it is better to write them down:

1. Rest frame Hamiltonian $\hat{\mathcal{H}}_0$: the observer moving with the fluid describes the *intrinsic fluid properties* using the Hamiltonian $\hat{\mathcal{H}}_0$,

$$\hat{\mathcal{H}}_0 = \sum_{i=1}^{N} \frac{\hat{\boldsymbol{p}}_i^2}{2m} + \hat{\mathcal{V}}. \tag{5.157}$$

Obviously, $\hat{\mathcal{H}}_0$ has no $\mathbf{v}$, and neither does the density operator $\exp(-\beta\hat{\mathcal{H}}_0)/Z_0$.

2. Moving-frame Hamiltonian $\hat{\mathcal{H}}_{\mathbf{v}}$: this Hamiltonian describes how the lab-frame observer describes the fluid, its intrinsic properties, and all the energy terms fluid motion introduces:

$$\hat{\mathcal{H}}_{\mathbf{v}} = \sum_{i=1}^{N} \frac{(\hat{\boldsymbol{p}}_i - m\mathbf{v})^2}{2m} + \hat{\mathcal{V}} = \hat{\mathcal{H}}_0 - \sum_{i=1}^{N} \hat{\boldsymbol{p}}_i \cdot \mathbf{v} + \frac{1}{2} Nmv^2 \tag{5.158}$$

$$= \hat{\mathcal{H}}_0 - \hat{\mathcal{P}} \cdot \mathbf{v} + \frac{1}{2} Nmv^2. \tag{5.159}$$

The energies shift by the amount $\frac{1}{2}Nmv^2$, as expected. There is also an extra coupling term between momentum and velocity, and this is really useful because it gives us access to the momentum operator[12]

$$\boxed{\nabla_{\mathbf{v}} \hat{\mathcal{H}}_{\mathbf{v}} = -\hat{\mathcal{P}} + Nm\mathbf{v}.} \tag{5.160}$$

Getting $\hat{\mathcal{P}}$ from $\hat{\mathcal{H}}_{\mathbf{v}}$ is a great relief because it spares us from computing coordinate-space gradients $\sum_{i=1}^{N} \nabla_i$.

Let us compute the expectation value of the momentum of the fluid in the moving frame,[13]

$$\langle\hat{\mathcal{P}}\rangle_{\mathbf{v}} := \mathrm{Tr}\left[\hat{\mathcal{P}}\frac{1}{Z_{\mathbf{v}}}e^{-\beta\hat{\mathcal{H}}_{\mathbf{v}}}\right] = \frac{1}{Z_{\mathbf{v}}}\mathrm{Tr}[\hat{\mathcal{P}}e^{-\beta\hat{\mathcal{H}}_{\mathbf{v}}}]. \tag{5.161}$$

The moving-frame partition function is

$$Z_{\mathbf{v}} := \mathrm{Tr}[e^{-\beta\hat{\mathcal{H}}_{\mathbf{v}}}], \tag{5.162}$$

so

$$\nabla_{\mathbf{v}} Z_{\mathbf{v}} = \mathrm{Tr}[(-\beta(\nabla_{\mathbf{v}}\hat{\mathcal{H}}_{\mathbf{v}}))e^{-\beta\hat{\mathcal{H}}_{\mathbf{v}}}] = \beta\,\mathrm{Tr}[\hat{\mathcal{P}}e^{-\beta\hat{\mathcal{H}}_{\mathbf{v}}}] - \beta Nm\mathbf{v}\,\mathrm{Tr}[e^{-\beta\hat{\mathcal{H}}_{\mathbf{v}}}] \tag{5.163}$$

$$= \beta Z_{\mathbf{v}}\langle\hat{\mathcal{P}}\rangle_{\mathbf{v}} - Nm\mathbf{v}Z_{\mathbf{v}} \tag{5.164}$$

---

[12] The velocity-space gradient $\nabla_{\mathbf{v}}$ (another notation is $\partial/\partial\mathbf{v}$) operates on velocity vectors in the same way that the coordinate-space gradient $\nabla \equiv \nabla_{\mathbf{r}}$ operates on position vectors.

[13] Thankfully, this is 'only' a Galilean transformation. Apparently, the question of how temperature transforms in a Lorentz transformation is still an open question [32].

and this gives

$$\langle \hat{\mathcal{P}} \rangle_{\mathbf{v}} = \frac{1}{\beta} \boldsymbol{\nabla}_{\mathbf{v}}(\ln Z_{\mathbf{v}}) - Nm\mathbf{v}. \tag{5.165}$$

The free energy in the moving frame, $F_{\mathbf{v}}$, is defined in the usual way,

$$F_{\mathbf{v}} := -\frac{1}{\beta} \ln Z_{\mathbf{v}}, \tag{5.166}$$

so we find the formula

$$\boxed{\langle \hat{\mathcal{P}} \rangle_{\mathbf{v}} = -\boldsymbol{\nabla}_{\mathbf{v}}F_{\mathbf{v}} + Nm\mathbf{v}.} \tag{5.167}$$

Based on experiments, we anticipate that only the normal fluid moves. There are $\rho_n \Omega$ normal fluid particles in volume Ω; therefore,

$$\langle \hat{\mathcal{P}} \rangle_{\mathbf{v}} \equiv \rho_n \Omega \mathbf{v} = -\boldsymbol{\nabla}_{\mathbf{v}}F_{\mathbf{v}} + Nm\mathbf{v}. \tag{5.168}$$

Take the divergence,

$$\boldsymbol{\nabla}_{\mathbf{v}} \cdot (\rho_n \Omega \mathbf{v}) = \Omega(\boldsymbol{\nabla}_{\mathbf{v}}\rho_n) \cdot \mathbf{v} + \rho_n \Omega = -\nabla_{\mathbf{v}}^2 F_{\mathbf{v}} + Nm. \tag{5.169}$$

The term $\Omega(\boldsymbol{\nabla}_{\mathbf{v}}\rho_n) \cdot \mathbf{v}$ is first or higher order in $\mathbf{v}$, so it does not contribute in the limit $\mathbf{v} \to 0$. The normal fluid density is

$$\rho_n = -\frac{1}{\Omega}\lim_{\mathbf{v}\to 0} \nabla_{\mathbf{v}}^2 F_{\mathbf{v}} + \frac{Nm}{\Omega}. \tag{5.170}$$

Since $(Nm)/\Omega = \rho$ is the total fluid density, the superfluid density must be

$$\boxed{\rho_s = \frac{1}{\Omega}\lim_{\mathbf{v}\to 0} \nabla_{\mathbf{v}}^2 F_{\mathbf{v}} = -\frac{1}{\Omega}\frac{1}{\beta}\lim_{\mathbf{v}\to 0} \nabla_{\mathbf{v}}^2 \ln Z_{\mathbf{v}}.} \tag{5.171}$$

We have found a microscopic definition for the superfluid density. Next, we need to figure out how to measure it in PIMC, and for that purpose we need to replace the moving frame with TBCs.

Remark: On the other hand,

$$Z_{\mathbf{v}} = e^{-\beta\frac{1}{2}Nmv^2} \int d\mathbf{x}\langle \mathbf{x}|e^{-\beta(\hat{\mathcal{H}}_0 - \hat{\mathcal{P}}\cdot\mathbf{v})}|P\mathbf{x}\rangle. \tag{5.172}$$

If we saw this expression in DMC, we would say the integrand is a Green's function with a momentum-dependent drift $\hat{D} = -\hat{\mathcal{P}} \cdot \mathbf{v}$. It is an interesting evolution, looping from positions $\mathbf{x}$ to permutations $P\mathbf{x}$.

Fluid rest frame with TBC

We showed earlier that TBC with the twist angle

$$\theta = -\frac{i}{\hbar}m\mathbf{v} \tag{5.173}$$

and PBCs is equivalent to the Galilean transformation between the fluid rest frame and the moving frame:

$$\boxed{Z_{\mathbf{v}} \equiv Z_0 \text{ and TBCs} \equiv Z_0 \text{ and phase factor and PBCs.}} \tag{5.174}$$

We obtain

$$Z_{\mathbf{v}} = Z_0|_{\text{TBC}} = \int d\mathbf{x}\langle\mathbf{x}|\epsilon^{-\beta\hat{\mathcal{H}}_0}|P\mathbf{x}\rangle|_{\text{TBC}} = \int d\mathbf{x}\sum_n e^{-\beta E_n}\phi_n^*(\mathbf{x})\phi_n(P\mathbf{x})|_{\text{TBC}} \tag{5.175}$$

$$= \int d\mathbf{x}\sum_n e^{-\beta E_n}e^{\frac{im}{\hbar}\mathbf{v}\cdot\mathbf{x}}\phi_n^*(\mathbf{x})e^{-\frac{im}{\hbar}\mathbf{v}\cdot P\mathbf{x}}\phi_n(P\mathbf{x})|_{\text{PBC}}, \tag{5.176}$$

where we used the convention adopted in this book, namely that $P\mathbf{x}$ implies a sum over permutations with the appropriate boson/fermion sign factor. The equation simplifies to

$$Z_{\mathbf{v}} = \int d\mathbf{x}e^{-\frac{im}{\hbar}\mathbf{v}\cdot(P\mathbf{x}-\mathbf{x})}\sum_n e^{-\beta E_n}\phi_n^*(\mathbf{x})\phi_n(P\mathbf{x})|_{\text{PBC}} \tag{5.177}$$

$$= \int d\mathbf{x}e^{-\frac{im}{\hbar}\mathbf{v}\cdot(P\mathbf{x}-\mathbf{x})}\langle\mathbf{x}|\epsilon^{-\beta\hat{\mathcal{H}}_0}|P\mathbf{x}\rangle|_{\text{PBC}} \tag{5.178}$$

$$= Z_0\left\langle e^{-\frac{im}{\hbar}\mathbf{v}\cdot(P\mathbf{x}-\mathbf{x})}\right\rangle_0|_{\text{PBC}}. \tag{5.179}$$

Only the phase factor contains $\mathbf{v}$, so the velocity-space gradients are easy to compute,

$$\boldsymbol{\nabla}_{\mathbf{v}}Z_{\mathbf{v}} = Z_0\left\langle -\frac{im}{\hbar}(P\mathbf{x}-\mathbf{x})e^{-\frac{im}{\hbar}\mathbf{v}\cdot(P\mathbf{x}-\mathbf{x})}\right\rangle_0|_{\text{PBC}} \tag{5.180}$$

$$\nabla_{\mathbf{v}}^2 Z_{\mathbf{v}} = Z_0\left\langle -\frac{m^2}{\hbar^2}(P\mathbf{x}-\mathbf{x})^2 e^{-\frac{im}{\hbar}\mathbf{v}\cdot(P\mathbf{x}-\mathbf{x})}\right\rangle_0|_{\text{PBC}}. \tag{5.181}$$

The small-$\mathbf{v}$ limits are

$$\lim_{\mathbf{v}\to 0} Z_{\mathbf{v}} = Z_0 \tag{5.182}$$

$$\lim_{\mathbf{v}\to 0} \boldsymbol{\nabla}_{\mathbf{v}}Z_{\mathbf{v}} = Z_0\left\langle -\frac{im}{\hbar}(P\mathbf{x}-\mathbf{x})\right\rangle_0|_{\text{PBC}} \tag{5.183}$$

$$\lim_{\mathbf{v}\to 0} \nabla_{\mathbf{v}}^2 Z_{\mathbf{v}} = Z_0\left\langle -\frac{m^2}{\hbar^2}(P\mathbf{x}-\mathbf{x})^2\right\rangle_0|_{\text{PBC}}. \tag{5.184}$$

We have

$$\nabla_{\mathbf{v}}^2 F_{\mathbf{v}} = -\frac{1}{\beta}\nabla_{\mathbf{v}}^2\ln Z_{\mathbf{v}} = -\frac{1}{\beta}\boldsymbol{\nabla}_{\mathbf{v}}\cdot\left(\frac{\boldsymbol{\nabla}_{\mathbf{v}}Z_{\mathbf{v}}}{Z_{\mathbf{v}}}\right) = -\frac{1}{\beta}\left[\frac{\nabla_{\mathbf{v}}^2 Z_{\mathbf{v}}}{Z_{\mathbf{v}}} - \left(\frac{\boldsymbol{\nabla}_{\mathbf{v}}Z_{\mathbf{v}}}{Z_{\mathbf{v}}}\right)^2\right], \tag{5.185}$$

From now on, we can drop the subscript 0. Using equation (5.171), the superfluid density is

$$\rho_s = -\frac{1}{\Omega}\lim_{\mathbf{v}\to 0}\nabla^2_{\mathbf{v}}\ F_{\mathbf{v}} = \frac{m^2}{\Omega\beta\hbar^2}[\langle(P\mathbf{x}-\mathbf{x})^2\rangle - \langle P\mathbf{x}-\mathbf{x}\rangle^2]|_{\text{PBC}}. \qquad (5.186)$$

The quantity in the brackets has the form of a *fluctuation* in statistical physics.[14] Let us define *dimensionless* winding and a winding number (*Pi* means permutations of the particle index *i*).

$$\boxed{\mathbf{W}: =\frac{1}{L}(P\mathbf{x}-\mathbf{x}) \equiv \frac{1}{L}\sum_{i=1}^{N}(\mathbf{x}_{Pi}-\mathbf{x}_i) \qquad \text{winding}} \qquad (5.187)$$

$$\boxed{W: =|\mathbf{W}| \qquad \text{winding number.}} \qquad (5.188)$$

The factor of $1/L$ is commonly added to cancel dimensions; L is the periodicity of the PBCs. Based on equation (5.186), the superfluid density is proportional to the fluctuation of the winding number.

Remark: Ceperley [12] defines winding for continuous paths $\mathbf{r}_i(\tau)$ as

$$\mathbf{W} = \sum_{i=1}^{N}\int_0^{\beta}d\tau\frac{d\mathbf{r}_i(\tau)}{d\tau}, \qquad (5.189)$$

which relates winding to a net path velocity.

Because of symmetry, isotropic fluids have $\langle\mathbf{W}\rangle = 0$. The connection between superfluid density and winding is

$$\boxed{\rho_s = \frac{m^2 L^2}{\Omega\beta\hbar^2}<\mathbf{W}^2>|\text{PBCs.}} \qquad (5.190)$$

In PIMC simulation in a 3D cubic box of sides L, we can average the three directions to get the superfluid density and the **superfluid fraction,**

$$\boxed{\rho_s = \frac{m^2 L^2}{\Omega 3\beta\hbar^2}<\mathbf{W}^2>|\text{PBCs, averaged over three directions}} \qquad (5.191)$$

$$\boxed{\frac{\rho_s}{\rho} = \frac{m L^2}{3\beta\hbar^2 N}<\mathbf{W}^2>|\text{PBCs, averaged over three directions.}} \qquad (5.192)$$

---

[14] The fluctuation of a stochastic variable A is commonly given by the variance,

$$\sigma_A: =\langle A^2\rangle - \langle A\rangle^2 \equiv \langle(A - \langle A\rangle)^2\rangle.$$

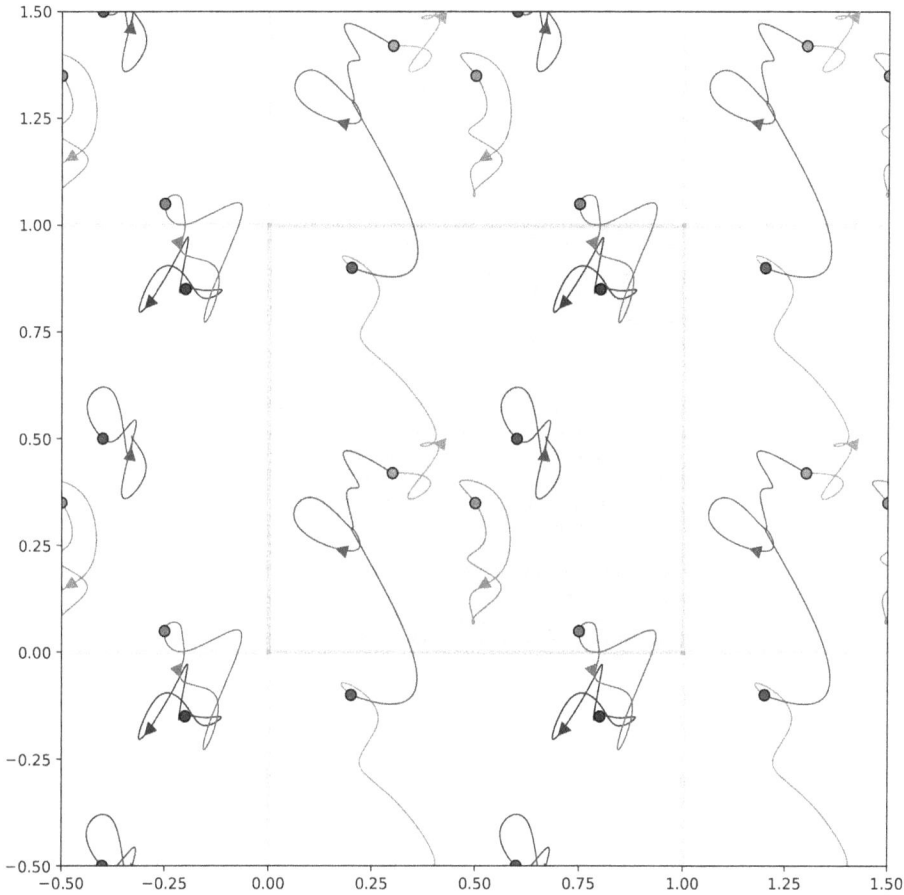

Figure 5.16. Six particles; two of them (red and blue) are in an exchange loop that winds across the whole simulation box. The arrows point in the direction of increasing imaginary time. Only the winding path is endless and has nonzero **W**; all closed loops give zero winding.

The beauty of the winding number is that it is a **topological property of the paths**. A loop with enough particles may reach all the way through the PBC simulation box, and figure 5.16 shows an example in a six-particle system.

The two-dimensional paths in figure 5.16 give $\mathbf{W} = (0, 1)$; the PBC box size was chosen to be $L = 1$. Figure 5.17 shows the same scenario separated in the x- and y-directions. If you draw a vertical line in the plots shown in figure 5.17 and count how many more times world lines cross the line from left to right compared with right to left, you get zero in the x-direction and unity in the y-direction. Winding is a **topological property of the paths: you get the same result no matter where you draw the vertical line**.

A world line that returns to its starting position has zero net displacement. Two world lines in an exchange loop swap positions in the x-direction, so that one world line has a net displacement of Δx while the other has $-\Delta x$, which leaves zero in the x-component of **W**. In the y-direction, the blue world line ends at the starting point

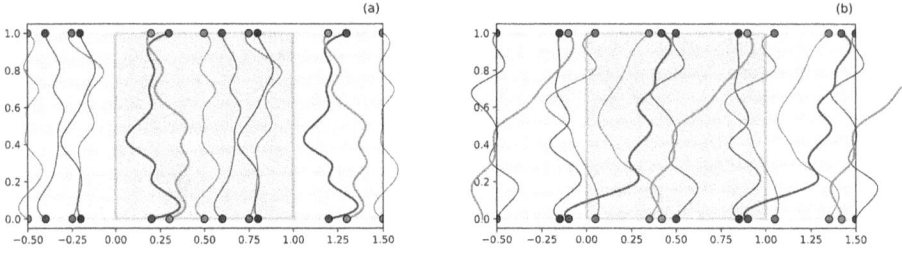

Figure 5.17. Paths shown in figure 5.16 viewed in the x- (a) and y-directions (b). In the x-direction, the red and blue paths (thick lines) look like ordinary exchange loops, while in the y-direction, they wind endlessly across the simulation box like stripes on a candy cane.

of the red world line, but the red world line ends at the starting point of the blue world line in the periodic image on the right! If you follow the blue–red exchange loop, the path has a net displacement of one in the simulation box.

The superfluid density formula, equation (5.192), suffers from a finite-size effect: the smaller L is, the more paths wind and are falsely assigned as belonging to the superfluid. The best one can do is to repeat the simulation with more particles to get a limiting value for the bulk ratio ρ_s/ρ. A finite-size scaling analysis shows that in 3D boson systems, the superfluid fraction should scale as [33]

$$\frac{\rho_s(t, L)}{\rho} \sim t^v, \ t: = \frac{T - T_c}{T_c}. \tag{5.193}$$

In PIMC, an ideal Bose gas can also produce a nonzero winding and superfluidity as a finite-size artifact.

5.6.1.5 PBCs and paths

Before continuing, let us see how PBCs affect paths in the partition function. Consider a simulation in a box with sides L and loops such as those in figure 5.16. Looking at paths within the box, one finds that paths crossing the boundaries continue from the other end of the box. If used directly, such jumps would play havoc in evaluating the vectors $\mathbf{x}_m - \mathbf{x}_{m-1}$.

Kinetic matrix elements

The partition function in the PA is

$$Z(\beta) = \int \prod_{m=1}^{M} d\mathbf{x}_m \times \prod_{m=1}^{M} \langle \mathbf{x}_m | e^{-\tau \hat{T}} | \mathbf{x}_{m-1} \rangle \times \prod_{m=1}^{M} e^{-\tau V(\mathbf{x}_m)} \tag{5.194}$$

$$= \int \prod_{m=1}^{M} d\mathbf{x}_m \times \prod_{m=1}^{M} (4\pi\lambda\tau)^{-3N/2} e^{-\frac{(\mathbf{x}_m - \mathbf{x}_{m-1})^2}{4\lambda\tau}} \times \prod_{m=1}^{M} e^{-\tau V(\mathbf{x}_m)}. \tag{5.195}$$

Under PBCs, the second line is also an approximation. In the DMC chapter, we solved the 1D particle-in-a-box kinetic matrix elements (see equations (4.292) and (4.293)) using the principles outlined in appendix B. In a 3D box with PBCs, the exact matrix elements are [12]

$$\langle \mathbf{x}_m | e^{-\tau \hat{T}} | \mathbf{x}_{m-1} \rangle = L^{-3N} \sum_{\mathbf{n} \in \mathbb{Z}^{3N}} e^{-\frac{4\pi^2 \lambda \tau}{L^2} \mathbf{n}^2 - i2\pi \mathbf{n} \cdot (\frac{\mathbf{x}_m - \mathbf{x}_{m-1}}{L})} \tag{5.196}$$

$$= L^{-3N} \vartheta \left(\frac{\mathbf{x}_m - \mathbf{x}_{m-1}}{L}, i\frac{4\pi \lambda \tau}{L^2} \right), \tag{5.197}$$

where ϑ is the Jacobi theta function. If

$$\lambda \tau / L^2 \ll 1, \tag{5.198}$$

then the $\mathbf{n}^2$ term is very close to a continuous Gaussian; the sum is approximately a Fourier transform of a Gaussian, which, as expected, gives the free-particle kinetic matrix elements.

The reason for mentioning the condition (5.198) here is that with the CA, one can use a larger τ than with the primitive action. A justified question is whether the condition is still satisfied, or should one use the accurate PBC kinetic matrix elements (5.197)? One usually makes the pragmatic decision that if satisfying the condition (5.198) is at risk, it is safer to keep τ smaller than necessary for the CA. The free-particle kinetic matrix elements are something one does not give up lightly.

Continuous paths

The spring term has the sum $\sum_{m=1}^{M}(\mathbf{x}_m - \mathbf{x}_{m-1})^2$. It is imperative to evaluate the sum by following *continuous paths*, just as we would do without PBCs. By 'continuous path,' we mean here a path from bead to bead that would become a continuous curve $\mathbf{x}(\tau)$ in the limit $M \to \infty$.

First, we need an anchor point, i.e. a point that uniquely defines the path we start to follow. A common choice is to anchor $\mathbf{x}_1$ and give the rest of the positions relative to it [12]:

$$\mathbf{x}_2 = \mathbf{x}_1 + \boldsymbol{\delta}_2, \ \mathbf{x}_3 = \mathbf{x}_2 + \boldsymbol{\delta}_3 = \mathbf{x}_1 + \boldsymbol{\delta}_2 + \boldsymbol{\delta}_3, \tag{5.199}$$

and so on,

$$\boldsymbol{\delta}_m := \mathbf{x}_m - \mathbf{x}_{m-1} \tag{5.200}$$

$$\mathbf{x}_m = \mathbf{r}_1 + \sum_{k=2}^{m} \boldsymbol{\delta}_k. \tag{5.201}$$

The sum is evaluated along continuous paths,

$$\sum_{i=1}^{M}(\mathbf{x}_m - \mathbf{x}_{m-1})^2 = \sum_{i=1}^{M}(\boldsymbol{\delta}_m)^2. \tag{5.202}$$

With small enough τ, there should not be any paths that jump across the box between slices $m-1$ and m, which is equivalent to saying that none of the displacements $\boldsymbol{\delta}_m$ gets a sudden jolt due to PBCs. Paths can still cross periodic boundaries, just 'not too fast.'

Minimum image convention

The minimum image convention is also widely used in VMC, DMC, and molecular dynamics simulations. A distance vector (x, y, z) is converted to the minimum image distance vector (x', y', z') using the transformation

$$
\begin{aligned}
x' &= x - L\,\mathrm{round}(x/L) \\
y' &= y - L\,\mathrm{round}(y/L) \\
z' &= z - L\,\mathrm{round}(z/L),
\end{aligned}
\tag{5.203}
$$

for a cubic simulation box with sides L.[15] The purpose of using the minimum image is to avoid counting potentials or other observables that are more influenced by the periodic image of the bead than the bead itself. This cuts the 'safe' distance to $L/2$; anything beyond that is closer to a periodic image. Of course, one should not count the interaction of a bead with itself either, but that is already a distance L apart.[16]

Should we wrap coordinates to the simulation box or not?

PBCs do not necessarily mean that all bead coordinates are explicitly folded back to the simulation box if they cross the periodic boundaries. One can let beads wander about and calculate the potential energy using the minimum image convention. This way, paths remain continuous. One can also use temporary coordinates folded to the simulation box in worm moves, especially in swap updates, to measure the density or for visualization purposes.

The question of why it is not always safe to deduce continuous-path coordinates from periodically folded coordinates is a subtle one. Unfolding a path is customarily done like this: start from a bead position in the PBC simulation box, calculate the minimum image vector in the direction of the next bead, and keep going in the minimum image direction. For long paths, adding displacements becomes numerically inaccurate. A worm head jumping far and crossing a PBC boundary is also ambiguous, although one should only allow small, local changes to take place. If the thermal wavelength is of the order of the size of the system, then you are in trouble. Eventually, this happens at some low temperature, since PIMC codes can handle only a relatively small number of particles N, and the size of the simulation box L is tied to the (target) density by $\rho = N/L^d$.

The use of folded coordinates, even for noninteracting particles, for which the Green's function is exact *in free space*, cannot be simulated accurately under PBCs at low temperatures—at least not without using a fairly large number of slices M. Recently, Spada, Giorgini, and Pilati [34] demonstrated how, with unfolded coordinates, even $M = 1$ is enough to exactly simulate noninteracting particles. They wrote the canonical partition function in PBCs in the form (in the notation used here)

---

[15] In Julia (and in C++), rounding can be done with round(), while in Fortran, the corresponding function is anint(). The function floor() does not preserve directions and should not be used to get minimum image directions.

[16] An exchange loop has two or more particles on the same time slice; hence, they do interact, and the loop has something that could be called 'loop self-interaction.'

$$Z_N = \frac{1}{N!}\sum_P \sum_{\mathbf{W}} \int d\mathbf{x} \langle \mathbf{x}; 0 | e^{-\beta\hat{\mathcal{H}}} | P\mathbf{x}; \mathbf{W}\rangle \tag{5.204}$$

$$= \frac{1}{N!}\sum_P \int \prod_{m=1}^{M} d\mathbf{x}_m \rho(\mathbf{x}_m, \mathbf{x}_{m+1}; \tau), \quad \mathbf{x}_M = P\mathbf{x}_0 + \mathbf{W}L. \tag{5.205}$$

Here, $\mathbf{W}$ is a dN-dimensional integer vector that enumerates the periodic images (not to be confused with winding). Paths start at $\mathbf{x}_0$, and after an imaginary time β, the path positions $\mathbf{x}_M$ are permutations of $\mathbf{x}_0$ plus the shift $\mathbf{W}L$ to the periodic image after M slices. The points $\mathbf{x}_0$ are inside the simulation box, while the rest of the path can reside in another periodic image, spanning the entire space.

Interaction and the superfluid transition temperature

As mentioned earlier, ideal bosons have BEC but no superfluidity. Long exchange loops indicate that the average interparticle distance is of the order of the thermal wavelength λ_T given in equation (5.27). Ideal bosons tend to be close to each other. There are empty spaces, and one is less likely to find another boson within the optimal λ_T distance for making long exchange paths. In contrast to ideal bosons, hard-sphere bosons and bosons with repulsive interactions fill the space more evenly and are therefore better suited for superfluidity [35]. In short, interactions can *increase* the superfluid transition temperature.

5.6.2 Bose–Einstein condensation

BEC is the macroscopic occupation of a single-particle state. The quantity central to BEC is the **one-body density matrix (OBDM)**,

$$\rho_1(\mathbf{r}', \mathbf{r}) := \langle \hat{\Psi}^\dagger(\mathbf{r}')\hat{\Psi}(\mathbf{r})\rangle = \frac{\text{Tr}\,[e^{-\beta\hat{\mathcal{H}}}\hat{\Psi}^\dagger(\mathbf{r}')\hat{\Psi}(\mathbf{r})]}{\text{Tr}\,[e^{-\beta\hat{\mathcal{H}}}]}, \tag{5.206}$$

where the field operator $\hat{\Psi}^\dagger(\mathbf{r})$ creates a particle at $\mathbf{r}$, and $\hat{\Psi}(\mathbf{r}')$ annihilates a particle at $\mathbf{r}'$. All necessary information is in the density matrix $\langle \mathbf{x}' | e^{-\beta\hat{\mathcal{H}}} | \mathbf{x}\rangle$, and the OBDM is found by integrating away all but one particle's coordinates,

$$\rho_1(\mathbf{r}', \mathbf{r}) = \frac{\Omega}{Z} \int d\mathbf{r}_2 ...d\mathbf{r}_N \langle \mathbf{r}', \mathbf{r}_2, ..., \mathbf{r}_N | e^{-\beta\hat{\mathcal{H}}} | \mathbf{r}, \mathbf{r}_2, ..., \mathbf{r}_N\rangle, \tag{5.207}$$

where Ω is the volume. Actually, the OBDM is $\rho_1(\mathbf{r}', \mathbf{r}; \beta)$, but for brevity, the temperature dependence is often left implicit. The Fourier transform of the OBDM is the **momentum distribution**,

$$n_{\mathbf{k}} = \int d\mathbf{r}d\mathbf{r}' e^{-i\mathbf{k}\cdot(\mathbf{r}-\mathbf{r}')} \rho_1(\mathbf{r}', \mathbf{r}). \tag{5.208}$$

In an isotropic system, the OBDM depends only on the distance $r := |\mathbf{r} - \mathbf{r}'|$,

$$\rho_1(r) = \frac{\Omega}{Z} \int d\mathbf{r}_2 ...d\mathbf{r}_N \langle \mathbf{x} | e^{-\beta\hat{\mathcal{H}}} | \mathbf{x}\rangle \delta(r - |\mathbf{r}_1 - \mathbf{r}_2|), \tag{5.209}$$

where Ω is the volume. If a system has BEC, the OBDM has **off-diagonal long-range order (ODLRO)**, meaning there is a finite expectation value associated with moving a particle from $\mathbf{r}'$ to a faraway position $\mathbf{r}$,

$$\lim_{|\mathbf{r}-\mathbf{r}'|\to\infty} \rho_1(\mathbf{r}', \mathbf{r}) \neq 0, \tag{5.210}$$

In the momentum distribution $n_{\mathbf{k}}$, this shows up as a sharp peak at wave vector $\mathbf{k} = 0$. In experiments on trapped bosons, the release of the trap lets most bosons fly away, while those with $k = 0$ stay where they were for a while and show up as a peak in the density [36].

With the occupation of the zero-momentum state n_0, the condensate fraction is

$$\frac{n_0}{n} = \frac{1}{N} \int d\mathbf{r} \lim_{|\mathbf{r}-\mathbf{r}'|\to\infty} \rho_1(\mathbf{r}', \mathbf{r}). \tag{5.211}$$

In finite systems, such as bosons in a harmonic trap, the limit $|\mathbf{r} - \mathbf{r}'| \to \infty$ is not meaningful. The OBDM contains information about single-particle states and their occupations. In the eigenvalue equation

$$\int d\mathbf{r} \rho_1(\mathbf{r}', \mathbf{r})\varphi_i(\mathbf{r}) = \lambda_i \varphi_i(\mathbf{r}'), \tag{5.212}$$

the $\varphi_i(\mathbf{r})$ are single-particle wave functions, and the eigenvalues λ_i are the occupations of the states.[17] The occupation of the zero-momentum state is n_0. In a generalized definition, ODLRO is present if one of the eigenvalues λ_i is of the order of N [37]. Using the eigenvalue equation, the spectral expansion of OBDM is

$$\rho_1(\mathbf{r}', \mathbf{r}) = \sum_i \lambda_i \varphi_i^*(\mathbf{r}')\varphi_i(\mathbf{r}), \tag{5.213}$$

where i runs through eigenstates, discrete or continuous.

Measuring OBDM in PIMC

The paths in the partition function are loops, so particles never jump from one point to another, and there is nothing in Z that gives the OBDM directly. However, the very operations of particle creation and annihilation are present in the worm algorithm (see section 5.4): the head and tail of a worm are $\hat{\Psi}^\dagger(\mathbf{r})$ and $\hat{\Psi}(\mathbf{r}')$.

In PIMC, the OBDM can be measured by accumulating data about the worm head and tail coordinates as $\mathbf{r}'$ and $\mathbf{r}$ in the same time slice.

In the canonical ensemble, the head and tail are never in the same slice, but one can do a 'fake close' of a worm by sampling a head at a chosen distance $\mathbf{r}$ from a tail,

---

[17] OBDM is self-adjoint and positive definite; hence, the occupations λ_i are positive, as they should be.

generating the path, and evaluating $\rho_1(\mathbf{r}) \propto \langle e^{\Delta S} \rangle$ for that sample. After this operation, the head is moved back and the generated path is erased.

The expectation value $\langle \rho_1(\mathbf{r}', \mathbf{r}) \rangle_{\text{worm}}$ is measured from paths in the 'worm partition function,' which is a combination of loops in Z and open worms in G,

$$Z_{\text{worm}} = Z + CG, \tag{5.214}$$

where worm opening and closing are governed by the parameter C that appeared in equations (5.99) and (5.105). The ratio of MC steps with loops (the 'Z-sector') to those with an open worm (the 'G-sector') is

$$\frac{N_G}{N_Z} = C\frac{G}{Z}, \tag{5.215}$$

and the normalized OBDM is

$$\rho_1(\mathbf{r}', \mathbf{r}) = \frac{1}{C} \langle \rho_1(\mathbf{r}', \mathbf{r}) \rangle_{\text{worm}}. \tag{5.216}$$

If C is small, there are fewer open worms, and less data is accumulated in $\langle \rho_1(\mathbf{r}', \mathbf{r}) \rangle_{\text{worm}}$. The automatically correct normalization makes it possible to reliably tell whether ODLRO is present, since it eliminates spurious cases of $\lim_{|\mathbf{r}-\mathbf{r}'| \to \infty} \rho_1(\mathbf{r}', \mathbf{r}) > 0$ caused by having the whole curve at a level that is slightly too high.

5.6.3 Energy estimators in PIMC

Our next task is to evaluate the finite-temperature expectation value of the energy $E(\beta)$ using the path integral expression for the partition function given in equation (5.10). There are many energy estimators, which all give the same $E(\beta)$ but a different variance. The frequently used ones are as follows.

- The thermodynamic energy estimator:

$$\boxed{E(\beta) = -\frac{1}{Z}\frac{\partial Z}{\partial \beta}.} \tag{5.217}$$

  We have access to Z, so we can also compute the expectation values of pressure, kinetic energy, potential energy, etc. using the standard expressions of thermodynamics.

- The direct or Hamiltonian energy estimator:

$$\boxed{E(\beta) = \frac{\text{Tr}[\hat{\mathcal{H}}e^{-\beta\hat{\mathcal{H}}}]}{\text{Tr}[e^{-\beta\hat{\mathcal{H}}}]}.} \tag{5.218}$$

- The virial estimator has no simple form.

We shall concentrate on the thermodynamic and virial estimators.

5.6.3.1 Thermodynamic energy estimator

The thermodynamic energy estimator $\langle E \rangle_{Th}$ has a serious shortcoming, related to the balance between two kinetic terms. For reference, a classical noninteracting monatomic gas has the kinetic energy

$$\langle T \rangle = \frac{3}{2} N k_B T = \frac{3N}{2\beta} \qquad \text{classical kinetic energy.} \tag{5.219}$$

The partition function of the primitive action is equation (5.32), repeated here:

$$Z(\beta) = \int \prod_{m=1}^{M} d\mathbf{x}_m \times \prod_{m=1}^{M} (4\pi\lambda\tau)^{-3N/2} e^{-\frac{(\mathbf{x}_m - \mathbf{x}_{m+1})^2}{4\lambda\tau}} \times \prod_{m=1}^{M} e^{-\tau V(\mathbf{x}_i)}. \tag{5.220}$$

Take the derivative w.r.t. β $(\tau: = \beta/M)$.

Factor by factor,

$$\frac{\partial}{\partial\beta} (4\pi\lambda\tau)^{-3NM/2} = -\frac{3NM}{2}(4\pi\lambda\tau)^{-3NM/2-1} 4\pi\lambda\frac{1}{M} = (4\pi\lambda\tau)^{-3NM/2} \times \left[-\frac{3N}{2\tau}\right] \tag{5.221}$$

$$\frac{\partial}{\partial\beta} e^{-\sum_{m=1}^{M}\frac{(\mathbf{x}_m - \mathbf{x}_{m+1})^2}{4\lambda\tau}} = e^{\sum_{m=1}^{M}-\frac{(\mathbf{x}_m - \mathbf{x}_{m+1})^2}{4\lambda\tau}} \times \left[\frac{1}{M}\sum_{m=1}^{M}\frac{(\mathbf{x}_m - \mathbf{x}_{m+1})^2}{4\lambda\tau^2}\right] \tag{5.222}$$

$$\frac{\partial}{\partial\beta} e^{-\tau\sum_{m=1}^{M} V(\mathbf{x}_i)} = e^{-\tau\sum_{m=1}^{M} V(\mathbf{x}_i)} \times \left[-\frac{1}{M}\sum_{m=1}^{M} V(\mathbf{x}_i)\right]. \tag{5.223}$$

Thus,

$$\frac{\partial Z}{\partial\beta} \approx -\frac{3N}{2\tau}Z(\beta) - \int d\mathbf{x}_1, \ldots, d\mathbf{x}_M \left[-\frac{1}{M}\sum_{m=1}^{M}\frac{(\mathbf{x}_m - \mathbf{x}_{m+1})^2}{4\lambda\tau^2}\right] e^{-\sum_{m=1}^{M} S_m^{\text{prim. approx.}}}$$
$$+ \int d\mathbf{x}_1, \ldots, d\mathbf{x}_M \left[-\frac{1}{M}\sum_{m=1}^{M} V(\mathbf{x}_i)\right] e^{-\sum_{m=1}^{M} S_m^{\text{prim. approx.}}}. \tag{5.224}$$

Thermodynamic energy estimator, primitive action Z (5.32)

$$E(\beta): = -\frac{1}{Z}\frac{\partial Z}{\partial\beta} \tag{5.225}$$

$$= \frac{3N}{2\tau} - \left\langle \frac{1}{M}\sum_{m=1}^{M}\frac{(\mathbf{x}_m - \mathbf{x}_{m+1})^2}{4\lambda\tau^2} \right\rangle + \left\langle \frac{1}{M}\sum_{m=1}^{M} V(\mathbf{x}_i) \right\rangle. \tag{5.226}$$

Figure 5.18. An initial configuration that gives a finite potential energy. Alas, the thermal estimator of the kinetic term is M times the classical kinetic energy, so *in the thermodynamic estimator, this is a very high total energy configuration.* Moreover, if you improve the accuracy of the PA by increasing M, the initial energy increases!

The first kinetic term is far too large:

$$\frac{3N}{2\tau} = \frac{3NM}{2\beta} = M \times \text{classical kinetic energy.} \qquad (5.227)$$

If you start a PIMC simulation with M identical time slices ($\mathbf{x}_m = \mathbf{x}_{m-1}$ for all m), as shown in figure 5.18, this *is* the initial value of the kinetic energy. Only after PIMC has run for a while do the world lines wiggle enough to make the second kinetic term, the 'spring energy,' nonzero, and the excess kinetic energy cancels out. To be fair, a straight world-line configuration is, in fact, extremely rare. All imaginary-time loops have collapsed to points, so we would be initializing the simulation with a configuration that is appropriate only at a very high temperature.

At large M (small τ), the kinetic terms fluctuate a lot [38, 39], and we conclude that, in PIMC, the combination (PA + thermodynamic energy estimator) is poor even for noninteracting particles. The CA (5.134) suffers from the same thermodynamic kinetic energy problem as the primitive action; the only differences are the variable time step and effective potentials. For M slices, CA has $\beta = M\tau/3$, and after a calculation similar to PA, one finds:

Thermodynamic energy estimator, CA Z (5.134)

$$E(\beta): = -\frac{1}{Z}\frac{\partial Z}{\partial \beta} = -\frac{1}{Z}\frac{\partial Z}{\partial \tau}\frac{3}{M} \qquad (5.228)$$

$$= \frac{3NM}{2\beta} - \left\langle \frac{1}{\beta}\sum_{m=1}^{M}\frac{(\mathbf{x}_m - \mathbf{x}_{m+1})^2}{4\lambda\tau t_m} \right\rangle + \left\langle \frac{\partial}{\partial \beta}\sum_{m=1}^{M}(\tau \tilde{V}_m(\mathbf{x};\tau)) \right\rangle. \qquad (5.229)$$

5.6.3.2 Virial energy estimator

Here, the rationale is that the potential energy typically fluctuates less than the kinetic energy. On a basic level, the virial theorem in 1D gives the total energy in the form

$$E = \underbrace{\frac{1}{2}x\frac{\partial V}{\partial x}}_{\text{virial term}} + V(x). \qquad (5.230)$$

In classical theory, the virial term is proportional to the work done by the force $-\frac{\partial V}{\partial x}$ over a distance x. In this form, the virial term is not translationally invariant, so it is applicable only for bound systems.[18]

Herman, Bruskin, and Berne used the virial theorem to derive an improved PIMC energy estimator [38]. One form of the virial energy estimator (3D, PA) is [3],

$$E(\beta) = \left\langle \frac{3N}{2\beta} + \frac{1}{M}\sum_{m=1}^{M}\sum_{j=1}^{N}\frac{1}{2}(\mathbf{r}_{j,m} - \mathbf{r}_j^*) \cdot \frac{\partial V(\mathbf{x}_m)}{\partial \mathbf{r}_{j,m}} + \frac{1}{M}\sum_{m=1}^{M}V(\mathbf{x}_m) \right\rangle. \qquad (5.231)$$

We will come back to the derivation soon, but I would just like to point out a few things. There is no large-M problem, as the first term is the classical kinetic energy without any M-dependence. The virial energy estimator uses a pivot, or a **reference point** $\mathbf{r}_j^*$, for each particle j. The reference point serves as a 'representative point' for the path of particle j. Here, $\mathbf{r}_j^*$ indicates that each particle j has its own reference point, the same point for every imaginary-time slice. Paths in Z are loops (ring polymers), so a logical choice would be to use the loop centroid reference point,

$$\mathbf{r}_j^* = \frac{1}{M}\sum_{m=1}^{M}\mathbf{r}_j \qquad \text{loop centroid.} \qquad (5.232)$$

This is adequate if loops have M beads, as in high-temperature systems or systems with distinguishable particles.

---

[18] Obviously, one cannot compute the kinetic energy from the potential energy for ideal particles while strictly keeping $V(x) = 0$, but it is possible with a confinement potential and Cauchy's boundary conditions [40].

At low temperatures, identical particles' paths can go through loop fusion and form longer and longer exchange loops. In this case, one should revise the definition of a reference point because the loop centroid $\mathbf{r}_j^*$ in equation (5.232) is designed for particle j and rather poorly represents the paths of particles j, k, l. .. in an exchange loop. In this case, a better reference point is a comoving centroid, where the reference point of particle j also depends on the imaginary-time slice m. A good choice is:

Comoving centroid
  Averaging window K, often $K = M$

$$\mathbf{r}_{j,\,m}^* = \frac{1}{2K}\sum_{k=0}^{K-1}(\mathbf{r}_{j,m+k} + \mathbf{r}_{j,m-k}),$$

(5.233)

---

where the window parameter K determines how many previous and next slices are averaged. With a comoving centroid, the virial term is

$$\frac{1}{M}\sum_{m=1}^{M}\sum_{j=1}^{N}\frac{1}{2}(\mathbf{r}_{j,m} - \mathbf{r}_{j,\,m}^*) \cdot \frac{\partial V(\mathbf{x}_m)}{\partial \mathbf{r}_{j,m}} \equiv \frac{1}{M}\sum_{m=1}^{M}\frac{1}{2}(\mathbf{x}_m - \mathbf{x}_m^*) \cdot \frac{\partial V(\mathbf{x}_m)}{\partial \mathbf{r}_m}.$$

(5.234)

This term vanishes if $K = 1$ because then every bead is its own reference, $\mathbf{r}_{j,m} = \mathbf{r}_{j,\,m}^*$. As indicated, a window of $K = M$ is commonly used in PIMC.

Derivation of the virial estimator

Let us work out the CA case under PBCs. Among the routine algebra, there are a few points that make it worth reviewing. For extra reading, I recommend the PhD thesis of Rota [41], where the virial energy estimator is derived using a slightly different notation.

The thermodynamic energy fluctuations are caused by two terms in equation (5.229):

$$\left\langle \frac{3NM}{2\beta} - \frac{1}{\beta}\sum_{m=1}^{M}\frac{(\mathbf{x}_m - \mathbf{x}_{m+1})^2}{4\lambda\tau t_m} \right\rangle.$$

(5.235)

There is not much we can do to the first term *per se*, but we can check whether the second term, the spring term, can be evaluated in a different manner with the aid of the virial operator,

$$(\mathbf{x}_m - \mathbf{x}_m^*) \cdot \frac{\partial}{\partial \mathbf{x}_m}.$$

(5.236)

Virials have a classical foundation, but let us see why the virial operator appears in the path integrals. The bead positions in Z fluctuate from slice to slice, but we know that $\mathbf{x}_m$ is most probably close to the reference point $\mathbf{x}_m^*$—that is what we had in mind when we constructed it. A good reference point is a rough estimate of the particle position, based on where it was before and after (in imaginary time). The position $\mathbf{x}_m$ is translated to $\mathbf{x}_m^*$ by the translation operator,[19]

$$|\mathbf{x}_m\rangle = e^{-(\mathbf{x}_m - \mathbf{x}_m^*)\cdot \frac{\partial}{\partial \mathbf{x}_m^*}}|\mathbf{x}_m^*\rangle, \tag{5.237}$$

so now we see where the virial operator comes from. The projection operator that lets paths go through $\mathbf{x}_m$ can now be written as

$$|\mathbf{x}_m\rangle\langle\mathbf{x}_m| = e^{-(\mathbf{x}_m - \mathbf{x}_m^*)\cdot \frac{\partial}{\partial \mathbf{x}_m^*}}|\mathbf{x}_m^*\rangle\langle\mathbf{x}_m^*|e^{(\mathbf{x}_m - \mathbf{x}_m^*)\cdot \frac{\partial}{\partial \mathbf{x}_m^*}}. \tag{5.238}$$

In the language of the double-slit experiment, the projection operator $|\mathbf{x}_m\rangle\langle\mathbf{x}_m|$ is a screen with a slit at $\mathbf{x}_m$. Equation (5.238) tells us that we get the same result if we first shift all paths by an amount $\mathbf{x}_m^* - \mathbf{x}_m$, put a screen with a slit at $\mathbf{x}_m^*$, and finally move the paths back to their original positions. Paths that would have gone through the slit at $\mathbf{x}_m$ have gone through the slit at $\mathbf{x}_m^*$.

In MC, replacing $|\mathbf{x}_m\rangle\langle\mathbf{x}_m|$ with $|\mathbf{x}_m^*\rangle\langle\mathbf{x}_m^*|$ is essentially importance sampling. The point $\mathbf{x}_m$ is a random point (integration $\int d\mathbf{x}_m$), while $\mathbf{x}_m^*$ is the best-guess position for the comoving reference

$$\mathbf{x}_m^* = \mathbf{x}_m^*(\mathbf{x}_m, \mathbf{x}_{m-1}, \mathbf{x}_{m+1}, \ldots, \mathbf{x}_{m-K}, \mathbf{x}_{m+K}). \tag{5.239}$$

Awareness of the environment is the key to a successful guess for $\mathbf{x}_m^*$.

To see where the spring term comes from, consider the diagonal part of the density matrix, i.e. the density that integrates to Z. From equation (5.134), we see that it is given by

$$\rho(\mathbf{x}_1, \ldots, \mathbf{x}_M; \beta = M\tau/3) := \prod_{n=1}^{M} \langle\mathbf{x}_n|e^{-t_n\tau\hat{T}}|\mathbf{x}_{n+1}\rangle \prod_{n=1}^{M} e^{-\tau\tilde{V}_n(\mathbf{x}_n; \tau)} \tag{5.240}$$

$$= \prod_{n=1}^{M} (4\pi\lambda\tau t_n)^{-3N/2} e^{-\frac{(\mathbf{x}_n - \mathbf{x}_{n+1})^2}{4\lambda\tau t_n}} \times \prod_{n=1}^{M} e^{-\tau\tilde{V}_n(\mathbf{x}_n; \tau)}. \tag{5.241}$$

The latter form relies on the commonly adopted free-particle approximation to the kinetic matrix elements. Yes, just as in DMC, the large energy fluctuations are caused by the free-particle (imaginary time) motion detached from the potentials. As expected, the remedy is to replace the free-particle energy with something that uses information drawn from the potential, in this case, the gradient in the virial operator.

The question of what reference point should be used under PBCs is a delicate one, since we have to make sure the quantities we would like to measure are well defined. Winding $\mathbf{W}$, defined in equation (5.187), measures the net displacement of paths, but a path displacement is only defined relative to some point—motion is also relative in imaginary time. We need a *fixed* reference point on the path. As indicated, the

---

[19] In 1D, a translation is $|x\rangle = e^{-i(x-x*)\hat{p}_x/\hbar}|x*\rangle$, where $\hat{p}_x = -i\hbar\frac{\partial}{\partial x}$ is the momentum operator.

comoving centroid $\mathbf{x}_m^*$ moves with m, so we cannot use it, at least not from the beginning.

As another example, detecting particle exchange is based on counting beads in a loop. We start from a bead and go through the loop until we end up in the same bead, counting how many imaginary-time steps are taken: M steps for a no-exchange loop, $2M$ steps for a two-particle exchange loop, and so on. The fact that we can start from a bead and return *to the same bead* is not self-evident, because imaginary time is cyclic. If none of the beads in a loop are uniquely labeled, there is no telling how many steps it takes to traverse a loop. A single *fixed* bead is enough to correctly account for particle exchange. An obvious choice is to **take slice 1 positions** $\mathbf{x}_1$ as our reference point. The points $\mathbf{x}_1$ uniquely define which path belongs to what particle. After the energy expectation value has been worked out, we can safely reconsider the choice of $\mathbf{x}_1$ as a reference point.

Since $\mathbf{x}_1$ is a fixed reference point, we refrain from integrating over it in order not to mix slice indices. Therefore, we apply the fixed-reference virial operator,

$$\sum_{m=2}^{M} (\mathbf{x}_m - \mathbf{x}_1) \cdot \frac{\partial}{\partial \mathbf{x}_m} \tag{5.242}$$

to the diagonal part of the density matrix $\rho(\mathbf{x}_1, ..., \mathbf{x}_M; \beta)$ and integrate over all coordinates except $\mathbf{x}_1$. The trick is to use partial integration:[20]

$$\boxed{\begin{aligned}
&\text{PA: } \beta = M\tau, \text{ CA: } \beta = M\tau/3 \\
&G: = -\int \prod_{m=2}^{M} d\mathbf{x}_m \sum_{m=2}^{M} (\mathbf{x}_m - \mathbf{x}_1) \cdot \frac{\partial}{\partial \mathbf{x}_m} \rho(\mathbf{x}_1, ..., \mathbf{x}_M; \beta) \\
&= \int \prod_{m=2}^{M} d\mathbf{x}_m \sum_{m=2}^{M} \left[\frac{\partial}{\partial \mathbf{x}_m} \cdot (\mathbf{x}_m - \mathbf{x}_1) \right] \rho(\mathbf{x}_1, ..., \mathbf{x}_M; \beta).
\end{aligned}} \tag{5.243}$$

As noted, the formula applies to both PA and CA with a minor change in the relation between β and τ. The divergence on the right-hand side of equation (5.243) is simple to evaluate; in 3D,

$$\sum_{m=2}^{M} \frac{\partial}{\partial \mathbf{x}_m} \cdot (\mathbf{x}_m - \mathbf{x}_1) = \sum_{m=2}^{M} (3N - 0) = 3N(M-1), \tag{5.244}$$

so

$$G = 3N(M-1) \int \prod_{m=2}^{M} d\mathbf{x}_m \rho(\mathbf{x}_1, ..., \mathbf{x}_M; \beta). \tag{5.245}$$

The factor of $3N(M-1)$ merely yields the number of free coordinates after fixing $\mathbf{x}_1$. The gradient on the left-hand side can be evaluated once the action has been

---

[20] Surface terms are negligible for small τ, which we should not forget even in the case of noninteracting particles.

specified. After some algebra, the left-hand side of equation (5.243) for the CA becomes (see appendix E)

$$
\begin{aligned}
G = \int \prod_{m=2}^{M} d\mathbf{x}_m \rho(\mathbf{x}_1, \ldots, \mathbf{x}_M; \beta) &\left\{ 2\sum_{m=1}^{M} \frac{(\mathbf{x}_m - \mathbf{x}_{m+1})^2}{4\lambda\tau t_m} \right. \\
&+ 2(\mathbf{x}_{M+1} - \mathbf{x}_1) \cdot \frac{(\mathbf{x}_M - \mathbf{x}_{M+1})}{4\lambda\tau t_M} \\
&\left. + \tau\sum_{m=1}^{M} (\mathbf{x}_m - \mathbf{x}_1) \cdot \frac{\partial}{\partial \mathbf{x}_m} \tilde{V}_m(\mathbf{x}_m; \tau) \right\}.
\end{aligned}
\tag{5.246}
$$

Equate the two expressions for G and divide both sides by 2β,

$$
\begin{aligned}
&\frac{3N(M-1)}{2\beta} \int \prod_{m=2}^{M} d\mathbf{x}_m \rho(\mathbf{x}_1, \ldots, \mathbf{x}_M; \beta) \\
&= \int \prod_{m=2}^{M} d\mathbf{x}_m \rho(\mathbf{x}_1, \ldots, \mathbf{x}_M; \beta) \left\{ \frac{1}{\beta}\sum_{m=1}^{M} \frac{(\mathbf{x}_m - \mathbf{x}_{m+1})^2}{4\lambda\tau t_m} + \frac{1}{\beta}(\mathbf{x}_{M+1} - \mathbf{x}_1) \cdot \frac{(\mathbf{x}_M - \mathbf{x}_{M+1})}{4\lambda\tau t_M} \right. \\
&\left. + \frac{\tau}{2\beta}\sum_{m=1}^{M} (\mathbf{x}_m - \mathbf{x}_1) \cdot \frac{\partial}{\partial \mathbf{x}_m} \tilde{V}_m(\mathbf{x}_m; \tau) \right\}.
\end{aligned}
\tag{5.247}
$$

It is now safe to integrate over $\mathbf{x}_1$ and read the expectation values,

$$
\begin{aligned}
\frac{3N(M-1)}{2\beta} = &\left\langle \frac{1}{\beta}\sum_{m=1}^{M} \frac{(\mathbf{x}_m - \mathbf{x}_{m+1})^2}{4\lambda\tau t_m} + \frac{1}{\beta}(\mathbf{x}_{M+1} - \mathbf{x}_1) \cdot \frac{(\mathbf{x}_M - \mathbf{x}_{M+1})}{4\lambda\tau t_M} \right. \\
&\left. + \frac{\tau}{2\beta}\sum_{m=1}^{M} (\mathbf{x}_m - \mathbf{x}_1) \cdot \frac{\partial}{\partial \mathbf{x}_m} \tilde{V}_m(\mathbf{x}_m; \tau) \right\rangle.
\end{aligned}
\tag{5.248}
$$

Rearrange and add a potential term to both sides,

$$
\begin{aligned}
&\left\langle \frac{3NM}{2\beta} - \frac{1}{\beta}\sum_{m=1}^{M} \frac{(\mathbf{x}_m - \mathbf{x}_{m+1})^2}{4\lambda\tau t_m} + \frac{\partial}{\partial\beta}\sum_{m=1}^{M} \tau\tilde{V}_m(\mathbf{x}; \tau) \right\rangle \\
&= \left\langle \frac{3N}{2\beta} + \frac{1}{\beta}(\mathbf{x}_{M+1} - \mathbf{x}_1) \cdot \frac{(\mathbf{x}_M - \mathbf{x}_{M+1})}{4\lambda\tau t_M} \right. \\
&\left. + \frac{\partial}{\partial\beta}\sum_{m=1}^{M} \tau\tilde{V}_m(\mathbf{x}; \tau) + \frac{\tau}{2\beta}\sum_{m=1}^{M} (\mathbf{x}_m - \mathbf{x}_1) \cdot \frac{\partial}{\partial \mathbf{x}_m} \tilde{V}_m(\mathbf{x}_m; \tau) \right\rangle.
\end{aligned}
\tag{5.249}
$$

The left-hand side is the thermodynamic energy estimator in equation (5.229), so we have found a new energy estimator,

$$
\begin{aligned}
E(\beta) = &\left\langle \frac{3N}{2\beta} + \frac{1}{\beta}(\mathbf{x}_{M+1} - \mathbf{x}_1) \cdot \frac{(\mathbf{x}_M - \mathbf{x}_{M+1})}{4\lambda\tau t_M} \right. \\
&\left. + \frac{\partial}{\partial\beta}\sum_{m=1}^{M} \tau\tilde{V}_m(\mathbf{x}; \tau) + \frac{\tau}{2\beta}\sum_{m=1}^{M} (\mathbf{x}_m - \mathbf{x}_1) \cdot \frac{\partial}{\partial \mathbf{x}_m} \tilde{V}_m(\mathbf{x}_m; \tau) \right\rangle.
\end{aligned}
\tag{5.250}
$$

The spring term is gone, along with the other M-dependent term, so the virial operator did its job.

As mentioned earlier, it is time to reconsider the reference point choice $\mathbf{x}_1$. The coordinate $\mathbf{x}_1$ is in no way special, and to further reduce fluctuations, we can cycle it through all M slices and take the average. If $\mathbf{x}_1$ cycles to $\mathbf{x}_m$, then the coordinate $\mathbf{x}_i$ cycles to $\mathbf{x}_{i+m-1}$ for any i, hence

$$
\left\langle \frac{1}{\beta}(\mathbf{x}_{M+1} - \mathbf{x}_1) \cdot \frac{(\mathbf{x}_M - \mathbf{x}_{M+1})}{4\lambda\tau t_M} \right\rangle
$$
$$
= \left\langle \frac{1}{M\beta}\sum_{m=1}^{M}(\mathbf{x}_{M+m} - \mathbf{x}_m) \cdot \frac{(\mathbf{x}_{M+m-1} - \mathbf{x}_{M+m})}{4\lambda\tau t_{M+m-1}} \right\rangle.
$$

(5.251)

The potential term in equation (5.250) with $\mathbf{x}_1$ already has a sum over m, so we can average $\mathbf{x}_1$ in any way we like, and since we found that $\mathbf{x}_m$ is close to the comoving centroid $\mathbf{x}_m^*$, we get a smaller fluctuation with

$$
\left\langle \frac{1}{2\beta}\sum_{m=1}^{M}(\mathbf{x}_m - \mathbf{x}_1) \cdot \frac{\partial}{\partial\mathbf{x}_m}[\tau\tilde{V}_m(\mathbf{x}_m; \tau)] \right\rangle
$$
$$
= \left\langle \frac{1}{2\beta}\sum_{m=1}^{M}(\mathbf{x}_m - \mathbf{x}_m^*) \cdot \frac{\partial}{\partial\mathbf{x}_m}[\tau\tilde{V}_m(\mathbf{x}_m; \tau)] \right\rangle.
$$

(5.252)

The final result is:

Virial energy estimator; CA; comoving centroid $\mathbf{x}_m^*$

$$
E(\beta) = \left\langle \frac{3N}{2\beta} + \frac{1}{M\beta}\sum_{m=1}^{M}(\mathbf{x}_{M+m} - \mathbf{x}_m) \cdot \frac{(\mathbf{x}_{M+m-1} - \mathbf{x}_{M+m})}{4\lambda\tau t_{M+m-1}} \right.
$$
$$
\left. + \frac{\partial}{\partial\beta}\sum_{m=1}^{M}\tau\tilde{V}_m(\mathbf{x}_m; \tau) + \frac{1}{2\beta}\sum_{m=1}^{M}(\mathbf{x}_m - \mathbf{x}_m^*) \cdot \frac{\partial}{\partial\mathbf{x}_m}[\tau\tilde{V}_m(\mathbf{x}_m; \tau)] \right\rangle.
$$

(5.253)

The first term is the classical kinetic energy.

The second term vanishes if there is no exchange, $\mathbf{x}_{M+m} = \mathbf{x}_m$, so it is an *exchange correction*. The vector $\mathbf{x}_{M+m} - \mathbf{x}_m$ can be longer than the PBC simulation box because it is between distant slices and the coordinates are on a continuous path.

The last energy term is the virial term, whose fluctuations depend on the magnitude of the difference $(\mathbf{x}_m - \mathbf{x}_m^*)$ and the steepness of the potential gradients. It is often this term that fluctuates the most, so the choice of reference point $\mathbf{x}_m^*$ has a large impact.

In CA, $\tilde{V}_m(\mathbf{x}_m; \tau)$ contains $|\frac{\partial}{\partial \mathbf{x}_m} V(\mathbf{x}_m)|^2|$, and we need the gradient for that, too. While doable, it is a rather tedious task. One way to avoid it is to resort to numerical differentiation, but not in coordinate space. For that, we need to look at coordinate scaling.

Virial estimator via coordinate scaling

Another way to derive the virial energy is based on *coordinate scaling* [42]. The physical premise is that loops in Z expand as the temperature decreases. If inverse temperatures β and β' are infinitesimally close, the difference in $Z(\beta)$ and $Z(\beta')$ can be accurately described by coordinate scaling. The strategy is to find the thermodynamic energy estimator for $E(\beta')$ and set $\beta' = \beta$ (undo the coordinate scaling) in order to find an energy estimator for $E(\beta)$. This turns out to be the virial estimator.

Coordinate scaling is well defined only with respect to a reference point, such as $\mathbf{x}_1$ or the comoving centroid $\mathbf{x}_m^*$ in equation (5.233). To avoid problems in exchange and PBCs, we start with the fixed reference point $\mathbf{x}_1$ and define scaled coordinates on slice m as follows:

$$\mathbf{x}_m^s := \mathbf{x}_m + s(\mathbf{x}_m - \mathbf{x}_1). \tag{5.254}$$

A small positive scaling s pulls $\mathbf{x}_m$ away from $\mathbf{x}_1$. Such a loop expansion is a bit lopsided, but we can improve upon it later. Scaling the fixed point $\mathbf{x}_1$ itself makes no difference, $\mathbf{x}_1^s = \mathbf{x}_1$. The CA potential in scaled coordinates is

$$\begin{aligned}
\tilde{V}_m(\mathbf{x}_m^s; \tau) &= \tilde{V}(\mathbf{x}_m + s(\mathbf{x}_m - \mathbf{x}_1); \tau) \\
&= \tilde{V}_m(\mathbf{x}_m; \tau) + s(\mathbf{x}_m - \mathbf{x}_1) \cdot \frac{\partial}{\partial \mathbf{x}_m} \tilde{V}_m(\mathbf{x}_m; \tau) + \mathcal{O}(s^2).
\end{aligned} \tag{5.255}$$

The highlighted factor is the virial operator applied to $\tilde{V}$, so now we begin to see what coordinate scaling has to do with the virial estimator. Coordinate scaling mimics the effects on paths if the temperature changes from β to β', so we must have $s = s(\beta, \beta')$. Coordinate scaling does not change the path topology, so it is better to keep $\beta' \approx \beta$ and require that

$$s \ll 1 \tag{5.256}$$

$$\lim_{\beta' \to \beta} s = 0. \tag{5.257}$$

For this reason, it is enough to keep terms linear in s. Another reason for the limitation $s \ll 1$ is that we would otherwise need to include higher derivatives of the potential in equation (5.255), something one does not usually want.

The scaled-coordinate partition function is, by definition, $Z(\beta')$,

$$Z(\beta') = \int d\mathbf{x}^s \langle \mathbf{x}^s | e^{-\beta'\hat{\mathcal{H}}} | P\mathbf{x}^s \rangle = \int \prod_{m=1}^{M} d\mathbf{x}_m^s \prod_{m=1}^{M} \langle \mathbf{x}_m^s | e^{-\tau'\hat{\mathcal{H}}} | \mathbf{x}_{m+1}^s \rangle. \tag{5.258}$$

As before, we keep $\mathbf{x}_1$ as a fixed reference point and consider the function

$$H(\mathbf{x}_1; \beta') := \int \prod_{m=2}^{M} d\mathbf{x}_m^s \prod_{m=1}^{M} \langle \mathbf{x}_m^s | e^{-\tau'\hat{\mathcal{H}}} | \mathbf{x}_{m+1}^s \rangle \tag{5.259}$$

$$
= \int \prod_{m=2}^{M} d\mathbf{x}_m^s \prod_{m=1}^{M} (4\pi\lambda t_m \tau')^{-3N/2} \times e^{-\sum_{m=1}^{M} \frac{(\mathbf{x}_m^s - \mathbf{x}_{m+1}^s)^2}{4\lambda t_m \tau}}
$$

$$
\times e^{-\sum_{m=1}^{M} \tau' \tilde{V}_m(\mathbf{x}_m^s; \tau')},
\tag{5.260}
$$

where the latter form is for CA and the usual free-particle kinetic matrix elements. Scaling changes the $3NM$-dimensional volume element according to[21]

$$
\prod_{m=2}^{M} d\mathbf{x}_m^s = \prod_{m=2}^{M} d\mathbf{x}_m (1+s)^{3N} = (1+s)^{3N(M-1)} \prod_{m=2}^{M} d\mathbf{x}_m.
\tag{5.261}
$$

To the first order in s, the sum of squared bead distances scales as

$$
\sum_{m=1}^{M} t_m^{-1} (\mathbf{x}_m^s - \mathbf{x}_{m+1}^s)^2 = \sum_{m=2}^{M-1} t_m^{-1} (\mathbf{x}_m^s - \mathbf{x}_{m+1}^s)^2 + t_1^{-1}(\mathbf{x}_1 - \mathbf{x}_2^s)^2 + t_M^{-1}(\mathbf{x}_M^s - \mathbf{x}_{M+1})^2
$$

$$
= (1+2s)\sum_{m=1}^{M} t_m^{-1}(\mathbf{x}_m - \mathbf{x}_{m+1})^2 + 2s t_M^{-1}(\mathbf{x}_M - \mathbf{x}_{M+1}) \cdot (\mathbf{x}_{M+1} - \mathbf{x}_1).
\tag{5.262}
$$

Notice that $\mathbf{x}_1$ and $\mathbf{x}_{M+1}$ (which equals $\mathbf{x}_1$ or its permutation) *do not* scale. So far, we have found that

$$
H(\mathbf{x}_1; \beta') = (1+s)^{3N(M-1)} \int \prod_{m=2}^{M} d\mathbf{x}_m \prod_{m=1}^{M} (4\pi\lambda t_m \tau')^{-3N/2}
$$

$$
\times e^{-\sum_{m=1}^{M} \frac{1+2s}{4\lambda t_m \tau}(\mathbf{x}_m - \mathbf{x}_{m+1})^2} \times e^{\frac{2s}{4\lambda t_M \tau}(\mathbf{x}_M - \mathbf{x}_{M+1})\cdot(\mathbf{x}_{M+1} - \mathbf{x}_1)}
$$

$$
\times e^{-\sum_{m=1}^{M} \tau'[\tilde{V}_m(\mathbf{x}_m; \tau') + s(\mathbf{x}_m - \mathbf{x}_1)\cdot \frac{\partial}{\partial \mathbf{x}_m}\tilde{V}_m(\mathbf{x}_m; \tau')]}.
\tag{5.263}
$$

The next task is to relate this expression to the thermodynamic energy estimator. Compute the derivative of $H(\mathbf{x}_1; \beta')$ w.r.t. β' at $\beta' = \beta$. For brevity, let

$$
s' := \frac{\partial s}{\partial \beta'}\Big|_{\beta'=\beta}.
\tag{5.264}
$$

We find

$$
\frac{\partial H(\mathbf{x}_1; \beta')}{\partial \beta'}\Big|_{\beta'=\beta} = 3N(M-1)s' H(\mathbf{x}_1; \beta) + \frac{3NM}{2\beta} H(\mathbf{x}_1; \beta)
$$

$$
+ \int \prod_{m=2}^{M} d\mathbf{x}_m \prod_{m=1}^{M} (4\pi\lambda t_m \tau)^{-3N/2} \times e^{-\sum_{m=1}^{M} \frac{(\mathbf{x}_m - \mathbf{x}_{m+1})^2}{4\lambda t_m \tau}} \times e^{-\sum_{m=1}^{M} \tau \tilde{V}_m(\mathbf{x}_m; \tau)}
$$

$$
\times \left\{ \left(\frac{1}{\beta} - 2s'\right)\sum_{m=1}^{M} \frac{(\mathbf{x}_m - \mathbf{x}_{m+1})^2}{4\lambda t_m \tau} + 2s'\frac{(\mathbf{x}_M - \mathbf{x}_{M+1})}{4\lambda t_M \tau} \cdot (\mathbf{x}_{M+1} - \mathbf{x}_1) \right.
$$

$$
\left. - \sum_{m=1}^{M} \frac{\partial}{\partial \beta}(\tau \tilde{V}_m(\mathbf{x}_m; \tau)) - s'\tau \sum_{m=1}^{M} (\mathbf{x}_m - \mathbf{x}_1) \cdot \frac{\partial}{\partial \mathbf{x}_m} V_m(\mathbf{x}_m; \tau) \right\}.
\tag{5.265}
$$

---

[21] Each dimension gives $x^s = x + s(x - x_1) \Rightarrow dx^s = (1+s)dx$.

Next, integrate over $\mathbf{x}_1$ and read expectation values. Since

$$\int d\mathbf{x}_1 H(\mathbf{x}_1; \beta) = Z(\beta), \tag{5.266}$$

we get the energy estimator

$$E(\beta) = \left\langle 3Ns' + 3NM\left(s' - \frac{1}{2\beta}\right) \right.$$

$$-\left(\frac{1}{\beta} - 2s'\right)\sum_{m=1}^{M} \frac{(\mathbf{x}_m - \mathbf{x}_{m+1})^2}{4\lambda t_m \tau} - 2s'\frac{(\mathbf{x}_M - \mathbf{x}_{M+1})}{4\lambda t_M \tau} \cdot (\mathbf{x}_{M+1} - \mathbf{x}_1) \tag{5.267}$$

$$\left. + \frac{\partial}{\partial\beta}\sum_{m=1}^{M} \tau \tilde{V}_m(\mathbf{x}_m; \tau) + s'\sum_{m=1}^{M} \tau(\mathbf{x}_m - \mathbf{x}_1) \cdot \frac{\partial}{\partial\mathbf{x}_m} V_m(\mathbf{x}_m; \tau) \right\rangle.$$

The highlighted M-dependent energy term and the spring term vanish by choosing

$$\boxed{s' = \frac{1}{2\beta} \qquad \text{(one possible choice)},} \tag{5.268}$$

and we recover the virial energy estimator in equation (5.250),

$$E(\beta) = \left\langle \frac{3N}{2\beta} + \frac{1}{\beta}\frac{(\mathbf{x}_M - \mathbf{x}_{M+1})}{4\lambda t_M \tau} \cdot (\mathbf{x}_{M+1} - \mathbf{x}_1) \right.$$

$$\left. + \frac{\partial}{\partial\beta}\sum_{m=1}^{M} \tau \tilde{V}_m(\mathbf{x}_m; \tau) + \frac{1}{2\beta}\sum_{m=1}^{M} \tau(\mathbf{x}_m - \mathbf{x}_1) \cdot \frac{\partial}{\partial\mathbf{x}_m} V_m(\mathbf{x}_m; \tau) \right\rangle. \tag{5.269}$$

Averaging over $\mathbf{x}_1$ and using the comoving centroid reference leads to equation (5.253). Although not needed here, the scaling

$$s = \sqrt{\frac{\beta'}{\beta}} - 1 \tag{5.270}$$

has the desired properties: $s' = 1/(2\beta)$ and scaling is positive if $\beta' > \beta$; that is, loops in Z expand if the temperature is lowered.

Virial energy estimator in a compact form

The expansion

$$\tilde{V}_m(\mathbf{x}_m^s; \tau') = \tilde{V}_m(\mathbf{x}_m + s(\mathbf{x}_m - \mathbf{x}_m^*); \tau') \tag{5.271}$$

$$= \tilde{V}_m(\mathbf{x}_m; \tau') + s(\mathbf{x}_m - \mathbf{x}_m^*) \cdot \frac{\partial}{\partial\mathbf{x}_m} \tilde{V}_m(\mathbf{x}_m; \tau') + \mathcal{O}(s^2) \tag{5.272}$$

gives a useful identity: using coordinate scaling with $s' = 1/(2\beta)$,

$$\frac{\partial}{\partial\beta'}(\beta' \tilde{V}_m(\mathbf{x}_m^s; \tau'))|_{\beta'=\beta} \equiv \frac{\partial}{\partial\beta}(\beta\tilde{V}_m(\mathbf{x}_m; \tau)) + \frac{1}{2}(\mathbf{x}_m - \mathbf{x}_m^*) \cdot \frac{\partial}{\partial\mathbf{x}_m} \tilde{V}_m(\mathbf{x}_m; \tau). \tag{5.273}$$

The virial energy estimator in equation (5.253) can now be written in a compact form,

$$E(\beta) = \left\langle \frac{3N}{2\beta} + \frac{1}{\beta}\frac{(\mathbf{x}_M - \mathbf{x}_{M+1})}{4\lambda t_M \tau} \cdot (\mathbf{x}_{M+1} - \mathbf{x}_1) + \sum_{m=1}^{M} \frac{\partial}{\partial \beta'}[\tau' \tilde{V}_m(\mathbf{x}_m^s; \tau')]|_{\beta'=\beta} \right\rangle. \qquad (5.274)$$

The constant-volume heat capacity,

$$C_V(T): = \left(\frac{\partial E}{\partial T}\right)_V = k_B \beta^2 \left[\frac{1}{Z}\frac{\partial^2 Z}{\partial \beta^2} - E^2\right], \qquad (5.275)$$

can be found using

$$C_V(T) = \langle \epsilon_V^2 \rangle - \langle \epsilon_V \rangle^2 - \langle \epsilon'_V \rangle, \qquad (5.276)$$

where

$$\epsilon_V: = \frac{3N}{2\beta} + \frac{1}{\beta}\frac{(\mathbf{x}_M - \mathbf{x}_{M+1})}{4\lambda t_M \tau} \cdot (\mathbf{x}_{M+1} - \mathbf{x}_1) + \sum_{m=1}^{M} \frac{\partial}{\partial \beta'}[\beta' \tilde{V}_m(\mathbf{x}_m^s; \tau')]|_{\beta'=\beta} \quad (5.277)$$

$$\epsilon'_V: = -\frac{3N}{2\beta^2} - \frac{1}{\beta^2}\frac{(\mathbf{x}_M - \mathbf{x}_{M+1})}{4\lambda t_M \tau} \cdot (\mathbf{x}_{M+1} - \mathbf{x}_1) + \sum_{m=1}^{M} \frac{\partial^2}{\partial \beta'^2}[\beta' \tilde{V}_m(\mathbf{x}_m^s; \tau')]|_{\beta'=\beta}. \quad (5.278)$$

so that $E(\beta) = \langle \epsilon_V \rangle$. The derivatives can be computed analytically, or using finite differences as in [42], or through automatic differentiation (AD). Notably, the derivatives are not taken from noisy stochastic data but from known functions.

5.6.4 Fermion PIMC and the sign problem

We have deliberately avoided discussing fermion PIMC because it suffers from the sign problem, which, to date, has no satisfactory solution. The sign problem occurs due to the near cancellation of contributions to observables from even and odd permutations. To elaborate further: consider two identical fermions, such as spin-polarized electrons. We cannot tell them apart; hence, the case with no exchange and the case with exchange are equally relevant in the partition function. The latter, however, has a minus sign, so in PIMC we evaluate Z by adding contributions that partly cancel. For a small number of fermions, the cancellation is tolerable, but for a few tens of fermions, the signal-to-noise ratio is already quite low.

Troyer and Wiese [43] studied the Ising model and concluded that the sign problem is the origin of its NP-hardness; consequently, a generic solution of the sign problem would also solve all problems in the complexity class NP in polynomial time. We seem to be out of luck, but maybe a generic solution is not needed for accurate fermion results?

In DMC, we pushed the sign problem aside by introducing a trial wave function that imposed approximate ground-state nodes on the imaginary-time-evolved wave function. However, finite-temperature systems are in a mixed state (a linear combination of many low-energy states), so the nodal surface changes with temperature. The obvious route would be to introduce a trial density matrix, but

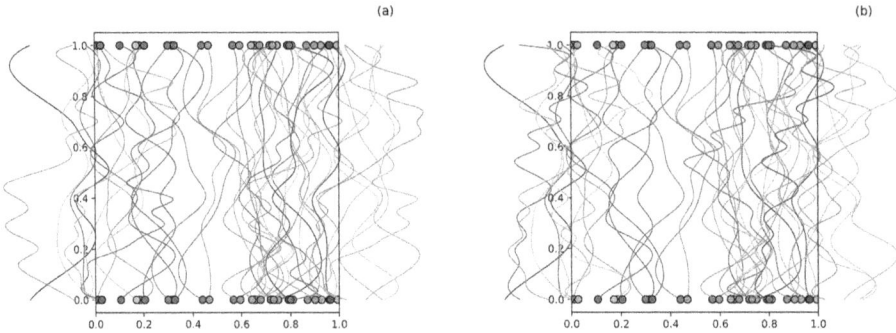

Figure 5.19. Panel (a) has 30 single-fermion paths and four two-electron exchange paths, so the permutation factor is $(-1)^4 = 1$. Panel (b) has 28 single-fermion paths and five two-electron exchange paths, so the permutation factor is $(-1)^5 = -1$. Many of the paths are identical, which means that in both cases, the measured quantities $\hat{\mathcal{A}}$ have almost equal magnitudes, so their contributions to $\langle\hat{\mathcal{A}}\rangle$ nearly cancel.

we are reluctant to give up the freedom to apply PIMC to many kinds of systems without tailoring the algorithm over and over again.

Let us start with a bird's-eye view of the sign problem. Figure 5.19 shows two possible 38-fermion paths. If you cannot tell which paths make a positive contribution and which ones make a negative contribution, you begin to understand why PIMC has so much trouble with many-fermion systems. PIMC has no trouble giving an expectation value for a measured quantity, but the error bars are large.

The sign problem is only a numerical issue introduced by some inaccuracy in the approach. In MC, it is the way in which a perfectly well-defined expectation value is shrouded in the mist of values fluctuating around it. There is no 'built-in' sign problem in the finite-temperature expectation values,

$$\langle\hat{\mathcal{A}}\rangle = \text{Tr}(\hat{\rho}\hat{\mathcal{A}}) = \frac{\sum_i e^{-\beta E_i}\langle i|\hat{\mathcal{A}}|i\rangle}{\sum_i e^{-\beta E_i}}. \tag{5.279}$$

Here the weights $e^{-\beta E_i}$ are manifestly positive, so *there is no sign problem*. Alas, solving $\hat{\mathcal{H}}|i\rangle = E_i|i\rangle$ is an exponentially hard problem because the number of degrees of freedom scales exponentially with system size. Exact diagonalization methods require calculations with exponentially large matrices, making them impractical beyond a few tens of particles. The computing time increases exponentially with the number of particles. A question one should ask is: just *how* accurately do we need to know the solution to $\hat{H}|i\rangle = E_i|i\rangle$ in order to avoid the sign problem?

The density matrix of two identical noninteracting fermions in 3D free space is a sum over permutations,

$$\rho_0(\mathbf{x}', \mathbf{x}; \tau) = \sum_P (-1)^P \langle\mathbf{x}'|e^{-\tau\hat{T}}|P\mathbf{x}\rangle, \tag{5.280}$$

which can also be written as a matrix (see section 4.7.1):

$$\rho_0(\mathbf{x}', \mathbf{x}; \tau): = \langle \mathbf{r}'_1, \mathbf{r}'_2 | e^{-\tau \hat{T}} | \mathbf{r}_1, \mathbf{r}_2 \rangle - \langle \mathbf{r}'_2, \mathbf{r}'_1 | e^{-\tau \hat{T}} | \mathbf{r}_1, \mathbf{r}_2 \rangle \tag{5.281}$$

$$= \langle \mathbf{r}'_1 | e^{-\tau \hat{T}} | \mathbf{r}_1 \rangle \langle \mathbf{r}'_2 | e^{-\tau \hat{T}} | \mathbf{r}_2 \rangle - \langle \mathbf{r}'_2 | e^{-\tau \hat{T}} | \mathbf{r}_1 \rangle \langle \mathbf{r}'_1 | e^{-\tau \hat{T}} | \mathbf{r}_2 \rangle \tag{5.282}$$

$$= \begin{vmatrix} \langle \mathbf{r}'_1 | e^{-\tau \hat{T}} | \mathbf{r}_1 \rangle & \langle \mathbf{r}'_2 | e^{-\tau \hat{T}} | \mathbf{r}_1 \rangle \\ \langle \mathbf{r}'_1 | e^{-\tau \hat{T}} | \mathbf{r}_2 \rangle & \langle \mathbf{r}'_2 | e^{-\tau \hat{T}} | \mathbf{r}_2 \rangle \end{vmatrix} \tag{5.283}$$

$$\overset{\text{free space}}{=} (4\pi \lambda \tau)^{-3} \begin{vmatrix} e^{-\frac{(\mathbf{r}'_1 - \mathbf{r}_1)^2}{4\lambda \tau}} & e^{-\frac{(\mathbf{r}'_2 - \mathbf{r}_1)^2}{4\lambda \tau}} \\ e^{-\frac{(\mathbf{r}'_1 - \mathbf{r}_2)^2}{4\lambda \tau}} & e^{-\frac{(\mathbf{r}'_2 - \mathbf{r}_2)^2}{4\lambda \tau}} \end{vmatrix} . \tag{5.284}$$

For a 100 fermion system, the sign begins to look almost random. Chin [44] argues that the more free-fermion propagations are used, the worse the sign problem gets, so one could 'delay' the onset of problems by using a high-order propagator. For example, the fourth-order propagator in equation (4.216) has three free-fermion propagations:

$$e^{-\tau(\hat{T} + \hat{V})} \approx e^{-v_0 \tau \hat{V}} e^{-t_1 \tau \hat{T}} e^{-v_1 \tau \hat{W}} e^{-t_1 \tau \hat{T}} e^{-v_1 \tau \hat{W}} e^{-t_1 \tau \hat{T}} e^{-v_0 \tau \hat{V}}, \tag{5.285}$$

but τ can be markedly larger than in the primitive or symmetrized primitive splittings. To prove the point, he solves the problem of up to 20 spin-polarized electrons in a two-dimensional (2D) circular, parabolic quantum dot.

5.6.4.1 The problem of alternating signs in a fermion sum

It would be nice to see how fermion permutations show up analytically, without obscuring the view with MC. Specifically, would it be a good idea to organize the canonical partition function into terms with no exchange, one exchange, two exchanges, and so on?

The canonical partition function of noninteracting particles can be computed recursively.

Recursive formula for the partition function of N identical noninteracting particles
Input: single-particle partition function $Z_1(\beta)$.
The upper sign is for bosons and the lower is for fermions.

$$Z_0(\beta): = 1 \tag{5.286}$$

$$Z_N(\beta) = \frac{1}{N} \sum_{n=1}^{N} (\pm 1)^{n+1} Z_1(n\beta) Z_{N-n}(\beta). \tag{5.287}$$

The fermion recursion begins with

$$Z_1(\beta): = \sum_n e^{-\beta \epsilon_n}, \text{ single-particle energies } e_n \qquad (5.288)$$

$$Z_2(\beta) = \frac{1}{2}(Z_1(\beta)^2 - Z_1(2\beta)) \qquad (5.289)$$

$$Z_3(\beta) = \frac{1}{3}(\underbrace{Z_1(\beta)^3}_{\text{no exchange}} - 3\underbrace{Z_1(2\beta)Z_1(\beta)}_{\text{one exchange}} + 2\underbrace{Z_1(3\beta)}_{\text{two exchanges}}). \qquad (5.290)$$

The recursion formula has been derived many times in the past, e.g. by Borrmann and Franke [45]. Recently, Barghathi, Yu, and del Maestro [46] derived a general framework that can be applied to the case of degenerate energy levels and to the computation of quantities such as specific heat and magnetic susceptibility. Noninteracting bosons and fermions in the canonical ensemble are a relevant limiting case in experiments on low-density atomic gases, where the number of particles is fixed by design.

As a side note, the recursion relation also offers insight into phase transitions in finite systems. The grand canonical partition function, equation (5.97), repeated here:

$$\mathcal{Z} = \sum_{N=0}^{\infty} e^{\mu N} Z_N: = \sum_{N=0}^{\infty} z^N Z_N, \qquad (5.291)$$

shows that $\mathcal{Z}$ is a polynomial of fugacity z. Yang and Lee [47] showed that the grand canonical partition function can be written as a function of its zeros in the complex fugacity plane and that phase transitions occur when these so-called Lee-Yang zeros accumulate and cross the real axis in the thermodynamical limit. Fisher [48] noticed that similar logic can be applied to the canonical partition function, which is a function of just β. Studying $Z_N(\beta)$ for complex β, one can find out where the singularities of the partition function appear in the limit $N \to \infty$. The so-called Fisher zeroes lie in the complex plane off the real β axis, but as N increases, the zeros close in on the real axis. In the wake of experiments on the BEC of trapped bosons in the early 2000s, Mülken et al [49] showed that phase transitions in macroscopic systems can be identified from the properties of the distribution of zeros of Z_N.

Why is this Z_3 the correct result? Insert Z_1:

$$Z_3(\beta) = \frac{1}{3}\left(\underbrace{\sum_{n,m,p} e^{-\beta(\epsilon_n + \epsilon_m + \epsilon_p)}}_{Z_1(\beta)^3} - 3\underbrace{\sum_{n,m} e^{-\beta\epsilon_n - 2\beta\epsilon_m}}_{Z_1(2\beta)Z_1(\beta)} + 2\underbrace{\sum_n e^{-3\beta\epsilon_n}}_{Z_1(3\beta)} \right). \qquad (5.292)$$

The first sum

$$\sum_{n,m,p} e^{-\beta(\epsilon_n + \epsilon_m + \epsilon_p)} \qquad (5.293)$$

sums over states m, n, p (ignoring the Pauli exclusion principle), so we should subtract cases with two identical single-particle states, that is, $m = n$, $m = p$, or $n = p$,

$$-3\sum_{n,m}e^{-\beta\epsilon_n - 2\beta\epsilon_m}. \tag{5.294}$$

This subtracts the state $m = n = p$ three times, which is twice too many, so add it back:

$$+2\sum_{n}e^{-3\beta\epsilon_n}. \tag{5.295}$$

Mathematically, the recursion formula is exact; and there is no randomness or statistical uncertainty; thus, the computation of, say, Z_{100} should be a breeze—but for fermions, it is not. We will soon see that the three terms in Z_3 have nearly perfect cancellation, which leads to numerical difficulties in finding their sum.

The one-body partition function $Z_1(\beta)$ is the same for bosons and fermions, so particle statistics depends on the signs. In a 3D HO potential $m\omega^2 r^2$,

$$Z_1(\beta) = \left(2\sinh(\tfrac{1}{2}\hbar\omega\beta)\right)^{-3} = (e^{\frac{1}{2}\hbar\omega\beta} - e^{-\frac{1}{2}\hbar\omega\beta})^{-3} \tag{5.296}$$

$$Z_2(\beta) = \frac{1}{2}\left((e^{\frac{1}{2}\hbar\omega\beta} - e^{-\frac{1}{2}\hbar\omega\beta})^{-6} - (e^{\frac{1}{2}\hbar\omega(2\beta)} - e^{-\frac{1}{2}\hbar\omega(2\beta)})^{-3}\right) \tag{5.297}$$

$$\begin{aligned}Z_3(\beta) = \frac{1}{3}\Big(&(e^{\frac{1}{2}\hbar\omega\beta} - e^{-\frac{1}{2}\hbar\omega\beta})^{-9} \\ &- 3(e^{\frac{1}{2}\hbar\omega(2\beta)} - e^{-\frac{1}{2}\hbar\omega(2\beta)})^{-3}(e^{\frac{1}{2}\hbar\omega\beta} - e^{-\frac{1}{2}\hbar\omega\beta})^{-3} \\ &+ 2(e^{\frac{1}{2}\hbar\omega(3\beta)} - e^{-\frac{1}{2}\hbar\omega(3\beta)})^{-3}\Big).\end{aligned} \tag{5.298}$$

The fermions Z_N become numerically troublesome at low temperatures, which shows as an unphysical negative partition function. Again, cancellation of terms is to blame. For example, in units of $\hbar\omega = 1$ and $k_B = 1$, the three terms in $Z_3(T = 0.01)$ are of the order of 10^{-196}. The small number itself is not a problem; after all, we take the logarithm of the free energy. The problem is that the sum of the three terms is of the order of 10^{-282}, so about 85 digits cancel perfectly! The signal is there, but it is deeply hidden.

In units of $\hbar\omega = 1$ and $k_B = 1$, the partition functions Z_N of particles in a 3D harmonic trap at selected temperatures are as follows:

$$\begin{aligned}&T = 1.0: \\ &Z_1 = 0.883\ 402 \\ &Z_2 = 0.390\ 199 - 0.038\ 507 = 0.351\ 692 \\ &Z_3 = 0.229\ 802 - 0.068\ 035 + 0.008\ 632 = 0.170\ 399\end{aligned} \tag{5.299}$$

$T = 0.01$:

$$Z_1 = 7.1750 \times 10^{-66}$$
$$Z_2 = 2.574 \times 10^{-131} - 2.574 \times 10^{-131} = 5.745\,508 \times 10^{-174} \tag{5.300}$$
$$Z_3 = 1.23\,1294 \times 10^{-196} - 3.693\,883 \times 10^{-196} + 2.462\,589 \times 10^{-196} = 3.067\,171 \times 10^{-282}$$

$T = 0.001$:

$$Z_1 = 3.616\,405 \times 10^{-652}$$
$$Z_2 = 6.539\,195 \times 10^{-1304} - 6.539\,195 \times 10^{-1304} = 1.991\,561 \times 10^{-1737} \tag{5.301}$$
$$Z_3 = 1.576\,559 \times 10^{-1955} - 4.729\,6766 \times 10^{-1955} + 3.153\,118 \times 10^{-1955} = 7.311\,709 \times 10^{-2823}.$$

Adding up the printed decimals for $T = 0.01$ and $T = 0.001$ gives exactly zero. In Z_3, as many as 85 significant numbers cancel at $T = 0.01$, and at $T = 0.001$, a huge 868 significant numbers cancel. However, Z_3 gives a positive result after all cancellations. The numbers were obtained using the Python package mpmath with a precision of mp.prec=3000 (2000 is not enough).

With symbolic math, it is easy to find the expression for, say, Z_{10}, but if one plugs in the numerical values, the cancellation of permutation terms is huge. Decreasing temperature and increasing N have the same effect. If the numerical accuracy of $Z_3(\beta)$ is sufficient for, say, $T > 0.1$, then $Z_{18}(\beta)$ will be accessible for $T > 0.6$, but $Z_{90}(\beta)$ will only be accessible for $T > 3$.

We know how to compute $Z_{100}(\beta)$ and even the grand canonical $\mathcal{Z}(\beta)$ for noninteracting fermions using other means, so the problem lies in the bookkeeping of paths and permutations. That is exactly the point: there must be a better way to deal with fermions in PIMC.

The field is open for inventions: here are some possible approaches:

- **Restricted PIMC (RPIMC)** [50] is the oldest on this list, but there are many recent variants, e.g. **configuration PIMC (CPIMC)**. Nodes are fixed using a trial density matrix, and paths are not allowed to cross nodal surfaces. This method uses the determinants of free-particle density matrices, which makes it accurate at high T. In jellium, the original version of RPIMC does not quite reproduce exact results at high density ($r_s < 1$).
- **Direct PIMC (DPIMC)** is accurate but computationally extremely costly. See, for example, reference [51].
- **CPIMC** [52] is excellent at high density and has no time-step error (for another example of *continuous-time QMC*, see section 5.8).
- **Permutation Blocking PIMC (PB-PIMC)** [53] uses antisymmetrized high-temperature density matrices for pairs or small groups of particles. PB-PIMC is formally exact; no fixed-node approximation is used.

5.7 Practical suggestions for testing a PIMC code

The following three test systems each serve a different purpose and help you to pinpoint problems.

5.7.1 First test system

Start with a small system that you can solve exactly for both distinguishable particles and bosons. For code testing purposes, the PIMC computation should be very fast.

The 1D HO with two noninteracting particles is a perfect choice. First, there is no interparticle potential, so exact solutions can be found easily. Secondly, the harmonic confining potential keeps the system together, and there is no need for PBC complications.

Close to $T = 0$, both bosons are approximately in the single-particle ground state, so the density is close to $|\phi_0(x)|^2$, a Gaussian around $x = 0$. If the density looks fine, the path sampling may be correct, and you can proceed to compute the energy.

The single-particle HO energy states are $\epsilon_n = (n + \frac{1}{2})\hbar\omega$; in suitable units, $\epsilon_n = n + \frac{1}{2}$. The partition function of two distinguishable particles is

$$Z_2(\beta) = \sum_{n_1=0}^{\infty}\sum_{n_2=0}^{\infty} e^{-\beta[(n_1 + \frac{1}{2})+(n_2 + \frac{1}{2})]} = \left[\sum_{n=0}^{\infty} e^{-\beta(n+\frac{1}{2})}\right]^2 = \left[e^{-\beta/2}\sum_{n=0}^{\infty} e^{-\beta n}\right]^2 \quad (5.302)$$

$$= \left[e^{-\beta/2}\frac{1}{1 - e^{-\beta}}\right]^2 = \left[\frac{1}{e^{\beta/2} - e^{-\beta/2}}\right]^2 :\, =Z_1(\beta)^2, \quad (5.303)$$

and the energy per particle is

$$E_2/N = \frac{1}{2\tanh(\beta/2)} \qquad \text{distinguishable particles.} \quad (5.304)$$

The boson case can be found, for example, using the recursion relation,

$$Z_2(\beta) = \frac{1}{2}(Z_1(\beta)^2 + Z_1(2\beta)) \quad (5.305)$$

$$E_2/N = \frac{1}{2}\left[1 + \frac{e^{-\beta}}{1 - e^{-\beta}} + 2\frac{e^{-2\beta}}{1 - e^{-2\beta}}\right] \qquad \text{bosons.} \quad (5.306)$$

5.7.2 Second test system

The next extension is PBCs. For testing, I suggest you use the **3D gas of noninteracting distinguishable particles or bosons in the canonical ensemble**. The former is good for debugging worm moves without swaps (and with bisection swap disabled, if it has one). The boson system can be used to find problems with paths extending throughout the periodic box. A worm should be able to close a path from a head in the simulation box to a tail in a nearby box; failing to do so shows up as zero winding, even in small systems (noninteracting bosons have no superfluidity in the thermodynamic limit).

Wrap all bead positions in an L-sided simulation box. Keeping both wrapped and unwrapped positions causes extra work in synchronization, and in most codes, positions are stored as wrapped.

I recommend you take a look at [34] and the way its authors test their code for noninteracting bosons, especially if you are prepared to use unwrapped positions.

Next, change all bead-to-bead distance calculations to use the minimum image convention. After that, choose a very large L and check that the HO system still works as expected.

The exact few-particle canonical result can be computed, for example, using the recursion relation for $Z_N(\beta)$. On the other hand, with sufficiently many particles ($N = 256$ or more), the canonical results should be close to their grand canonical counterparts, which are easy to find analytically. The simulation box can be chosen to correspond to the grand canonical BEC critical temperature $T_c = 1$, meaning that

$$T_c = \frac{2\pi\hbar^2}{mk_B}\zeta(3/2)^{-2/3}(N/V)^{2/3} = 4\pi\lambda\zeta(3/2)^{-2/3}(N/V)^{2/3} := 1, \qquad (5.307)$$

where the Riemann zeta function $\zeta(3/2) \approx 2.612\,375\,348\,685\,4883$ and $k_B = 1$ (energy in units of temperature). From this, we can solve

$$N/V = (4\pi\lambda)^{-3/2}\zeta(3/2) := N/L^3 \qquad (5.308)$$

$$\Leftrightarrow L = (4\pi\lambda)^{1/2}(N/\zeta(3/2))^{1/3}. \qquad (5.309)$$

This choice ensures that a large-N boson simulation shows signs of BEC near $T = 1$.

In less than three dimensions, you can use, for example,

$$L = 10N^{1/dim}. \qquad (5.310)$$

The simulation box must be large enough to prevent world lines from passing through the PBC simulation box in imaginary time τ. Such an event would play havoc with the kinetic energy terms, caused by jumps of size L in the bead-to-bead distances, and the code would return negative kinetic energy.

The grand canonical (internal) energy of a noninteracting particle in 3D is

$$U = Nk_B \begin{cases} \dfrac{3}{2}\dfrac{\zeta(5/2)}{\zeta(3/2)}\dfrac{T^{5/2}}{T_c^{3/2}} \approx 0.771\dfrac{T^{5/2}}{T_c^{3/2}}, & T < T_c \\ \dfrac{3}{2}\dfrac{g_{5/2}(z)}{g_{3/2}(z)}T, & T > T_c \end{cases}. \qquad (5.311)$$

Here, $\zeta(5/2) \approx 1.341\,487\,257\,25$, and the so-called Bose integral (also called the polylogarithm function $\mathrm{Li}_p(z)$) is

$$g_p(z) = \frac{1}{\Gamma(p)}\int_0^\infty dx\frac{x^{p-1}}{z^{-1}e^x - 1} \overset{\text{sometimes}}{=} \sum_{l=1}^\infty \frac{z^l}{l^p}, \qquad (5.312)$$

which can be accurately computed by, for example, the Python package mpmath function polylog. Notice that $g_p(1) \equiv \zeta(p)$, so U is continuous at T_c. In addition to the explicit temperature dependence, the fugacity z also depends on T,

$$z := e^{\beta\mu} \qquad \text{fugacity}, \qquad (5.313)$$

which gives $z = 1$ below T_c (where the chemical potential $\mu = 0^-$, negative but very close to zero). Above T_c, the chemical potential $\mu(T)$ has to be solved using the condition

$$g_{3/2}(z) = \lambda_T^3 (N/V), \qquad (5.314)$$

where the thermal wavelength λ_T is given in equation (5.27). In our dimensionless units, fugacity can be solved as zeros of the function

$$f(z): = T^{3/2} g_{3/2}(z) - \zeta(3/2). \qquad (5.315)$$

Remember that $\mu < 0$; otherwise, the boson single-particle state occupation number $1/(e^{-\beta(\epsilon_n - \mu)} - 1)$ becomes negative for the ground state. Pressure is related to the energy density,

$$P = \frac{2}{3} \frac{U}{V}, \qquad (5.316)$$

and the constant-volume heat capacity can easily be calculated using $C_V = (\partial U / \partial T)_V$.

5.7.3 Third test system

Once you are happy with your implementation of PBCs and identical particles and at least the one-body potential of the HO works reasonably well, you can add a particle–particle pair interaction. Interactions slow down PIMC because move acceptance drops. A reasonable system to play with is **liquid He^4**. There is plenty of information about its energy versus T, specific heat, superfluid density, Bose–Einstein condensate fraction, etc.—both experimentally and from PIMC. Liquid He^4 PIMC calculations can be found in many articles, for example, in [13, 54] and many more.

This is not a light computation, so I suggest you start with, say, $N = 16$. Only 16 atoms do not make much of a liquid, so the energetics will be off, and the superfluid density will show a noticeable finite-size smearing near the superfluid transition temperature of $T_\lambda \approx 2.17$ K. You may also want to take a look at liquid He^4 confined in nanopores [55] or a dilute Bose gas [56].

Considering that a 64-atom He^4 liquid can be computed using a desktop computer and some patience (overnight with a reasonably fast code), the dramatic difficulty of computing just a 38-atom *fermion* liquid He^3 system shows how badly the fermion sign problem hits PIMC. Dornheim, Moldabekov, Vorberger, and Militzer published such fermion PIMC results in 2022, after some 10^5 CPU hours on a supercomputer [57].[22]

---

[22] Followed by computations of the properties of a warm dense electron gas [58], warm dense matter [59], and many more.

5.7.4 About the sample PIMC code

The sample PIMC code is intended to show how PIMC could be coded in Julia. The code is not very effective or robust, but it has worm moves, a few measurements, and the three test systems mentioned earlier. Only the **canonical ensemble** has been implemented, although the grand canonical ensemble is mentioned here and there for future work.

At 4000+ lines, the code is a bit lengthy, but it is split into a few modules, each performing a well-defined task:

- PIMC_Common.jl makes the basic choices, i.e. which of the three test systems to compute and what parameters to try. I have decided to fix the particle number N with compiler optimization in mind. The number of slices M is calculated based on a target τ because the latter is more critical to the accuracy of the splitting of $e^{-\beta\hat{\mathcal{H}}}$.
- PIMC_Systems.jl contains definitions of the potentials in each implemented system and exact energies (if available).
- PIMC_main.jl is the main program. It calls the appropriate initialization function for the chosen system. Here, I choose what PIMC moves to use and how frequently, what to measure and how frequently, and what to report (on screen) and how frequently. PIMC_Structs.jl defines most structs. Although Julia is not an object-oriented language, I find it more manageable to keep things structured. The naming t_something reminds me that these are 'types.' The most important ones are:
  - t_pimc holds information about the PIMC parameters that are not hard-wired constants, the one-particle and pair potentials of the system, and the PIMC moves, measurements, and reports I decided to use in PIMC_main.jl.
  - t_beads, where the initial bead positions on each slice are stored, along with information about their slices. Beads only move in space; they are kept fixed in their initial slice.
  - t_links represents the interconnectivity of beads on paths. It is an array implementation of a linked list, which stores information about what a bead is next to. Beads are identified using an integer identity, and *links*, which is of type t_links, knows that the bead after bead *id* is links.next[*id*], and the previous one is links.prev[*id*].
  - t_move is the type for PIMC moves, such as the bisection move and the worm move. Moves can be freely added in PIMC_main.jl using the function add_move![23]
  - t_measurement is the type for various measurements, such as energy, radial distribution function, etc. Similar to moves, measurements are added using add_measurement!

---

[23] The exclamation mark following a function name, such as add_move!, is just a reminder that the function changes its arguments.

- t_report is, as you already guessed, the type for reports, very much like t_moves and t_measurement. I have the habit of displaying information snapshots on screen, although 'serious' data is stored on disk.

- PIMC_Moves has the basic moves: bead move, rigid move (translation), bisection move, and worm move (including open, close, wiggle, and swap).

- PIMC_Measurements has a few measurements, such as the superfluid fraction and the radial distribution function. The action-dependent energy estimators are in the action modules.

- PIMC_Reports produces screen reports for a few quantities, mainly for debugging and testing purposes to see whether the simulation is performing well.

- PIMC_Primitive_Action.jl and PIMC_Chin_Action.jl are the actions and action-specific energy estimators, respectively. Both contain the thermodynamical and virial energy estimators derived in this book. The kinetic action and the interaction are also defined here. The interaction is computed as an update to the existing value, which is initialized in PIMC_main.jl and refreshed now and then to avoid error accumulation.

- PIMC_Utilities.jl contains generic help functions, which are independent of the defined types.

- QMC_Statistics.jl is the same statistics-collecting routine as that used in DMC, where you can add data that is collected into blocks. Error estimation is based on the assumption that each block is statistically independent, but policing that is up to you, and there is no correlation time estimation. There are registered Julia packages that can do the same tasks as QMC_Statistics.jl, but I introduced this module to make the code more self-contained.

The potential energy tail correction (Vtail in the code) is relevant for fluids, where $\lim_{r \to \infty} g(r) \to 1$. Its purpose is to correct for the PBC cutoff distance at half the box width $L/2$ (beyond that distance, nearby periodic images spoil correlations). In the code, the potential energy evaluation is limited to a distance of $L/2$, where the potential is often nonzero. One possibility is to force the potential to go smoothly to zero at $L/2$; for small N, this may shift the potential minimum toward zero, causing the potential energy to end up far too high.[24] Smooth cutoffs are more important in DMC, where the whole simulation flow relies on the energetics.

The PIMC code writes numerous output files, mainly useful for plotting the data while running the code, which, by the way, has no stopping criterion. Personally, I work on a Linux workstation, and my on-the-fly plotting is done using *gnuplot*. The default graphics output of gnuplot may be less sophisticated than the graphical

---

[24] Although it is not used in the code, a smooth cutoff can be found in PIMC_Systems in the liquid He part so that you can experiment with it. For $N = 16$, it leads to a really inadequate potential energy and $g(r)$ that is visibly wrong. The lesson is that smoothing should not change anything near the potential minimum.

output produced by `matplotlib`, but it is so simple to plot computed data and just press 'e' on the keyboard to reread the data to see where it is going. The data files also contain the raw block averages, in case you want to get an idea of how well a quantity has been sampled or to run error analysis in postproduction.

The 'serious' data is stored in an HDF5 file (suffix `.h5`), one file for each PIMC simulation. HDF5 files can be read using, for example, Julia or Python, and in Linux, the contents can be dumped on screen using h5dump. The data is divided into GROUP and DATASET, such as the thermal energy estimator GROUP "E_th", the energy DATASET "E", and the error DATASET "std_E". The actual data is in DATA:

```
GROUP "/" {
 GROUP "E_th" {
 DATASET "E" {
 DATATYPE H5T_IEEE_F64LE
 DATASPACE SCALAR
 DATA {
 (0): -8.19343
 }
 }
 DATASET "std_E" {
 DATATYPE H5T_IEEE_F64LE
 DATASPACE SCALAR
 DATA {
 (0): 0.322425
 }
 . . .
```

The code `analyze_hdf5.py` reads and plots a few liquid ^{4}He results, stored in the files `results.He_liquid_chin_*` in the same directory. Further refinement of what to plot and in what color is achieved using `idlist` (for example, `idlist = [ ['tau0.01','16','olive'],['tau0.01','32','-green']]`) and by filtering out results that do not have $\tau = 0.01$ and particle numbers of 16 or 32. The colors are standard `matplotlib` colors. I used this Python script during computation; therefore, clicking on figure 1 rereads the chosen HDF5 data files and refreshes the plots.

Test on liquid ^{4}He

The test is run in the canonical ensemble. The density is interpolated from experimental data, which shows a maximum at the superfluid transition temperature. A more self-consistent way would be to find the saturated vapor pressure (SVP) conditions by locating the density at zero pressure.

Figure 5.20 shows the radial distribution function $g(r)$ computed using the PIMC code with the CA and $\tau = 0.01$, compared with experimental data. The radial distribution function is a sensitive error indicator. If $g(r)$ is completely wrong, it

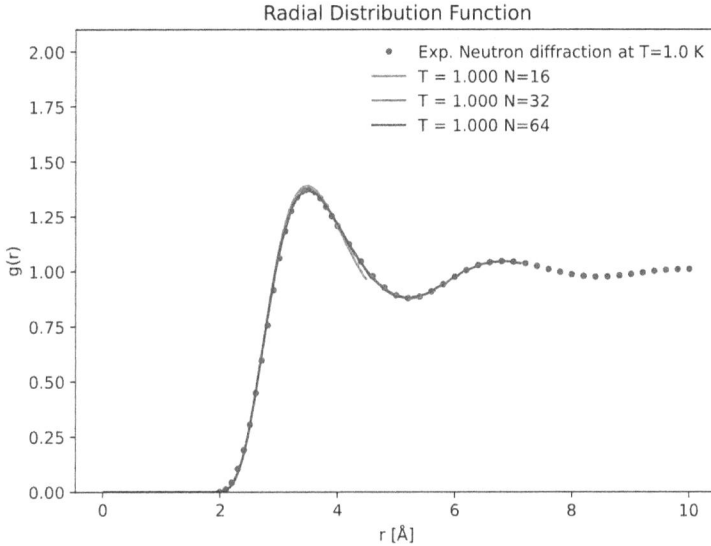

Figure 5.20. The radial distribution function of liquid ^{4}He at $T = 1$ K. The solid lines are PIMC results, and the markers show neutron diffraction data provided by Svensson *et al* [60]. Using only 16 atoms cuts off $g(r)$ as early as $L/2 = 4.5$ Å, and the curve misses all but the nearest-neighbor maximum. With 64 atoms, the cutoff is at $L/2 = 7.16$ Å, which already reproduces the experimental result quite accurately.

signals that there is a mistake in the PIMC moves; either a coding error or the detailed balance condition is broken. The problem may also be due to an accuracy issue; for example, increasing τ to, say, $\tau = 0.05$ shows almost instantly in PIMC near the leg of the $g(r)$ curve, indicating that the approximate Green's function has begun to fail and has allowed atoms to get too close to each other.

The potential and total energies of liquid ^{4}He are shown in figures 5.21 and 5.22, respectively. The centroid-reference virial estimator with window M used in the code performs relatively better at higher temperatures, where there are fewer multiparticle exchange loops. The potential energy is noticeably more accurately estimated than in the thermal energy estimator, a fact that led us to use the virial estimator in the first place.

Figure 5.23 shows the superfluid density, measured using the same PIMC calculation as that used for the energies and the radial distribution function. Taking into account the finite-size effects, it would be possible to approximate the superfluid transition temperature. The low-temperature points are clearly inaccurate, and the error estimates are unreliable. This indicates sampling issues at low T. Winding is a global, topological property, and sampling different windings may cause trouble; if one is not careful, it is easy to miss superfluidity altogether. The worm algorithm swap moves are perfectly capable of producing long exchange loops that pass through periodic boxes, so one has to pay attention to the worm parameters. It is also possible to propose swap moves that more frequently try swaps with beads that are further away from the worm head, keeping in mind that

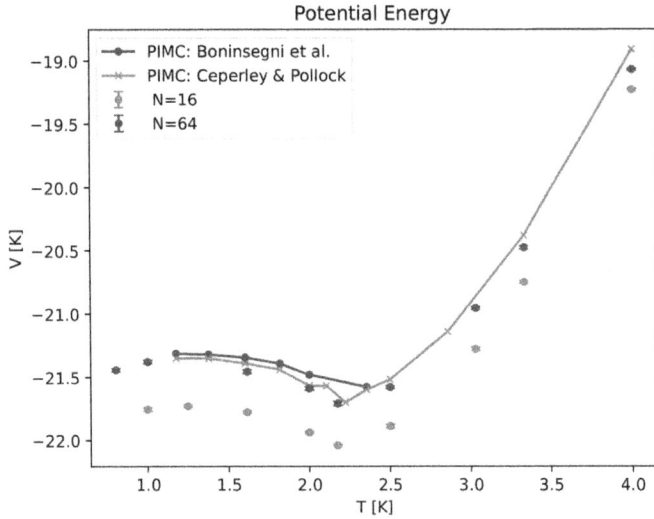

Figure 5.21. The potential energy of liquid ^{4}He as a function of temperature. The solid red curve with crosses represents the PIMC result from [20], and the blue curve with circles represents the PIMC result from [61]. As expected for the small cutoff distance, the $N = 16$ result exhibits substantial deviation.

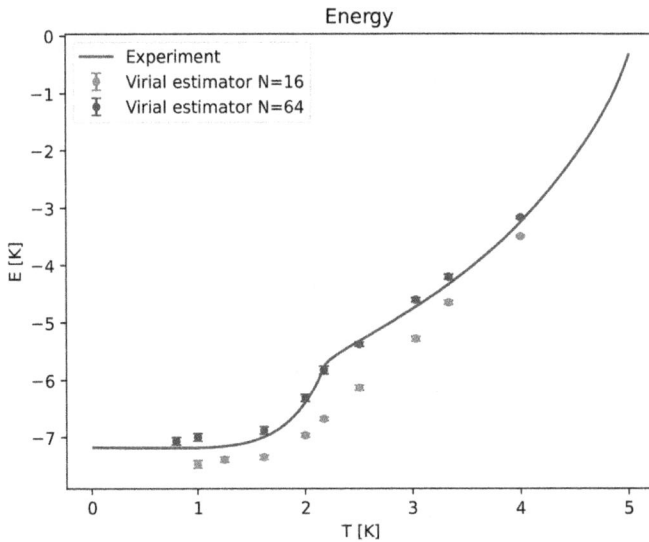

Figure 5.22. The total energy of liquid ^{4}He as a function of temperature. The blue curve represents the experimental energy, integrated from the specific heat data from [62]. Points with error bars represent the PIMC virial estimator results.

such nonsymmetric proposals have to be compensated to maintain detailed balance (as per Metropolis–Hastings). One can also experiment with larger τ to see whether ρ_s is calculated accurately using fewer slices.

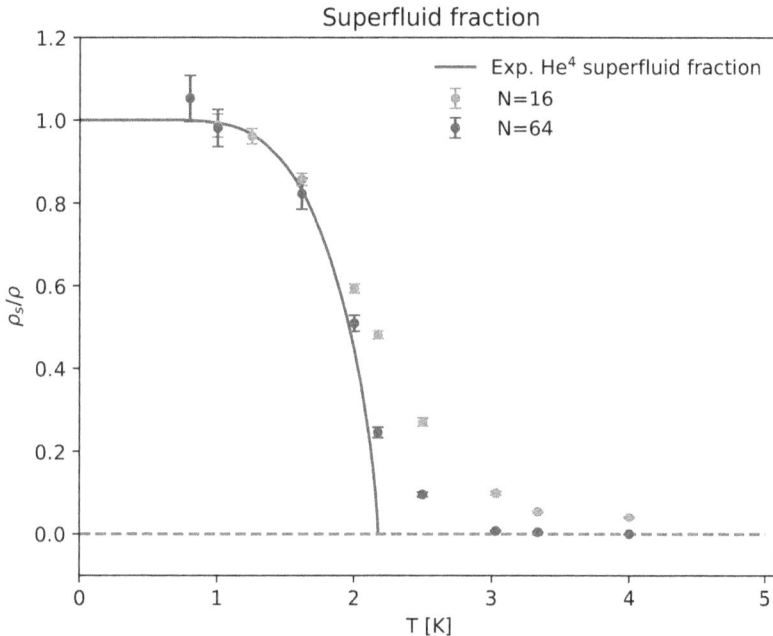

Figure 5.23. The superfluid density of liquid ^{4}He as a function of temperature. The blue curve represents the experimental ρ_s [62]. Finite-size effects are apparent around the superfluid transition in PIMC. Sampling winding at low temperatures can be ineffective, so it is a good idea to plot the ρ_s samples to see whether winding is stuck on one value over many MC cycles.

5.8 Stochastic series expansion

QMC is about sampling the possible configurations in a quantum system, and in this respect, the path integrals of discrete systems are a bit more tractable than continuous real-space path integrals. The method introduced here has *no time-step error* from splitting the operator $e^{-\beta\hat{\mathcal{H}}}$, and such approaches generally go under the name of **continuous-time QMC algorithms**. The continuous-time splitting of $e^{-\beta\hat{\mathcal{H}}}$ is by no means tied to discrete systems. For example, one can always define $\hat{\mathcal{H}} = \hat{\mathcal{H}}_0 + \hat{\mathcal{H}}'$ and write a perturbation expansion for $e^{-\beta\hat{\mathcal{H}}}$ using the soluble Hamiltonian $\hat{\mathcal{H}}_0$ and perturbation $\hat{\mathcal{H}}'$ (see appendix G).

The **stochastic series expansion (SSE)** method

$$Z = \mathrm{Tr}[e^{-\beta\hat{\mathcal{H}}}] = \sum_s \sum_{n=0}^{\infty} \frac{(-\beta)^n}{n!} \langle s|\hat{\mathcal{H}}^n|s\rangle \qquad \text{SSE}, \qquad (5.317)$$

was invented by A. Sandvik and J. Kurkijärvi [63] as an extension to Handscomb's method. SSE can be used for lattice spin systems, such as Ising and Hubbard models, and the expansion itself is generally valid.

Consider the **Hubbard model** Hamiltonian with nearest-neighbor jumps t, on-site interaction U, and chemical potential μ (in the grand canonical ensemble),

$$\hat{\mathcal{H}} = \sum_{\langle ij \rangle} \left[-t(\hat{a}_i^\dagger \hat{a}_j + \hat{a}_j^\dagger \hat{a}_i) \right] + \sum_i \left[-\mu \hat{n}_i + \frac{U}{2} \hat{n}_i(\hat{n}_i - 1) \right], \qquad (5.318)$$

where the creation and annihilation operators $\hat{a}$ obey either boson or fermion commutation rules and $\hat{n}$ is the single-site particle-number operator. The discreteness in a real-space system results from particle–particle Coulomb repulsions. The interactions approximately localize particles to sites, which in turn can be described by Wannier wave functions. Once the transition probabilities have been worked out, one finds that the essential physics is captured by electrons hopping from site to site with the hopping parameter t. The origin of the U-term is the on-site Coulomb repulsion: two electrons with opposite spins occupying the same site add an amount U to the energy. A negative U, on the other hand, would encourage electrons to pair, as in superconductivity. The Hubbard model has been a treasure trove for studies of magnetism and gained importance with the advent of ultracold boson lattices.

For SSE, the Hubbard Hamiltonian is expressed as a sum over bond operators,[25]

$$\hat{\mathcal{H}} = - \sum_i \hat{\mathcal{K}}_i$$

$$= - \sum_{\langle ij \rangle} \left[\underbrace{t(\hat{a}_i^\dagger \hat{a}_j + \hat{a}_j^\dagger \hat{a}_i)}_{\text{off-diagonal}} + \underbrace{\frac{\mu}{Z}(\hat{n}_i + \hat{n}_j) - \frac{U}{2Z}[\hat{n}_i(\hat{n}_i - 1) + \hat{n}_j(\hat{n}_j - 1)]}_{\text{diagonal}} \right], \qquad (5.319)$$

and the partition function includes the products of bond operators,

$$Z = \sum_s \sum_{n=0}^{\beta} \frac{\beta^n}{n!} \sum_{i_1, \dots, i_n} \langle s | \hat{\mathcal{K}}_{i_1} \dots \hat{\mathcal{K}}_{i_n} | s \rangle. \qquad (5.320)$$

The idea is to sample the combinations of bond operators and the end states $|s\rangle$. One way to sample the stochastic series is through a **directed loop update**, which is a global worm update [64]. A few notions about the directed loop update are:

- The number of operators is controlled by the temperature.
- Only diagonal operators can be inserted/removed locally.
- It is easier to deal with constant M operators; use identity operators $\mathbb{1}$ as placeholders for diagonal operators that actually appear in $\hat{\mathcal{H}}$ (see figure 5.24).

The implementation is as follows:
- **Start**: Pick any reasonable state $|s\rangle$ and insert M identity operators $\mathbb{1}$ (see figure 5.25).

- **Diagonal update**: Change identity operators to diagonal ones or vice versa with the following acceptance probabilities:

---

[25] The coordination number Z is the system-dependent number of nearest neighbors.

Example: 8 sites   / Update

12324211 ——→ 12324211
⬚   ▪
12324211 ←—— 12324211

⬚ Identity operator
▪ Diagonal operator   Site occupations

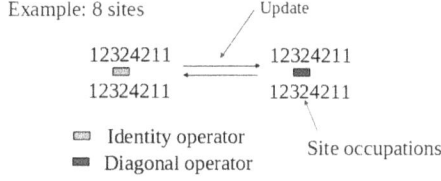

Figure 5.24. Identity operators and diagonal operators are interchangeable.

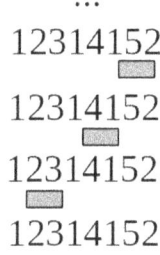

$$\cdots$$

$$12314152$$
⬚

$$12314152$$
⬚

$$12314152$$
⬚

$$12314152$$

Figure 5.25. Identity operators (gray boxes) are placeholders for future diagonal operators.

$$A_{ins}(\mathbb{1} \to \hat{\mathcal{K}}_d) = \min\left[\frac{\langle s|\hat{\mathcal{K}}_d|s\rangle \beta N_{\text{bonds}}}{M-n},\ 1\right] \tag{5.321}$$

$$A_{rem}(\hat{\mathcal{K}}_d \to \mathbb{1}) = \min\left[\frac{M-n+1}{\langle s|\hat{\mathcal{K}}_d|s\rangle \beta N_{\text{bonds}}},\ 1\right]. \tag{5.322}$$

Derivation: Use the detailed balance condition for $I \to \hat{\mathcal{K}}_d$ and $\hat{\mathcal{K}}_d \to I$:

$$\mathbb{1}:\quad M-n+1 \ \leftrightarrow\ M-n$$
$$\hat{\mathcal{K}}_d:\quad\quad n-1 \ \leftrightarrow\ n$$

$$\frac{P(S \to S')}{P(S' \to S)} = \frac{W(S')}{W(S)} = \frac{\langle s'|\beta\hat{\mathcal{K}}_d|s'\rangle}{\langle s|\mathbb{1}|s\rangle} = \beta\langle s'|\hat{\mathcal{K}}_d|s'\rangle = \frac{T(S \to S')A_{ins}(\mathbb{1} \to \hat{\mathcal{K}}_d)}{T(S' \to S)A_{rem}(\hat{\mathcal{K}}_d \to \mathbb{1})}, \tag{5.323}$$

where

$$T(S \to S') = \text{probability of picking a } \mathbb{1} = \frac{M-n+1}{N_{bonds}} \tag{5.324}$$

$$T(S' \to S) = \text{probability of picking the specific } \hat{\mathcal{K}}_d = 1, \tag{5.325}$$

so that

$$\frac{A_{ins}(\mathbb{1} \to \hat{\mathcal{K}}_d)}{A_{rem}(\hat{\mathcal{K}}_d \to \mathbb{1})} = \frac{\langle s|\hat{\mathcal{K}}_d|s\rangle \beta N_{bonds}}{M-n+1}, \tag{5.326}$$

and use, for example, the Metropolis solution for acceptance probabilities $P_{ins}(1 \rightarrow \hat{\mathcal{K}}_d)$ and $P_{rem}(\hat{\mathcal{K}}_d \rightarrow 1)$. Here, operations must be *exactly reversible*, and one must take care that adding a particle and removing a particle are exactly opposite operations. Nominally, insertion is carried out using the operator $n - 1 \rightarrow n$, and removal takes place via $n \rightarrow n - 1$; however, in practice, n is the *current* number of $\hat{\mathcal{K}}_d$s, so insertion is instead $n \rightarrow n + 1$. Therefore, the insert acceptance has $M - n$, not $M - n + 1$.

Let us consider a Bose–Hubbard model simulation. A suitable update that can make a boson jump sites is the boson add/remove loop update $(+1)/(-1)$. As shown in figure 5.26, this update can change the type of an operator but not the total number of operators. The number of bosons can change, so the update naturally samples the grand canonical ensemble; the canonical ensemble needs more restrictions. The end state in Z is sampled automatically (see figure 5.27).

The update propagation rule

Knowing the in-leg, how do we pick the out-leg so that Z is sampled? Again, it is up to detailed balance to choose the out-leg. The weights are matrix elements,[26]

$$W(S) = \langle s_1 | \hat{\mathcal{K}} | s_2 \rangle, \ W(S') = \langle s' | \hat{\mathcal{K}}' | s' \rangle \qquad (5.327)$$

and we can create probability tables, denoted by $P(\text{update, in-leg,out-leg})$, such as

$$P(S \rightarrow S') = P(+1; i, j), \ P(S \rightarrow S') = P(-1; i, j) \text{ or } P(+1; j, i). \qquad (5.328)$$

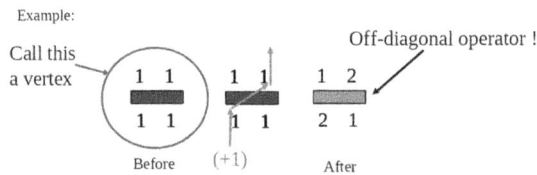

Figure 5.26. A $(+1)$ off-diagonal update changes a diagonal vertex to an off-diagonal one.

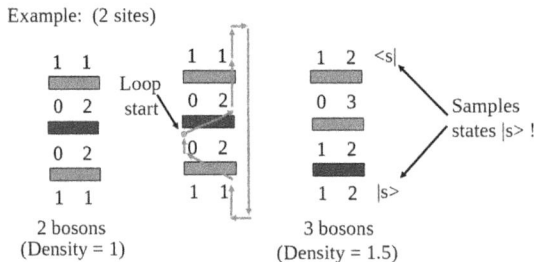

Figure 5.27. A full update loop can change the number of particles and sample the end states.

---

[26] The matrix elements can be evaluated using known rules for creation and annihilation operators.

One possible choice that satisfies the detailed balance condition is the so-called *heat bath*,

$$P(S \to S') = \frac{W(S')}{\sum\limits_{S} W(S)}, \qquad (5.329)$$

which assumes positive weights. A loop update may 'bounce,' meaning that the up-going update adds a boson, while a down-going one removes a boson. This is not desirable, because it leaves the vertex unchanged. Bounces can be made rarer by making a choice other than the heat bath for the acceptance $P(S \to S')$ (while keeping the detailed balance condition valid).

Positive weights $W(S)$ can be ensured by adding a constant C to the Hamiltonian,

$$\hat{\mathcal{H}} = -\sum_{\langle ij \rangle} \left[\underbrace{t(\hat{a}_i^\dagger \hat{a}_j + \hat{a}_j^\dagger \hat{a}_i)}_{\text{off-diagonal}} + \underbrace{C + \frac{\mu}{Z}(\hat{n}_i + \hat{n}_j) - \frac{U}{2Z}[\hat{n}_i(\hat{n}_i - 1) + \hat{n}_j(\hat{n}_j - 1)]}_{\text{diagonal}} \right],$$

- $\langle s | \text{off-diagonal} | s' \rangle \geqslant 0$, so it is already fine.
- $\langle s | \text{diagonal} | s \rangle \geqslant 0$ if we choose a maximum allowed occupation n_{max} and a large enough C to match it.
- If n_{max} turns out to be too low, start anew with a larger one.

An example of a full loop update is shown in figure 5.28. The loop continues until it reaches the starting point (i.e. the worm closes). This usually happens soon enough because the number of paths the loop update can take is limited.

Estimators

Similar to real-space PIMC simulations, observables in SSE are measured using estimators. For example, the thermodynamical energy estimator is

$$\langle E \rangle = -\frac{1}{Z}\frac{\partial Z}{\partial \beta} = -\frac{1}{\beta}\frac{1}{Z}\sum_{s}\sum_{n=0}^{\infty}\frac{n\beta^n}{n!}\sum_{i_1,\ldots,i_n} \langle s | \hat{\mathcal{K}}_{i_1} \ldots \hat{\mathcal{K}}_{i_n} | s \rangle = -\frac{\langle n \rangle}{\beta},$$

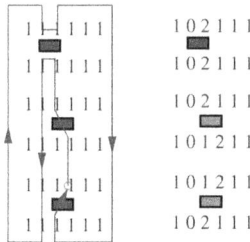

Figure 5.28. Another example of a full loop update. The start/finish is indicated by the red circle. In this particular loop update, two off-diagonal updates appear, and the number of bosons remains at six. The update does not bounce at the uppermost operator, because the in-leg is not the same as the out-leg.

where $\langle n \rangle$ is the average number of operators (vertices). The ground-state energy is a finite, fixed number, so $\langle n \rangle$ increases with β. Therefore, low-temperature simulations have more terms to sample, which makes them slow. The energy fluctuates around $\langle E \rangle$, so there is a finite maximum number of terms M that is reached during any finite-time simulation. Density and compressibility are simple to measure, and the superfluid density ρ_s is measured via the winding number, which is now the net amount of hopping across a boundary. A detailed study of a boson lattice can be found in [65].

References

[1] Feynman R P and Hibbs A R 1965 Quantum Mechanics and Path Integrals *International Series in Pure and Applied Physics* (New York: McGraw-Hill)

[2] Lévy P 1954 Le Mouvement Brownien *Mémor. Sci. Math.* **126** 1–81

[3] Tuckerman M E 2023 *Statistical Mechanics: Theory and Molecular Simulation* 2nd edn (Oxford: Oxford University Press)

[4] Pollock E L and Ceperley D M 1984 Simulation of quantum many-body systems by path-integral methods *Phys. Rev.* B **30** 2555–68

[5] Whitfield T W and Martyna G J 2007 Low variance energy estimators for systems of quantum Drude oscillators: treating harmonic path integrals with large separations of time scales *J. Chem. Phys.* **126** 074104

[6] Prokof'ev N, Svistunov B and Tupitsyn I 1998 Exact, complete, and universal continuous-time worldline Monte Carlo approach to the statistics of discrete quantum systems *J. Exp. Theor. Phys.* **87** 310–21

[7] Boninsegni M, Prokof'ev N and Svistunov B 2006 Worm algorithm for continuous-space path integral Monte Carlo simulations *Phys. Rev. Lett.* **96** 070601

[8] Herdman C M, Rommal A and Del Maestro A 2014 Quantum Monte Carlo measurement of the chemical potential of ^{4}He *Phys. Rev.* B **89** 224502

[9] Dornheim T, Bonitz M, Moldabekov Z, Schwalbe S, Tolias P and Vorberger J 2025 Chemical potential of the warm dense electron gas from *ab initio* path integral Monte Carlo simulations *Phys. Rev.* B **111** 115149

[10] Chin S A and Chen C R 2002 Gradient symplectic algorithms for solving the Schrödinger equation with time-dependent potentials *J. Chem. Phys.* **117** 1409–15

[11] Lindoy L P, Huang G S and Jordan M J T 2018 Path integrals with higher order actions: application to realistic chemical systems *J. Chem. Phys.* **148** 074106

[12] Ceperley D M 1995 Path integrals in the theory of condensed helium *Rev. Mod. Phys.* **67** 279–355

[13] Sakkos K, Casulleras J and Boronat J 2009 High order Chin actions in path integral Monte Carlo *J. Chem. Phys.* **130** 204109

[14] Leggett A J 1999 Superfluidity *Rev. Mod. Phys.* **71** S318–23

[15] London F 1938 On the Bose-Einstein condensation *Phys. Rev.* **54** 947–54

[16] London. F 1938 The λ -phenomenon of liquid helium and the Bose–Einstein egeneracy *Nature* **141** 643–4

[17] Landau L 1941 Theory of the superfluidity of helium II *Phys. Rev.* **60** 356–8

[18] Feynman R P 1953 Atomic theory of the λ transition in helium *Phys. Rev.* **91** 1291–301

[19] Feynman R P 1954 Atomic theory of the two-fluid model of liquid helium *Phys. Rev.* **94** 262–77

[20] Pollock E L and Ceperley D M 1987 Path-integral computation of superfluid densities *Phys. Rev.* B **36** 8343–52

[21] Mareschal. M 2021 The early years of quantum Monte Carlo (2): finite-temperature simulations *Eur. Phys. J.* H **46** 26

[22] Sindzingre P, Klein M L and Ceperley D M 1989 Path-integral Monte Carlo study of low-temperature ^{4}He clusters *Phys. Rev. Lett.* **63** 1601–4

[23] Toennies J P, Vilesov A F and Whaley K B 2001 Superfluid helium droplets: an ultracold nanolaboratory *Phys. Today* **54** 31–7

[24] Zillich R E, Paesani F, Kwon Y and Whaley K B 2005 Path integral methods for rotating molecules in superfluids *J. Chem. Phys.* **123** 114301

[25] Leggett A J 1970 Can a solid be 'superfluid'? *Phys. Rev. Lett.* **25** 1543–6

[26] Kim E and Chan M H W 2004 Probable observation of a supersolid helium phase *Nature* **427** 225–7

[27] Léonard J, Morales A, Zupancic P, Esslinger T and Donner T 2017 Supersolid formation in a quantum gas breaking a continuous translational symmetry *Nature* **543** 87–90

[28] Aikawa K, Frisch A, Mark M, Baier S, Rietzler A, Grimm R and Ferlaino F 2012 Bose-Einstein condensation of erbium *Phys. Rev. Lett.* **108** 210401

[29] Casotti E, Poli E, Klaus L, Litvinov A, Ulm C, Politi C, Mark M J, Bland T and Ferlaino F 2024 Observation of vortices in a dipolar supersolid *Nature* **635** 327–31

[30] Fisher M E, Barber M N and Jasnow D 1973 Helicity modulus, superfluidity, and scaling in isotropic systems *Phys. Rev.* A **8** 1111–24

[31] Rousseau V G 2014 Superfluid density in continuous and discrete spaces: avoiding misconceptions *Phys. Rev.* B **90** 134503

[32] Farías C, Pinto V A and Moya P S 2017 What is the temperature of a moving body? *Sci. Rep.* **7** 17657

[33] Pollock E L and Runge K J 1992 Finite-size-scaling analysis of a simulation of the ^{4}He superfluid transition *Phys. Rev.* B **46** 3535–9

[34] Spada G, Giorgini S and Pilati S 2022 Path-integral Monte Carlo worm algorithm for Bose systems with periodic boundary conditions *Condens. Matter* **7** 30

[35] Grüter P, Ceperley D and Laloë F 1997 Critical temperature of Bose-Einstein condensation of hard-sphere gases *Phys. Rev. Lett.* **79** 3549–52

[36] Anderson M H, Ensher J R, Matthews M R, Wieman C E and Cornell E A 1995 Observation of Bose-Einstein condensation in a dilute atomic vapor *Science* **269** 198–201

[37] Pethick C J and Smith H 2008 *Bose-Einstein Condensation in Dilute Gases* 2nd edn (Cambridge: Cambridge University Press)

[38] Herman M F, Bruskin E J and Berne B J 1982 On path integral Monte Carlo simulations *J. Chem. Phys.* **76** 5150–5

[39] Janke W and Sauer T 1997 Optimal energy estimation in path-integral Monte Carlo simulations *J. Chem. Phys.* **107** 5821–39

[40] Cabrera-Trujillo R and Vendrell O 2020 On the virial theorem for a particle in a box: accounting for Cauchyas boundary condition *Am. J. Phys.* **88** 1103–8

[41] Rota R 2011 Path integral Monte Carlo and Bose-Einstein condensation in quantum fluids and solids *PhD Thesis* Universitat Politècnica de Catalunya

[42] Yamamoto T M 2005 Path-integral virial estimator based on the scaling of fluctuation coordinates: application to quantum clusters with fourth-order propagators *J. Chem. Phys.* **123** 104101

[43] Troyer M and Wiese U-J 2005 Computational complexity and fundamental limitations to fermionic quantum Monte Carlo simulations *Phys. Rev. Lett.* **94** 170201

[44] Chin S A 2015 High-order path-integral Monte Carlo methods for solving quantum dot problems *Phys. Rev.* E **91** 031301

[45] Borrmann P and Franke G 1993 Recursion formulas for quantum statistical partition functions *J. Chem. Phys.* **98** 2484–5

[46] Barghathi H, Yu J and Del Maestro A 2020 Theory of noninteracting fermions and bosons in the canonical ensemble *Phys. Rev. Res.* **2** 043206

[47] Yang C N and Lee T D 1952 Statistical theory of equations of state and phase transitions. I. Theory of condensation *Phys. Rev.* **87** 404–9

[48] Fisher. M E 1965 The nature of critical points *Lectures in Theoretical Physics* **vol 7C** ed W E Brittin (Boulder, CO: University of Colorado Press) pp 1–159

[49] Mülken O, Borrmann P, Harting J and Stamerjohanns H 2001 Classification of phase transitions of finite Bose-Einstein condensates in power-law traps by Fisher zeros *Phys. Rev.* A **64** 013611

[50] Ceperley D M 1991 Fermion nodes *J. Stat. Phys.* **63** 1237–67

[51] Dornheim T 2021 Fermion sign problem in path integral Monte Carlo simulations: grand-canonical ensemble *J. Phys. A: Math. Theor.* **54** 335001

[52] Schoof T, Groth S, Vorberger J and Bonitz M 2015 *Ab initio* thermodynamic results for the degenerate electron gas at finite temperature *Phys. Rev. Lett.* **115** 130402

[53] Dornheim T, Groth S, Filinov A and Bonitz M 2015 Permutation blocking path integral Monte Carlo: a highly efficient approach to the simulation of strongly degenerate non-ideal fermions *New J. Phys.* **17** 073017

[54] Morresi T and Garberoglio G 2025 Revisiting the properties of superfluid and normal liquid ^{4}He using *ab initio* potentials *J. Low Temp. Phys.* **219** 103–22

[55] Vranje L, Markić š and Glyde H R 2015 Superfluidity, BEC, and dimensions of liquid ^{4}He in nanopores *Phys. Rev.* B **92** 064510

[56] Spada G, Pilati S and Giorgini S 2022 Thermodynamics of a dilute Bose gas: a path-integral Monte Carlo study *Phys. Rev.* A **105** 013325

[57] Dornheim T, Moldabekov Z A, Vorberger J and Militzer B 2022 Path integral Monte Carlo approach to the structural properties and collective excitations of liquid ^{3}He without fixed nodes *Sci. Rep.* **12** 708

[58] Dornheim T, Bonitz M, Moldabekov Z A, Schwalbe S, Tolias P and Vorberger J 2025 Chemical potential of the warm dense electron gas from *ab initio* path integral Monte Carlo simulations *Phys. Rev.* B **111** 115149

[59] Dornheim T, Moldabekov Z A, Schwalbe S and Vorberger J 2025 Direct free energy calculation from *ab initio* path integral Monte Carlo simulations of warm dense matter *Phys. Rev.* B **111** L041114

[60] Svensson E C, Sears V F, Woods A D B and Martel P 1980 Neutron-diffraction study of the static structure factor and pair correlations in liquid ^{4}He *Phys. Rev.* B **21** 3638–51

[61] Boninsegni M, Prokof'ev N V and Svistunov B V 2006 Worm algorithm and diagrammatic Monte Carlo: a new approach to continuous-space path integral Monte Carlo simulations *Phys. Rev.* E **74** 036701

[62] Donnelly R J and Barenghi. C F 1998 The observed properties of liquid helium at the saturated vapor pressure *J. Phys. Chem. Ref. Data* **27** 1217–74

[63] Sandvik A W and Kurkijärvi J 1991 Quantum Monte Carlo simulation method for spin systems *Phys. Rev.* B **43** 5950–61

[64] Syljuåsen O F and Sandvik A W 2002 Quantum Monte Carlo with directed loops *Phys. Rev.* E **66** 046701

[65] Syljuåsen O F 2003 Directed loop updates for quantum lattice models *Phys. Rev.* E **67** 046701

IOP Publishing

A Practical Course on Quantum Monte Carlo

Vesa Apaja

Chapter 6

Path integral ground-state Monte Carlo

It may appear a bit convoluted to work out a finite-temperature path integral theory and then realize that similar ideas can be used to find the ground state. The need for ground-state path integrals was born from the fact that path integral Monte Carlo (PIMC) cannot handle near-zero temperatures. The ground-state version goes by the names **ground-state path integral Monte Carlo (GS-PIMC)** or **path integral ground state (PIGS)**. For a recent review of the ground-state method, see [1].

Coming from PIMC, first forget the notion that imaginary time is cyclic and that it is related to temperature. The imaginary-time evolution, calculated in practice using diffusion Monte Carlo (DMC), is

$$\left|\Psi(\tau)\right\rangle = e^{-\tau\hat{\mathcal{H}}}\left|\varphi_T\right\rangle, \tag{6.1}$$

where τ is just an arbitrary imaginary-time step. This evolution was shown to represent a projection that projects out the ground state for sufficiently large τ, and the ground state can be made stable by subtracting (a guessed) E_0 from $\hat{\mathcal{H}}$. Splitting the evolution operator $e^{-\tau\hat{\mathcal{H}}}$ into M more manageable short-time evolutions gives

$$\left|\Psi(\tau)\right\rangle = e^{-\tau/M\hat{\mathcal{H}}}e^{-\tau/M\hat{\mathcal{H}}}. ..e^{-\tau/M\hat{\mathcal{H}}}|\Psi(0)\rangle. \tag{6.2}$$

You may imagine the evolution as follows (see figure 6.1). The trial state $|\varphi_T\rangle$ is the end of a chain, and you hold the chain in your hand. The chain hangs down, and the ground state is the floor. If your chain is long enough, the lower end of the chain will reach the floor, and you can start sampling the ground state by moving the chain. You now realize that you can sample the ground state, i.e. the floor, by holding one end of the chain in each hand and lowering your hands closer to the floor. You now have a *path* from one $|\varphi_T\rangle$ to the other, and you are evaluating an expectation value [2]:

doi:10.1088/978-0-7503-6310-5ch6

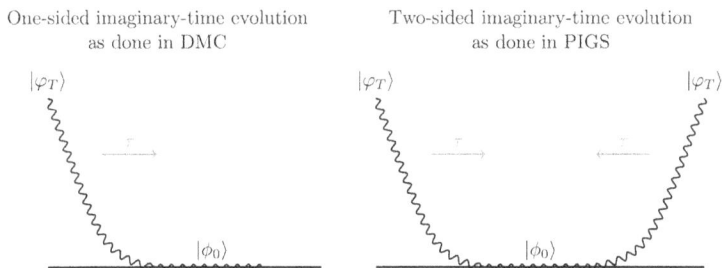

One-sided imaginary-time evolution
as done in DMC

Two-sided imaginary-time evolution
as done in PIGS

Figure 6.1. Schematic of imaginary-time evolution in DMC and in PIGS.

$$\langle \hat{O}(\tau) \rangle = \frac{\left\langle \varphi_T \middle| e^{-\tau\hat{\mathcal{H}}} \hat{O} e^{-\tau\hat{\mathcal{H}}} \middle| \varphi_T \right\rangle}{\left\langle \varphi_T \middle| e^{-\tau\hat{\mathcal{H}}} e^{-\tau\hat{\mathcal{H}}} \middle| \varphi_T \right\rangle}. \tag{6.3}$$

If τ is large enough, this is a ground-state expectation value. DMC can project to as large a value of τ as one needs, but the number of slices one can use in equation (6.3) is finite. Therefore, one must either use a reasonably good trial wave function or at least use a high-order splitting of $e^{-\tau\hat{\mathcal{H}}}$ to keep the number of slices small. Indeed, in [3], it is shown that for bosons, even $\varphi_T = 1$ (i.e. no correlations at all) can be used to find liquid and solid He^4 ground-state properties with the aid of a fourth-order Suzuki–Chin propagator. This demonstrates the extraordinary **robustness** of PIGS.

Now that the trial wave function made a comeback, one can apply the fixed-node approximation for fermions, **FN-PIGS**, as is done in [4] for pure ^{3}He and ^{3}He–^{4}He mixtures.

The practical implementation of PIGS is similar to that of PIMC, except that the 'end beads' are sampled from $\varphi_T(\mathbf{x})$ using, e.g. Metropolis sampling. Only the midslices are presumably sampling the ground state, but for checking purposes, it is a good idea to evaluate the energy in every slice to ascertain whether the imaginary-time propagation is long enough to reach the ground state. The potential and total energies should plateau in the midslices and make a curve similar to the schematic one shown in figure 6.1. A V-shaped curve indicates more slices or a better trial wave function is needed.

In PIGS, it is possible to overextend beyond the safe validity regime of a propagator. There is no real reason to use the same small time step everywhere just to keep the error in the propagator below a certain limit. PIGS is so robust that one can take a relatively long step in imaginary time, as long as the step gives a lower energy. In reference [5], the Chin–Suzuki action is shown to give a surprisingly good update for the trial wave function after just one fourth-order τ propagation! Similar to equation (5) in reference [5], we can use the Chin action Green's function in equation (5.122) and the density matrix in equation (5.133) (with $M = 3$) to construct a new trial wave function as follows:

$$\psi_T^{\text{new}}(\mathbf{x}) = \int d\mathbf{x}_1 G_{\text{CA}}(\mathbf{x}, \mathbf{x}_1; \tau_1)\psi_T(\mathbf{x}_1) \tag{6.4}$$

$$= \int d\mathbf{x}_1 d\mathbf{x}_2 d\mathbf{x}_3 \langle \mathbf{x} | e^{-t_1 \eta \hat{T}} | \mathbf{x}_1 \rangle \langle \mathbf{x}_1 | e^{-t_2 \eta \hat{T}} | \mathbf{x}_2 \rangle \langle \mathbf{x}_2 | e^{-t_3 \eta \hat{T}} | \mathbf{x}_3 \rangle$$
$$\times e^{-\eta \tilde{V}_1(\mathbf{x}_2; \eta)} e^{-\eta \tilde{V}_1(\mathbf{x}_2; \eta)} e^{-\eta \tilde{V}_1(\mathbf{x}_1; \eta)} \psi_T(\mathbf{x}_1).$$

(6.5)

Notice that here, τ_1 is an *adjustable parameter*. We can insert $\psi_T^{\text{new}}(\mathbf{x})$ into $\psi_T(\mathbf{x})$ and repeat the process; each iteration introduces a new, adjustable τ_2, and so on.

References

[1] Yan Y and Blume D 2017 Path integral Monte Carlo ground state approach: formalism, implementation, and applications. *J. Phys.* B **50** 223001
[2] Sarsa A, Schmidt K E and Magro W R 2000 A path integral ground state method *J. Chem. Phys.* **113** 1366–71
[3] Rossi M, Nava M, Reatto L and Galli D E 2009 Exact ground state Monte Carlo method for Bosons without importance sampling *J. Chem. Phys.* **131** 154108
[4] Ujevic S, Zampronio V, de Abreu B R and Vitiello S A 2023 Properties of fermionic systems with the path-integral ground state method *SciPost Phys. Core* **6** 031
[5] Rota R, Casulleras J, Mazzanti F and Boronat J 2010 High-order time expansion path integral ground state *Phys. Rev.* E **81** 016707

IOP Publishing

A Practical Course on Quantum Monte Carlo

Vesa Apaja

Appendix A

Central limit theorem

The central limit theorem states that:

> *Given a distribution with a mean μ and a variance σ^2, the sampling distribution of the mean approaches a normal distribution with a mean μ and a variance σ^2/N as the sample size N increases.*

This is almost magic: when taking the mean value of variables x with *almost any* probability distribution function (PDF) $f(x)$, such that

$$z = \frac{x_1 + x_2 + \ldots + x_N}{N}, \tag{A.1}$$

the sampling distribution $g(z)$ of the mean value approaches a normal distribution! Even more amazingly, this even seems to happen for a rather small N.

What exactly is N and a 'sampling distribution'? In principle, the number of samples x_i is infinite, but we are not going to add them all up to get the mean. So we *sample the mean* by taking a finite-sized specimen of N points and calculating their mean to get an idea of the true mean.

Proof:

The probability distribution f has a mean and a variance, defined as

$$\mu := \int dx f(x) x \qquad \text{mean} \tag{A.2}$$

$$\sigma^2 := \int dx f(x)(x - \mu)^2 \qquad \text{variance.} \tag{A.3}$$

The N values x_i sampled from f are statistically independent (the value x_i does not depend on x_k if $i \neq k$), so their joint probability distribution of N values is the product $f(x_1)f(x_2) \ldots f(x_N)$. We wish to find the distribution $g(z)$ of the mean z given in equation (A.1). Such a constraint can be expressed in an integral using the Dirac delta function. So far, we have

$$g(z) = \int dx_1 dx_2 \ldots dx_N \underbrace{f(x_1)f(x_2) \ldots f(x_N)}_{\text{samples from the distribution } f} \underbrace{\delta(z - \frac{x_1 + x_2 + \ldots + x_N}{N})}_{\text{constraint that the mean is } z}. \quad (A.4)$$

As a Monte Carlo algorithm:

Sample N values x_i from the distribution f and call their mean z. Each value of z samples the distribution g.

Next, write the Dirac delta in integral form,

$$\delta(x) = \frac{1}{2\pi} \int dk e^{ikx}, \quad (A.5)$$

to get

$$g(z) = \frac{1}{2\pi} \int dk dx_1 dx_2 \ldots dx_N f(x_1)f(x_2) \ldots f(x_N) e^{ik\left(z - \frac{x_1 + x_2 + \ldots + x_N}{N}\right)}. \quad (A.6)$$

The x-integrals are separate and they are all the same integral,

$$g(z) = \frac{1}{2\pi} \int dk \int dx_1 f(x_1) e^{ik\left(\frac{z - x_1}{N}\right)} \ldots \int dx_N f(x_N) e^{ik\left(\frac{z - x_N}{N}\right)} \quad (A.7)$$

$$= \frac{1}{2\pi} \int dk \left[\int dx f(x) e^{ik\left(\frac{z - x}{N}\right)}\right]^N. \quad (A.8)$$

Since the values are close to the mean μ, add and subtract μ,

$$g(z) = \frac{1}{2\pi} \int dk \left(\int dx f(x) e^{ik\left(\frac{z - \mu + \mu - x}{N}\right)}\right)^N \quad (A.9)$$

$$= \frac{1}{2\pi} \int dk e^{ik(z - \mu)} \left(\int dx f(x) e^{ik\left(\frac{\mu - x}{N}\right)}\right)^N. \quad (A.10)$$

The integral in parentheses can be expanded in powers of the small deviation $\mu - x$,

$$\int dx f(x)\left(1 + \frac{ik(\mu - x)}{N} - \frac{k^2(\mu - x)^2}{2N^2} + \ldots\right) \quad (A.11)$$

$$= 1 + \frac{ik}{N}\underbrace{\left(\mu - \int dx f(x)x\right)}_{=0} - \frac{k^2}{2N^2}\underbrace{\int dx f(x)(\mu - x)^2}_{\sigma^2} + \ldots \quad (A.12)$$

$$= 1 - \frac{k^2 \sigma^2}{2N^2} + \dots, \tag{A.13}$$

using the definitions in equation (A.3). Insert this approximation into $g(z)$ and use the identity

$$\left(1 - \frac{x}{N}\right)^N \approx e^{-x}, \quad \text{for large } N. \tag{A.14}$$

The integral in $g(z)$ becomes

$$g(z) \approx \frac{1}{2\pi} \int dk\, e^{ik(z-\mu)} \left(1 - \frac{k^2 \sigma^2}{2N^2}\right)^N \approx \frac{1}{2\pi} \int dk\, e^{ik(z-\mu)} e^{-\frac{k\sigma^2}{2N}}, \tag{A.15}$$

The integral computes the Fourier transform of a Gaussian, so the result is:

CENTRAL LIMIT THEOREM

If

$$z := \frac{x_1 + x_2 + \dots + x_N}{N} \tag{A.16}$$

where x_is are sampled from a distribution with mean μ and variance σ^2, then for large N, the distribution of z is

$$g(z) \approx \frac{1}{\sqrt{2\pi}\,\sigma_z} e^{-\frac{(z-\mu)^2}{2\sigma_z^2}} \qquad \text{with } \sigma_z := \sigma/\sqrt{N},$$

a normal distribution with mean μ and variance $\sigma_z = \sigma^2/N$.

In other words: *for large N, the distribution of mean values is a normal distribution with a variance of σ^2/N.* This shows why Monte Carlo results converge as $1/\sqrt{N}$. Remarkably, the probability distribution $f(x)$ of values x is used only to find the two momenta μ and σ^2.

Figure A.1 illustrates how fast the distribution of the mean approaches the normal distribution. The underlying distribution $f(x)$ is box-shaped; it takes a value of one between -0.5 and 0.5 and zero elsewhere. In other words, $x \in U[-0.5, 0.5]$. The average of two values, $z = (x_1 + x_2)/2$, is limited to $-1 \leqslant z \leqslant 1$, so the tails of the normal distribution are beyond reach. Still, the triangular shape already resembles *Normal(z)*. The average of 12 numbers is so close to *Normal(z)* that it is hard to see the difference in the figure. Still, we know that the values are limited to $-6 \leqslant z \leqslant 6$, so the distribution in the tails is zero, unlike a normal distribution. Such a tail error is common to all distributions $f(x)$ that have a limited domain.

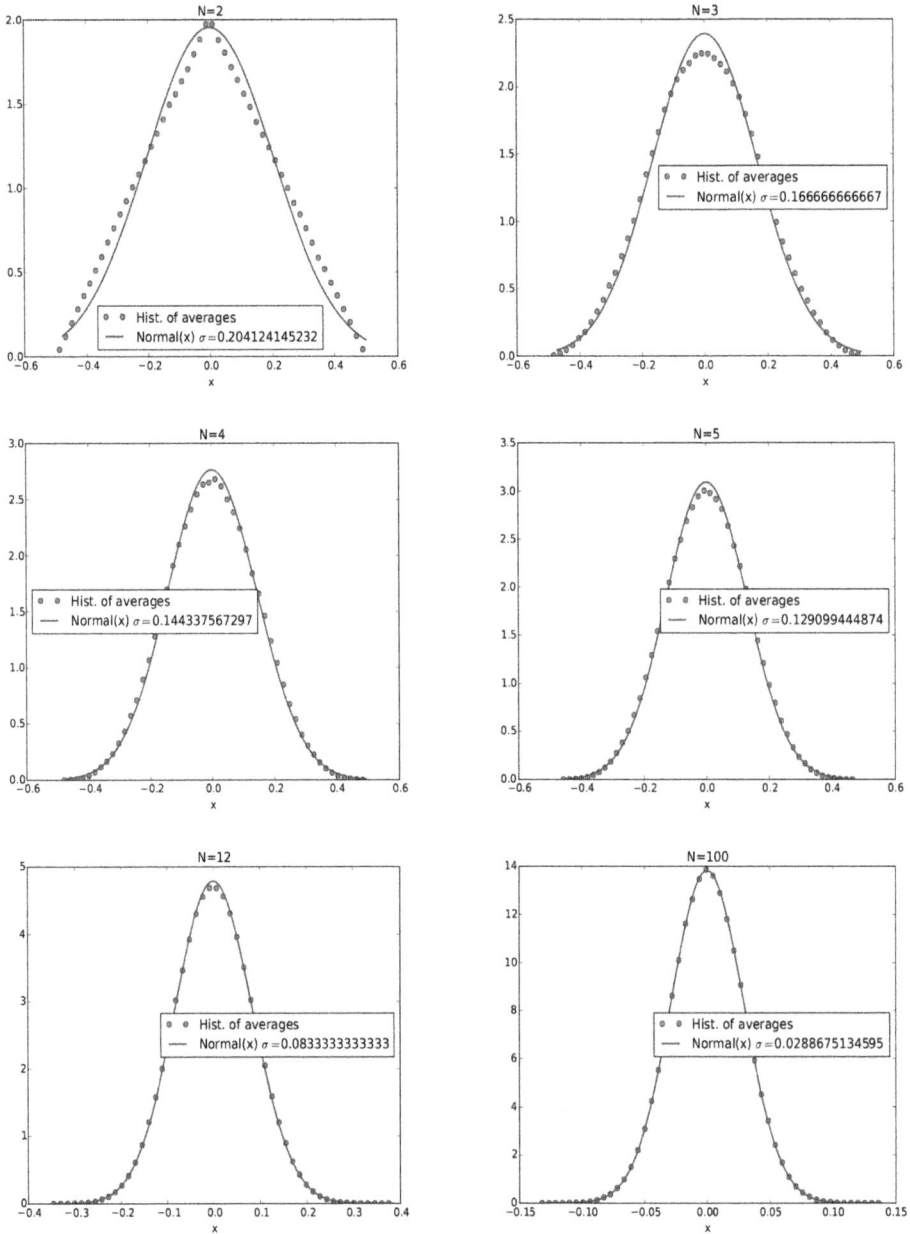

Figure A.1. Distribution of the mean value of N random numbers $\in U[-0.5, 0.5]$ for $N = 2, 3, 4, 5, 12, 100$ (markers), compared with the normal distribution (solid line). The input distribution has a variance of $\sigma_0^2 = 1/12$, and the plotted normal distribution variance is σ_0^2/N.

A.1 Cauchy distribution

The central limit theorem is valid only if the 'seed distribution,' i.e. the distribution from which the values x_i are sampled, has a well-defined mean and variance. To see that this is not always the case, consider the standard Cauchy distribution

$$w(x) = \frac{1}{\pi(1 + x^2)}, \tag{A.17}$$

which looks a bit like the standard normal distribution (see figure A.2). It is a probability distribution,

$$\int_{-\infty}^{\infty} w(x) = 1, \tag{A.18}$$

with a (mathematical) mean value of

$$\mu: = \int_{-\infty}^{\infty} w(x)x = 0. \tag{A.19}$$

The reason for this is that the argument is antisymmetric, and the contributions from $x < 0$ and $x > 0$ cancel exactly. However, the separate integrals for $x < 0$ and $x > 0$ diverge. If one samples N points $x_1, ..., x_N$ from the Cauchy distribution and computes their average, it cannot converge to any value because the points are not chosen symmetrically about $x = 0$: *the Cauchy distribution has no statistical mean.* The variance does not exist at all:

$$\sigma^2: = \int_{-\infty}^{\infty} w(x)(x - \mu)^2 = \infty. \tag{A.20}$$

The nonexistence of these momenta pulls the carpet from under the central limit theorem. You *can* sample the Cauchy distribution, and with a great deal of wishful thinking, calculate the mean over a set of points. But no matter how many data points you collect, your statistical estimate of the mean is as 'accurate' as with just one sampled point! With every new point, you need to adjust the statistical mean, and it never converges to a definite value.

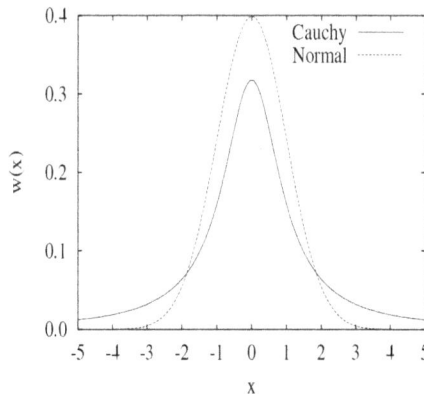

Figure A.2. The standard Cauchy probability distribution compared with the standard normal distribution. The Cauchy distribution looks benign, but it has some interesting statistical properties discussed in the main text.

A-5

One often uses the Cauchy distribution as a test distribution to see how sensitive a method is to the contributions made by the tails. As an aside, the distribution of the ratio of two normally distributed random variables with $\mu = 0$ is a Cauchy distribution; hence, it is also called the *normal ratio distribution*.

IOP Publishing

A Practical Course on Quantum Monte Carlo

Vesa Apaja

Appendix B

Diffusion matrix elements

The matrix elements $G_D(\mathbf{x}', \mathbf{x}; \tau)$ can also be derived directly using the eigenstates of $\hat{\mathcal{T}}$. These are plane waves with wave vectors $\mathbf{q}$,

$$\hat{\mathcal{T}}|\mathbf{q}\rangle = Dq^2|\mathbf{q}\rangle \tag{B.1}$$

$$\mathbf{q}: =(\mathbf{k}_1, \mathbf{k}_2, \ldots, \mathbf{k}_N), \quad q: =|\mathbf{q}|. \tag{B.2}$$

Here, $\mathbf{q}$ is a shorthand notation similar to $\mathbf{x}$. The operator $\hat{\mathcal{T}}$ is self-adjoint with real eigenvalues, so $e^{-\tau\hat{\mathcal{T}}}$ can be written as an *eigenfunction expansion* or *spectral expansion*,

$$e^{-\tau\hat{\mathcal{T}}} = \sum_{\mathbf{q}} e^{-\tau Dq^2}|\mathbf{q}\rangle\langle\mathbf{q}| \qquad \text{spectral expansion.} \tag{B.3}$$

Exponent operators are encountered so often that a word of warning is appropriate. For operators with *unbounded eigenvalues*, such as $\hat{\mathcal{T}}$, it is unsafe to expand the exponent operator as a Taylor series expansion $\exp(x) = 1 + x + x^2/2! + \cdots$,

$$e^{-\tau\hat{\mathcal{T}}} \stackrel{?}{=} 1 - \tau\hat{\mathcal{T}} + \frac{1}{2!}\tau^2\hat{\mathcal{T}}^2 - \ldots \qquad \text{take care!} \tag{B.4}$$

but

$$e^{-\tau\hat{\mathcal{H}}} = 1 - \tau\hat{\mathcal{H}} + \frac{1}{2!}\tau^2\hat{\mathcal{H}}^2 - \ldots \qquad \text{correct, if the series converges!} \tag{B.5}$$

Without delving into the intricacies of operator theory, the problem is due to domains and the convergence of the series. Consider particles initially in a confined region; there are no walls—they just happen to be there initially. We know diffusion will spread them out without limits. That is what $e^{-\tau\hat{\mathcal{T}}}$ does, but the right-hand side of equation (B.4) cannot move the particles outside the original region, no matter how many terms you add. It is as if there are invisible walls that keep them in, and the motion of the particles looks quite exotic. Instead of using the $\exp(x)$ Taylor

doi:10.1088/978-0-7503-6310-5ch8

series expansion, it is safe to use the spectral expansion. Be aware that even the spectral expansion is valid only for self-adjoint operators, such as $e^{-\tau \hat{T}}$.

Insert the spectral expansion (B.3) into the diffusion matrix elements,

$$\langle \mathbf{x}' | e^{-\tau \hat{T}} | \mathbf{x} \rangle = \sum_{\mathbf{q}} e^{-\tau D q^2} \langle \mathbf{x}' | \mathbf{q} \rangle \langle \mathbf{q} | \mathbf{x} \rangle. \tag{B.6}$$

Using the Heisenberg uncertainty relation, one can show that in the volume Ω,

$$\langle \mathbf{x} | \mathbf{q} \rangle = \frac{1}{\Omega^{N/2}} e^{-i\mathbf{q} \cdot \mathbf{x}}, \tag{B.7}$$

and using the orthogonality of plane waves, $\langle \mathbf{q}' | \mathbf{q} \rangle = \delta_{\mathbf{q}', \mathbf{q}}$, we get

$$\langle \mathbf{x}' | e^{-\tau \hat{T}} | \mathbf{x} \rangle = \frac{1}{\Omega^N} \sum_{\mathbf{q}} e^{-i\mathbf{q} \cdot (\mathbf{x}' - \mathbf{x})} e^{-\tau D q^2}. \tag{B.8}$$

If the possible values $\mathbf{q}$ form a continuum, then the sum becomes an integral,

$$\langle \mathbf{x}' | e^{-\tau \hat{T}} | \mathbf{x} \rangle = \frac{1}{\Omega^N} \int d\mathbf{q} \, e^{-i\mathbf{q} \cdot (\mathbf{x}' - \mathbf{x})} e^{-\tau D q^2} = \left[\frac{1}{\Omega} \int d\mathbf{k} \, e^{-i\mathbf{k} \cdot (\mathbf{r}' - \mathbf{r})} e^{-\tau D k^2} \right]^N. \tag{B.9}$$

The integral in the brackets is the d-dimensional Fourier transform of a Gaussian with variance $\sqrt{2D\tau}$, so we recover equation (4.39).

Appendix C

Rejection method

Let us derive one way to sample a distribution. If a distribution $w(x)$ with $x \in [0, 1]$ has an upper limit $\mathcal{M}$, then one has the identity

$$w(x) = \int_0^{\mathcal{M}} d\xi_2 \, \theta(\xi_2 - w(x)) = \int_0^1 d\xi_1 \int_0^{\mathcal{M}} d\xi_2 \, \delta(x - \xi_1)\theta(w(\xi_1) - \xi_2), \quad \text{(C.1)}$$

where $\theta(x)$ is the step function,

$$\theta(x) = \begin{cases} 1, & \text{if } x > 0 \\ 0, & \text{else} \end{cases}. \quad \text{(C.2)}$$

The first equality in equation (C.1) can be read as a strange way of expressing the two-dimensional area of a $1 \times w(x)$ rectangle as a longer $1 \times \mathcal{M}$ rectangle, where the part exceeding $w(x)$ is 'empty,' as illustrated in figure C.1. The second equality is essentially the Dirac delta function, which is used to replace the argument x in $w(x)$ with ξ_1.

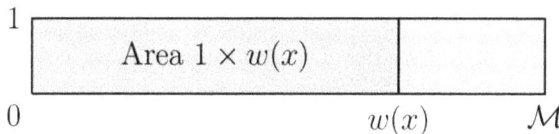

Figure C.1. Rejection method illustrated as an area computation. The whole area is integrated over, but the step function integrand is zero outside the blue area.

The double integral can be evaluated using Monte Carlo, and the resulting algorithm is known as the *rejection method*:

doi:10.1088/978-0-7503-6310-5ch9

Rejection method

Compute the maximum $\mathcal{M}$ of distribution $w(x)$.

1. Sample two random numbers, $\xi_1 \in U[0, 1]$ and $\xi_2 \in U[0, \mathcal{M}]$.
2. Evaluate $w(\xi_1)$.
3. If $\xi_2 < w(\xi_1)$, accept $x = \xi_1$, else reject it.
4. Return to step 1.

The identity (C.1) proves that the numbers x sample the distribution $w(x)$.

IOP Publishing

A Practical Course on Quantum Monte Carlo

Vesa Apaja

Appendix D

Updating Slater determinants

The *Slater matrix* D (often denoted by S) of single-particle states (orbitals) is defined as

$$D_{ij}: = \phi_j(\mathbf{r}_i), \tag{D.1}$$

where $\mathbf{r}_i$ is the coordinate of the ith electron. Recall that the order of the indices is: first, the electron index; then, the orbital index. The inverse matrix is

$$D^{-1} = \frac{C^T}{|D|}, \tag{D.2}$$

where the *cofactor* C has elements

$$C_{ij} = (-1)^{i+j} M_{ij}, \tag{D.3}$$

M_{ij} being the so-called i, j -minor, a Slater matrix with the row i and the column j left out. Looking at the definition (D.1), the row i has the coordinate of the ith electron, so C_{ij} **does not depend on** $\mathbf{r}_i$. This is a crucial point in gradient and Laplacian computation. The determinant $|D|$ can be solved using equation (D.2),

$$D^{-1}|D| = C^T \quad |D* \tag{D.4}$$

$$\Leftrightarrow DD^{-1}|D| = DC^T \tag{D.5}$$

$$\Leftrightarrow |D| = DC^T = \sum_j D_{ij}(C^T)_{ji} = \sum_j D_{ij} C_{ij} \quad \forall \, i, \tag{D.6}$$

and this gives the cofactor (Legendre) expansion formula

$$\boxed{|D| = \sum_j D_{ij} C_{ij} \quad \forall \, i,} \tag{D.7}$$

doi:10.1088/978-0-7503-6310-5ch10

and also

$$C^T = |D|D^{-1} \Leftrightarrow C = |D|(D^{-1})^T; \qquad \text{(D.8)}$$

therefore,

$$\boxed{C_{ij} = |D|(D^{-1})_{ji}.} \qquad \text{(D.9)}$$

Both are useful in updating matrices.

D.0.1 Drift and local energy of the Slater–Jastrow trial wave function

Fixed-node diffusion Monte Carlo (DMC) of electron systems often uses the Slater–Jastrow trial wave function,

$$\boxed{\varphi_T := e^J |D^\uparrow||D^\downarrow|,} \qquad \text{(D.10)}$$

and for importance sampling, we need to compute the drift and the local energy. The drift is the gradient

$$\boxed{\mathbf{F}_i = 2\frac{\nabla_i \varphi_T}{\varphi_T} = 2\frac{\nabla_i e^J}{e^J} + 2\frac{\nabla_i |D^\uparrow|}{|D^\uparrow|} + 2\frac{\nabla_i |D^\downarrow|}{|D^\downarrow|}.} \qquad \text{(D.11)}$$

The local energy is (in Hartree atomic units)

$$\boxed{E_L := \varphi_T^{-1} \hat{\mathcal{H}} \varphi_T = -\frac{1}{2}\varphi_T^{-1}\sum_i \nabla_i^2\ \varphi_T + V} \qquad \text{(D.12)}$$

$$\boxed{\begin{aligned} = -\frac{1}{2}&\left[\sum_i \frac{\nabla_i^2 e^J}{e^J} + \sum_{i \in ups} \frac{\nabla_i^2 |D^\uparrow|}{|D^\uparrow|} + \sum_{i \in downs} \frac{\nabla_i^2 |D^\downarrow|}{|D^\downarrow|} \right. \\ &\left. +2\sum_{i \in ups} \frac{\nabla_i e^J}{e^J} \cdot \frac{\nabla_i |D^\uparrow|}{|D^\uparrow|} + 2\sum_{i \in downs} \frac{\nabla_i e^J}{e^J} \cdot \frac{\nabla_i |D^\downarrow|}{|D^\downarrow|}\right] + V. \end{aligned}} \qquad \text{(D.13)}$$

There is no up–down term because the ith electron has either spin up or spin down.

D.0.2 Gradients of a Slater determinant

For the gradient ∇_i, we can use the fact that the cofactor C_{ij} has no $\mathbf{r}_i$. The gradient of a Slater determinant is

$$\frac{\nabla_i |D|}{|D|} = \frac{\nabla_i \sum_j D_{ij} C_{ij}}{\sum_j D_{ij} C_{ij}} \qquad \text{(D.14)}$$

$$= \frac{\sum_j (\boldsymbol{\nabla}_i D_{ij}) C_{ij}}{\sum_j D_{ij} C_{ij}} = \frac{\sum_j (\boldsymbol{\nabla}_i \phi_j(\mathbf{r}_i)) |\not{D}| (D^{-1})_{ji}}{\sum_j D_{ij} |\not{D}| (D^{-1})_{ji}} \tag{D.15}$$

$$= \frac{\sum_j (\boldsymbol{\nabla}_i \phi_j(\mathbf{r}_i))(D^{-1})_{ji}}{(DD^{-1})_{ii}}, \tag{D.16}$$

and $(DD^{-1})_{ii} = 1$, so

$$\boxed{\frac{\boldsymbol{\nabla}_i |D|}{|D|} = \sum_j (\boldsymbol{\nabla}_i \phi_j(\mathbf{r}_i))(D^{-1})_{ji}.} \tag{D.17}$$

Similarly, the Laplacian is

$$\boxed{\frac{\nabla_i^2 |D|}{|D|} = \sum_j (\nabla_i^2 \phi_j(\mathbf{r}_i))(D^{-1})_{ji}.} \tag{D.18}$$

The gradient and the Laplacian of single-particle states are simple to calculate analytically. We could recompute the inverse matrix $(D^{-1})_{ji}$ after each position update, but this becomes time-consuming for large N. Instead, we can update $(D^{-1})_{ji}$ after changing one coordinate.

D.0.3 Updating the inverse Slater matrix

Again, the cofactor C_{ij} does not change, so

$$C_{ij} \equiv C_{ij}^{\text{new}} \equiv C_{ij}^{\text{old}} \tag{D.19}$$

$$\Leftrightarrow C_{ij} = |D|^{\text{old}} (D_{\text{old}}^{-1})_{ji} = |D|^{\text{new}} (D_{\text{new}}^{-1})_{ji}, \tag{D.20}$$

and we get two formulas,

$$|D|^{\text{old}} = \sum_j D_{ij}^{\text{old}} C_{ij} \tag{D.21}$$

$$|D|^{\text{new}} = \sum_j D_{ij}^{\text{new}} C_{ij}. \tag{D.22}$$

In variational Monte Carlo (VMC), the move of the ith electron is accepted depending on the *square* of the ratio

$$\frac{\varphi_T^{\text{new}}}{\varphi_T^{\text{old}}} = \frac{(e^J)^{\text{new}} |D^\uparrow|^{\text{new}} |D^\downarrow|^{\text{new}}}{(e^J)^{\text{old}} |D^\uparrow|^{\text{old}} |D^\downarrow|^{\text{old}}} = \frac{(e^J)^{\text{new}} |D^\uparrow|^{\text{new}}}{(e^J)^{\text{old}} |D^\uparrow|^{\text{old}}}, \tag{D.23}$$

if the ith electron has spin up. In any case, only one determinant remains, and the determinant ratio is

$$\frac{|D|^{\text{new}}}{|D|^{\text{old}}} = \frac{\sum_j D_{ij}^{\text{new}} C_{ij}}{\sum_j D_{ij}^{\text{old}} C_{ij}} = \frac{\sum_j D_{ij}^{\text{new}} \,|\cancel{D}|^{\cancel{\text{old}}}\,(D_{\text{old}}^{-1})_{ji}}{\sum_j D_{ij}^{\text{old}} \,|\cancel{D}|^{\cancel{\text{old}}}\,(D_{\text{old}}^{-1})_{ji}} \tag{D.24}$$

$$= \frac{\sum_j D_{ij}^{\text{new}} (D_{\text{old}}^{-1})_{ji}}{\sum_j D_{ij}^{\text{old}} (D_{\text{old}}^{-1})_{ji}} = \frac{\sum_j D_{ij}^{\text{new}} (D_{\text{old}}^{-1})_{ji}}{(D^{\text{old}} (D_{\text{old}}^{-1}))_{ii}} = \frac{\sum_j D_{ij}^{\text{new}} (D_{\text{old}}^{-1})_{ji}}{1} \tag{D.25}$$

and the result is (remembering $D_{ij}^{\text{new}} = \phi_j(\mathbf{r}_i^{\text{new}})$)

$$\boxed{\frac{|D|^{\text{new}}}{|D|^{\text{old}}} = \sum_j D_{ij}^{\text{new}} (D_{\text{old}}^{-1})_{ji} = (D^{\text{new}} D_{\text{old}}^{-1})_{ii} \quad , \text{ if } i\text{:th electron moves}.} \tag{D.26}$$

This formula would not be much good if there were not a fast way of keeping the inverse matrix up-to-date.

Inverse matrix update Start from the matrix identity

$$(A + B)^{-1} = A^{-1} - (A + B)^{-1} B A^{-1}. \tag{D.27}$$

Proof:

$$(A + B)^{-1} = A^{-1} - (A + B)^{-1} B A^{-1} \;\; |(A + B)* \tag{D.28}$$

$$\Leftrightarrow (A + B)(A + B)^{-1} = (A + B) A^{-1} - (A + B)(A + B)^{-1} B A^{-1} \tag{D.29}$$

$$\Leftrightarrow (A + B)(A + B)^{-1} = (A + B) A^{-1} - B A^{-1} = I + B A^{-1} - B A^{-1} = I. \tag{D.30}$$

Write the identity for $A: = D_{\text{old}}$ and $B: = D_{\text{new}} - D_{\text{old}}$, so that $A + B = D_{\text{new}}$,

$$D_{\text{new}}^{-1} = D_{\text{old}}^{-1} - D_{\text{new}}^{-1} B D_{\text{old}}^{-1}. \tag{D.31}$$

In our case (use index k for positions and reserve i for the moved electron),

$$(D_{\text{new}})_{kj} = \phi_j(\mathbf{r}_k^{\text{new}}) = \phi_j(\mathbf{r}_k^{\text{old}}) + (\phi_j(\mathbf{r}_k^{\text{new}}) - \phi_j(\mathbf{r}_k^{\text{old}})) \tag{D.32}$$

$$= (D_{\text{old}})_{kj} + \underbrace{(\phi_j(\mathbf{r}_k^{\text{new}}) - \phi_j(\mathbf{r}_k^{\text{old}}))}_{B_{kj}}, \tag{D.33}$$

so we find that

$$(D_{\text{new}}^{-1})_{kj} = (D_{\text{old}}^{-1})_{kj} - \sum_{nm} (D_{\text{new}}^{-1})_{kn} B_{nm} (D_{\text{old}}^{-1})_{mj} \tag{D.34}$$

$$= (D_{\text{old}}^{-1})_{kj} - \sum_{nm} (D_{\text{new}}^{-1})_{kn} [\phi_m(\mathbf{r}_n^{\text{new}}) - \phi_m(\mathbf{r}_n^{\text{old}})](D_{\text{old}}^{-1})_{mj}. \tag{D.35}$$

Only $\mathbf{r}_i$ was updated, so the bracket vanishes unless $n = i$,

$$(D_{\text{new}}^{-1})_{kj} = (D_{\text{old}}^{-1})_{kj} - (D_{\text{new}}^{-1})_{ki} \sum_m [\phi_m(\mathbf{r}_i^{\text{new}}) - \phi_m(\mathbf{r}_i^{\text{old}})](D_{\text{old}}^{-1})_{mj}$$

$$= (D_{\text{old}}^{-1})_{kj} - (D_{\text{new}}^{-1})_{ki} \sum_m \underbrace{\phi_m(\mathbf{r}_i^{\text{new}})}_{(D_{\text{new}})_{im}} (D_{\text{old}}^{-1})_{mj} \tag{D.36}$$

$$+ (D_{\text{new}}^{-1})_{ki} \underbrace{\sum_m \underbrace{\phi_m(\mathbf{r}_i^{\text{old}})}_{(D_{\text{old}})_{im}} (D_{\text{old}}^{-1})_{mj}}_{(D_{\text{old}} D_{\text{old}}^{-1})_{ij} = \delta_{ij}}, \tag{D.37}$$

which gives

$$(D_{\text{new}}^{-1})_{kj} = (D_{\text{old}}^{-1})_{kj} - (D_{\text{new}}^{-1})_{ki} \sum_m (D_{\text{new}})_{im}(D_{\text{old}}^{-1})_{mj} + (D_{\text{new}}^{-1})_{ki}\delta_{ij}. \tag{D.38}$$

This can be written in the form

$$(D_{\text{new}}^{-1})_{kj} = (D_{\text{old}}^{-1})_{kj} - (D_{\text{new}}^{-1})_{ki} K_{ij} + (D_{\text{new}}^{-1})_{ki}\delta_{ij}, \tag{D.39}$$

where

$$K_{ij} := \sum_m (D_{\text{new}})_{im}(D_{\text{old}}^{-1})_{mj} = (D_{\text{new}} D_{\text{old}}^{-1})_{ij}. \tag{D.40}$$

Remember that i is the fixed index of the moved electron. Obviously, K_{ij} is a generalization of the sum in equation (D.26), where we used only the diagonals,

$$K_{ii} = \frac{|D|^{\text{new}}}{|D|^{\text{old}}} \quad \text{after } i\text{th electron move.} \tag{D.41}$$

We get two cases,

$$(D_{\text{new}}^{-1})_{kj} = (D_{\text{old}}^{-1})_{kj} - (D_{\text{new}}^{-1})_{kj} K_{ii} + (D_{\text{new}}^{-1})_{kj}, \quad \text{if } j = i \tag{D.42}$$

$$(D_{\text{new}}^{-1})_{kj} = (D_{\text{old}}^{-1})_{kj} - (D_{\text{new}}^{-1})_{ki} K_{ij}, \text{ if } j \neq i, \tag{D.43}$$

and the first equation can be immediately solved for $(D_{\text{new}}^{-1})_{ki}$,

$$\boxed{(D_{\text{new}}^{-1})_{kj} = \frac{(D_{\text{old}}^{-1})_{kj}}{K_{ii}} = \frac{(D_{\text{old}}^{-1})_{kj}}{\dfrac{|D|^{\text{new}}}{|D|^{\text{old}}}} \quad , \text{if } j = i,} \tag{D.44}$$

which can be inserted into the latter equation as $(D_{\text{new}}^{-1})_{ki}$ to get

$$
\begin{aligned}
(D_{\text{new}}^{-1})_{kj} &= (D_{\text{old}}^{-1})_{kj}\left(1 - \frac{K_{ij}}{K_{ii}}\right) \\
&= (D_{\text{old}}^{-1})_{kj}\left(1 - \frac{\sum\limits_{m}(D_{\text{new}})_{im}(D_{\text{old}}^{-1})_{mj}}{\dfrac{|D|^{\text{new}}}{|D|^{\text{old}}}}\right), \quad \text{if } j \neq i.
\end{aligned}
\tag{D.45}
$$

The inverse matrix update formulas are also very useful in writing efficient VMC and DMC codes.

Appendix E

Path integral Monte Carlo virial estimator

Let us evaluate the left-hand side of equation (5.243) for the Chin action (CA). The CA has (see equation (5.241))

$$\rho(\mathbf{x}_1, \ldots, \mathbf{x}_M; \beta) = \prod_{n=1}^{M} (4\pi\lambda\tau t_n)^{-3N/2} e^{-\frac{(\mathbf{x}_n - \mathbf{x}_{n+1})^2}{4\lambda\tau t_n}} \times \prod_{n=1}^{M} e^{-t_n\tau \tilde{V}_n(\mathbf{x}_n; \tau)}. \tag{E.1}$$

The last coordinate $\mathbf{x}_{M+1}$ is $\mathbf{x}_1$ for distinguishable particles and $P(\mathbf{x}_1)$ for identical particles. The coordinate $\mathbf{x}_m$ with $m \neq 1$ (the gradient of $\mathbf{x}_1$ is not needed) appears in the spring factor for indices $n = m$ and $n = m - 1$ and in the potential factor for $n = m$,

$$\frac{\partial}{\partial \mathbf{x}_m} \rho(\mathbf{x}_1, \ldots, \mathbf{x}_M; \beta) \tag{E.2}$$

$$= \rho(\mathbf{x}_1, \ldots, \mathbf{x}_M; \beta) \frac{\partial}{\partial \mathbf{x}_m} \left[-\frac{(\mathbf{x}_m - \mathbf{x}_{m+1})^2}{4\lambda\tau t_m} - \frac{(\mathbf{x}_{m-1} - \mathbf{x}_m)^2}{4\lambda\tau t_{m-1}} - \tau t_m \tilde{V}_m(\mathbf{x}_m; \tau) \right] \tag{E.3}$$

$$= \rho(\mathbf{x}_1, \ldots, \mathbf{x}_M; \beta) \left[-2\frac{(\mathbf{x}_m - \mathbf{x}_{m+1})}{4\lambda\tau t_m} + 2\frac{(\mathbf{x}_{m-1} - \mathbf{x}_m)}{4\lambda\tau t_{m-1}} - \frac{\partial}{\partial \mathbf{x}_m}[\tau t_m \tilde{V}_m(\mathbf{x}_m; \tau)] \right], \tag{E.4}$$

where the factors of two show the origin of the doubling of part of the kinetic energy. Insert this into the left-hand side of equation (5.243):

$$G = -\int \prod_{m=2}^{M} d\mathbf{x}_m \sum_{m=2}^{M} (\mathbf{x}_m - \mathbf{x}_1) \cdot \frac{\partial}{\partial \mathbf{x}_m} \rho(\mathbf{x}_1, \ldots, \mathbf{x}_M; \beta) \tag{E.5}$$

$$= -\int \prod_{m=1}^{M} d\mathbf{x}_m \rho(\mathbf{x}_1, \ldots, \mathbf{x}_M; \beta) \tag{E.6}$$

doi:10.1088/978-0-7503-6310-5ch11

$$\times \left\{ 2 \sum_{m=1}^{M} (\mathbf{x}_m - \mathbf{x}_1) \cdot \left[-\frac{(\mathbf{x}_m - \mathbf{x}_{m+1})}{4\lambda \tau t_m} + \frac{(\mathbf{x}_{m-1} - \mathbf{x}_m)}{4\lambda \tau t_{m-1}} \right] \right. \tag{E.7}$$

$$\left. - \sum_{m=1}^{M} (\mathbf{x}_m - \mathbf{x}_1) \cdot \frac{\partial}{\partial \mathbf{x}_m} [\tau t_m \tilde{V}_m(\mathbf{x}_m; \tau)] \right\}. \tag{E.8}$$

For convenience, sums start from $m = 1$; the first term is zero. The sums can be simplified (see the details below), and the result is

$$G = \int \prod_{m=2}^{M} d\mathbf{x}_m \rho(\mathbf{x}_1, \ldots, \mathbf{x}_M; \beta) \left\{ 2 \sum_{m=1}^{M} \frac{(\mathbf{x}_m - \mathbf{x}_{m-1})^2}{4\lambda \tau t_m} \right. \tag{E.9}$$

$$+ 2(\mathbf{x}_{M+1} - \mathbf{x}_1) \cdot \frac{(\mathbf{x}_M - \mathbf{x}_{M+1})}{4\lambda \tau t_M} \tag{E.10}$$

$$\left. + \sum_{m=1}^{M} (\mathbf{x}_m - \mathbf{x}_1) \cdot \frac{\partial}{\partial \mathbf{x}_m} [\tau t_m \tilde{V}_m(\mathbf{x}_m; \tau)] \right\}. \tag{E.11}$$

Details:

$$\sum_{m=1}^{M} (\mathbf{x}_m - \mathbf{x}_1) \cdot \left[-\frac{(\mathbf{x}_m - \mathbf{x}_{m+1})}{4\lambda \tau t_m} + \frac{(\mathbf{x}_{m-1} - \mathbf{x}_m)}{4\lambda \tau t_{m-1}} \right] \tag{E.12}$$

$$= -\sum_{m=1}^{M} (\mathbf{x}_m - \mathbf{x}_{m+1} + \mathbf{x}_{m+1} - \mathbf{x}_1) \cdot \frac{(\mathbf{x}_m - \mathbf{x}_{m+1})}{4\lambda \tau t_m} + \sum_{m=1}^{M} (\mathbf{x}_m - \mathbf{x}_1) \cdot \frac{(\mathbf{x}_{m-1} - \mathbf{x}_m)}{4\lambda \tau t_{m-1}} \tag{E.13}$$

$$= -\sum_{m=1}^{M} \frac{(\mathbf{x}_m - \mathbf{x}_{m+1})^2}{4\lambda \tau t_m}$$
$$- \sum_{m=1}^{M} (\mathbf{x}_{m+1} - \mathbf{x}_1) \cdot \underbrace{\frac{(\mathbf{x}_m - \mathbf{x}_{m+1})}{4\lambda \tau t_m}}_{:=\Delta_m} + \sum_{m=1}^{M} (\mathbf{x}_m - \mathbf{x}_1) \cdot \underbrace{\frac{(\mathbf{x}_{m-1} - \mathbf{x}_m)}{4\lambda \tau t_{m-1}}}_{\Delta_{m-1}} \tag{E.14}$$

$$= -\sum_{m=1}^{M} \frac{(\mathbf{x}_m - \mathbf{x}_{m+1})^2}{4\lambda \tau t_m} - \sum_{m=1}^{M-1} (\mathbf{x}_{m+1} - \mathbf{x}_1) \cdot \Delta_m + \sum_{m=1}^{M-1} (\mathbf{x}_{m+1} - \mathbf{x}_1) \cdot \Delta_m$$
$$- (\mathbf{x}_{M+1} - \mathbf{x}_1) \cdot \Delta_M + (\mathbf{x}_1 - \mathbf{x}_1) \cdot \Delta_0 \tag{E.15}$$

$$= -\sum_{m=1}^{M} \frac{(\mathbf{x}_m - \mathbf{x}_{m+1})^2}{4\lambda \tau t_m} - (\mathbf{x}_{M+1} - \mathbf{x}_1) \cdot \frac{(\mathbf{x}_M - \mathbf{x}_{M+1})}{4\lambda \tau t_M}. \tag{E.16}$$

To check this, see the SymPy code in `sympy_sum_x_simplicifation.py`.

IOP Publishing

A Practical Course on Quantum Monte Carlo

Vesa Apaja

Appendix F

Error estimation

F.1 Definitions

Consider evaluating a quantity A using N samples. The mean value over the sample —the so-called *sample average*—is

$$\bar{A}_N := \frac{1}{N}\sum_{i=1}^{N} A_i. \tag{F.1}$$

This is not the expectation value (the ensemble average or the mean) of A, marked $\langle A \rangle$, because the sample is finite. The expectation value is the sum over every possible value of A_i,

$$\langle A \rangle \equiv \sum_i p_i A_i, \tag{F.2}$$

where p_i is the probability of A_i. For a continuous distribution, this is

$$\langle A \rangle \equiv \int dx p(x) A(x). \tag{F.3}$$

The difference is that:
- $\bar{A}_N$ is a random variable that depends on the chosen sample.
- $\langle A \rangle$ is a fixed number.

According to the law of large numbers, the mean and the expectation value coincide for an infinitely large sample,

$$\lim_{N\to\infty} \bar{A}_N = \langle A \rangle. \tag{F.4}$$

doi:10.1088/978-0-7503-6310-5ch12   F-1   © IOP Publishing Ltd 2025. All rights,

F.2 Biased and unbiased estimators of variance

The variance σ^2 is the mean square deviation from the mean value. The standard deviation σ is the reported error. Monte Carlo (MC) results are given as

$$result = \underbrace{1.234}_{\text{mean value}} \pm \underbrace{0.002}_{\sigma}. \tag{F.5}$$

First, one must be clear about which mean value is used in the evaluation of σ^2. If we use the mean over N samples, $\bar{A}_N$, then we get the *biased estimator of the variance*:

$$\sigma^2_{biased} := \frac{1}{N}\sum_{i=1}^{N}(A_i - \bar{A}_N)^2 \quad \text{biased estimator of the variance.} \tag{F.6}$$

The *unbiased estimator of the variance* is the mean square deviation from the true expectation value,

$$\sigma^2_{unbiased} := \frac{1}{N}\sum_{i=1}^{N}(A_i - \langle A \rangle)^2 \quad \text{unbiased estimator of the variance.} \tag{F.7}$$

Obviously, the biased and unbiased estimators are related. Start from the definition of the biased estimator and add and subtract the constant $\langle A \rangle$:

$$\sigma^2_{biased} = \frac{1}{N}\sum_{i=1}^{N}(A_i - \bar{A}_N)^2 = \frac{1}{N}\sum_{i=1}^{N}[A_i - \langle A \rangle - (\bar{A}_N - \langle A \rangle)]^2$$

$$= \underbrace{\frac{1}{N}\sum_{i=1}^{N}[A_i - \langle A \rangle]^2}_{\sigma^2_{unbiased}} - 2\frac{1}{N}\sum_{i=1}^{N}(A_i - \langle A \rangle)\underbrace{(\bar{A}_N - \langle A \rangle)}_{no\ i} + \frac{1}{N}\sum_{i=1}^{N}\underbrace{(\bar{A}_N - \langle A \rangle)^2}_{no\ i} \tag{F.8}$$

$$= \sigma^2_{unbiased} - 2(\bar{A}_N - \langle A \rangle)\underbrace{\frac{1}{N}\sum_{i=1}^{N}(A_i - \langle A \rangle)}_{\bar{A}_N - \langle 1 \rangle} + (\bar{A}_N - \langle A \rangle)^2$$

$$= \sigma^2_{unbiased} - (\bar{A}_N - \langle A \rangle)^2. \tag{F.9}$$

In MC, statistical error estimation is based on the central limit theorem, equation (A.16).

We compute the average of the collected data, which has unbiased variance $\sigma^2_{unbiased}$. In the central limit theorem, the mean μ is over all values, so it equals the expectation value,

$$\mu \equiv \langle A \rangle. \tag{F.10}$$

Assuming the central limit theorem is valid for our data—and we must assume this, otherwise we do not have a clue about the statistical error—we have:

$$(\bar{A}_N - \langle A \rangle)^2 = (\bar{A}_N - \mu)^2 = \left[\frac{\sigma_{unbiased}}{\sqrt{N}} \right]^2 = \frac{\sigma_{unbiased}^2}{N}. \tag{F.11}$$

Inserting this into equation (F.9) gives the relation

$$\boxed{\sigma_{biased}^2 = \sigma_{unbiased}^2 \left[1 - \frac{1}{N} \right].} \tag{F.12}$$

This can be interpreted as a finite sample-size correction to the variance. To conclude, the bias-corrected MC standard deviation, **the error**, is

$$\boxed{\sigma = \frac{\sigma_{unbiased}}{\sqrt{N}} = (\bar{A}_N - \langle A \rangle)^2 = \sqrt{\frac{1}{N(N-1)} \sum_{i=1}^{N} (A_i - \bar{A}_N)^2}.} \tag{F.13}$$

A few notions about the result so far:
- Without bias correction, finite sampling underestimates variances
  $\Rightarrow$ susceptibilities (response functions) are underestimated by a factor of $1 - \frac{1}{N}$.
  One might argue: so what? Just collect more data and get a smaller $1/N$. True, but here comes the real bad news:
- N = the number of statistically independent (!) samples.
- Near criticality (e.g. near T_c in the Ising model), sampling may become inefficient $\Leftrightarrow$ Autocorrelation times increase $\Leftrightarrow$ more computational time is needed to increase N.
- N is related to the so-called *integrated autocorrelation time* τ_{int} by the relation (where τ_{int} is given in measurement time intervals)

$$N_{statistically\ independent\ samples} = \frac{N_{all\ samples}}{1 + 2\tau_{int}}; \tag{F.14}$$

therefore, if the integrated autocorrelation time increases dramatically for some reason, you need a lot more MC steps to reach the same accuracy with the same N.

In addition to correcting the sample bias, we need to address the question of **correlations** in samples.

F.3 Integrated autocorrelation time

The integrated autocorrelation time τ_{int} appeared in the discussion without derivation. Correlations are expected to die out exponentially, meaning that the measured values at MC times i and $i + t$ satisfy

$$\langle A_i A_{i+t} \rangle - \langle A \rangle^2 \propto e^{-t/\tau_{exp}}, \tag{F.15}$$

where τ_{exp} is the exponential correlation time, that is, the correlation decay rate specific to the quantity A. The index i vanishes due to the assumption of translational invariance in simulation time, meaning that we can consider an arbitrary simulation step i,

$$\langle A_i A_{i+t} \rangle = \langle A_0 A_{0+t} \rangle. \tag{F.16}$$

Naturally, this is valid only in equilibrium, i.e. after the thermalization steps. The normalized autocorrelation function of A is defined as

$$\phi(t) = \frac{\langle A_i A_{i+t} \rangle - \langle A \rangle^2}{\langle A^2 \rangle - \langle A \rangle^2} \; \forall \; i. \tag{F.17}$$

A better average is obtained by integrating—or rather summing—the autocorrelation function. The integrated autocorrelation time is defined as

$$\boxed{\tau_{int} := \frac{\sum\limits_{t=1}^{\infty}(\langle A_i A_{i+t} \rangle - \langle A \rangle^2)}{\langle A^2 \rangle - \langle A \rangle^2}.} \tag{F.18}$$

The infinite limit in the summation is impractical, and only a finite data window is available for the evaluation of τ_{int}.

So, how is τ_{int} related to the MC error estimate? The MC error estimate tells us how far the sample mean is from the true mean. We have the formula (F.13)

$$\sigma^2 = \frac{\sigma^2_{unbiased}}{N} = (\bar{A}_N - \langle A \rangle)^2, \tag{F.19}$$

so let us see how large this is *expected* to be on average and evaluate

$$\langle \sigma^2 \rangle = \langle (\bar{A}_N - \langle A \rangle)^2 \rangle. \tag{F.20}$$

There is a sum squared, so we anticipate that it can be expressed as a convolution, such as the integrated autocorrelation time in equation (F.18). Let us work out the details:

$$\langle \sigma^2 \rangle = \langle (\bar{A}_N - \langle A \rangle)^2 \rangle = \left\langle \left(\frac{1}{N} \sum_{i=1}^{N} (A_i - \langle A \rangle) \right)^2 \right\rangle \tag{F.21}$$

$$= \left\langle \left(\frac{1}{N} \sum_{i=1}^{N} (A_i - \langle A \rangle) \right) \left(\frac{1}{N} \sum_{j=1}^{N} (A_j - \langle A \rangle) \right) \right\rangle. \tag{F.22}$$

The idea is to assume that N is large, so that

$$\frac{1}{N} \sum_{i=1}^{N} A_i = \langle A \rangle + \text{small deviation}, \tag{F.23}$$

and see how significant the small deviation is in the square. An isolated A_i can be replaced with $\langle A \rangle$, but the square also has the terms $A_i A_j$. Separate the cases $j = i$ and $j < i$, $j > i$ (the latter two give the same result):

$$\langle \sigma^2 \rangle = \frac{1}{N^2} \sum_{i=1}^{N} \langle (A_i - \langle A \rangle)^2 \rangle + \frac{2}{N^2} \sum_{i=1}^{N} \sum_{j=i+1}^{N} \langle A_i A_j - \langle A \rangle^2 \rangle, \tag{F.24}$$

In the second sum, $j > i$, so it is useful to define $j = i + t$:

$$\langle \sigma^2 \rangle = \frac{1}{N^2} \sum_{i=1}^{N} \langle (A_i - \langle A \rangle)^2 \rangle + \frac{2}{N^2} \sum_{i=1}^{N} \sum_{t=1}^{N-i} (\langle A_i A_{i+t} \rangle - \langle A \rangle^2). \tag{F.25}$$

Assuming the t-sum converges long before $t = N - i$, we may extend the sum to infinity:

$$\langle \sigma^2 \rangle \approx \frac{1}{N^2} \sum_{i=1}^{N} \langle (A_i - \langle A \rangle)^2 \rangle + \frac{2}{N^2} \sum_{i=1}^{N} \sum_{t=1}^{\infty} (\langle A_i A_{i+t} \rangle - \langle A^2 \rangle) \tag{F.26}$$

$$= \frac{1}{N^2} \sum_{i=1}^{N} \langle (A_i - \langle A \rangle)^2 \rangle + \frac{2}{N} \tau_{int} (\langle A^2 \rangle - \langle A \rangle^2) \tag{F.27}$$

$$= \frac{\langle A^2 \rangle - \langle A \rangle^2}{N} (1 + 2\tau_{int}). \tag{F.28}$$

This means that the so-called *statistical inefficiency* given by the factor $(1 + 2\tau_{int})$ changes the number of statistically independent samples from all samples N to a smaller number, $N / (1 + 2\tau_{int})$.

Analyzing the autocorrelations in data gives valuable information about how efficient the MC sampling algorithm is, which is a relevant question in designing Markov chain MC. However, if the algorithm is not subject to change, then the discussion about autocorrelations may not be that fruitful, and there are more practical ways to estimate the error in MC results.

F.4 Block averaging

The averages of sufficiently large blocks of data are statistically independent.

We need to establish how large is 'sufficiently large.' Split the collected data, N values, into blocks of size n. Say there are N_b blocks,

$$N = N_b \times n, \tag{F.29}$$

so we work with N_b mean values B_i computed for each block i:

$$\underbrace{A_1 A_2 \ \ldots \ A_n}_{\text{mean } B_1} \underbrace{A_{n+1} A_{n+2} \ \ldots \ A_{2n}}_{\text{mean } B_2} \underbrace{A_{2n+1} \ldots}_{} \underbrace{\ldots A_N}_{\text{mean } B_{N_b}}. \tag{F.30}$$

If n is sufficiently large, spanning a simulation time longer than the integrated autocorrelation time τ_{int}, then the values B_i are *stochastically independent*. If this is the case, the error is given by the unbiased standard deviation,

$$\sigma = \sqrt{\frac{1}{N_b(N_b - 1)}\sum_{i=1}^{N_b}(B_i - \bar{B}_{N_b})^2}\,. \tag{F.31}$$

If the blocks cover the whole sample of N values, then of course, the average of the blocks is the total average, $\bar{B}_{N_b} = \bar{A}_N$. An equivalent, cleaner form is obtained by evaluating the square in the previous formula,

$$error = \sqrt{\frac{1}{N_b - 1}(\overline{(B^2)}_{N_b} - (\bar{B}_{N_b})^2)}\,. \tag{F.32}$$

This method can be used for both primary variables, such as energy, that are measured directly from MC simulation, and for secondary variables, such as the specific heat, computed from energy data.

F.5 Resampling methods

The original set of data is collected by taking samples of a quantity during the simulation, which is known as sampling. In resampling, one takes samples from the data that has been collected, so it is sampling from a sample.

F.5.1 Jackknife

Von Mises had already applied the jackknife resampling method in some form before the computing era (1930s), but the main work was done by Quenouille (1949) (who wanted to find a way to estimate bias) and Tukey (1958).
- Divide the data into N blocks with a block size $>\tau_{int}$
  $\Rightarrow$ blocks are independent
- Omit the first, 2nd, 3rd,...block from the sample of size N. The decimated set $A_{(i)}$ has the ith block omitted. For example, for $N = 5$:

Full set of blocks	1	2	3	4	5	average $\bar{A}_N$
averages of blocks	$\bar{a}_1$	$\bar{a}_2$	$\bar{a}_3$	$\bar{a}_4$	$\bar{a}_5$	
$A_{(1)}$		2	3	4	5	average $\bar{A}_{(1)}$
$A_{(2)}$	1		3	4	5	average $\bar{A}_{(2)}$
$A_{(3)}$	1	2		4	5	average $\bar{A}_{(3)}$
$A_{(4)}$	1	2	3		5	average $\bar{A}_{(4)}$
$A_{(5)}$	1	2	3	4		average $\bar{A}_{(5)}$.

- The standard deviation can be computed using

$$\sigma = \sqrt{\frac{N-1}{N}\sum_{i=1}^{N}(\bar{A}_{(i)} - \bar{A}_N)^2}. \tag{F.33}$$

The factor of $\frac{N-1}{N}$ can be made plausible by noticing that

$$\bar{A}_{(i)} = \frac{1}{N-1}\sum_{j(\neq i)}\bar{a}_j, \tag{F.34}$$

so one recovers the standard deviation of the block data,

$$\sigma = \sqrt{\frac{1}{N(N-1)}\sum_{i=1}^{N}(\bar{a}_i - \bar{A}_N)^2}. \tag{F.35}$$

The resampled sets in the jackknife method are not really new measurements and have less data than the original set, so their variance is smaller than the variance in the full data. The resampled sets deviate from the original set by only one data point, so in this sense, the jackknife method probes the near vicinity of the original set.

Do not use the jackknife method to estimate the median—it will fail miserably. The median—the middle value of the data—has what we call 'non-smooth statistics': a small change in the data set can cause a big change in the median. This can be considered a sort of non-differentiability. Another quantity that is non-smooth is the maximum value of data.

F.5.2 Bootstrap

Bootstrapping, in the form represented here, is a generalization of jackknife resampling, suggested by Efron (1979).

- Resample, with replacement, the existing data to get a new data set. Compute the statistics from this resampled set (figure F.1).

    Original sample: 1 2 3 4 5
    1st bootstrap replica: 1 1 2 5 4
    2nd bootstrap replica: 5 2 2 1 3
    3rd bootstrap replica: 4 1 1 3 4
    etc.

- About $N_{\text{boot}} = 1000$ replicas are often enough.
- Compute the error using

$$error = \sqrt{\frac{1}{N-1}[\overline{(A^2)}_N - (\bar{A}_N)^2]}. \tag{F.36}$$

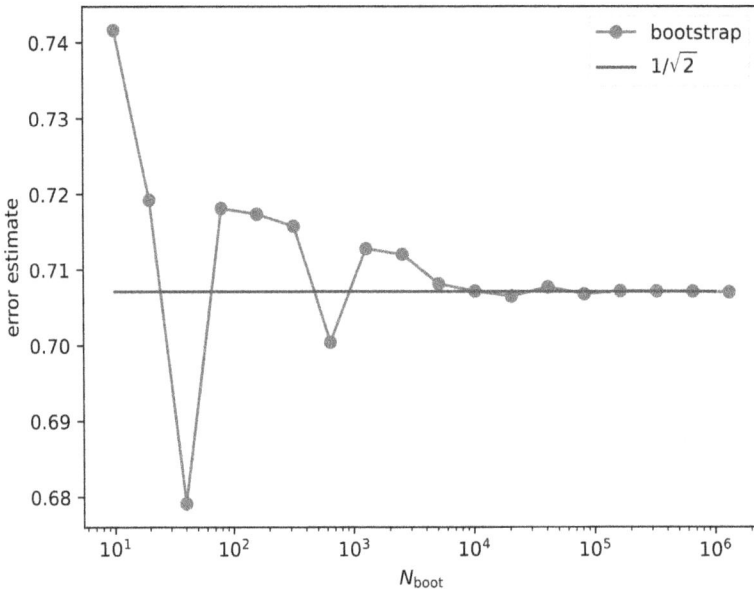

Figure F.1. The convergence of the bootstrap error estimate as a function of the number of bootstrap replicas. The original data set is {1,2,3,4,5}.

Do not divide by $(N_{boot} - 1)$.[1]
- You can block data to get independent samples, but it is not really necessary. Some blocking is usually done anyway because storing all the data on disk is uneconomical.

No theoretical calculations are needed; the method works no matter how complicated the chosen estimator (such as the variance) is. Just like the jackknife, the bootstrap can be used to simultaneously estimate bias, kurtosis, and other statistical properties of the data.

The bootstrap-resampled sets deviate from the original set more than the jackknife-resampled sets, so in this sense, the bootstrap method samples further away from the original set.

Jackknife or bootstrap ?
- Jackknife requires as many resampled sets as there are original data points ⇒ for over 100 data points, bootstrap becomes more effective.
- However, for bootstrap, you need about 1000 samples to obtain a decent estimate of the variance ⇒ jackknife seems to be a better choice for up to about 1000 data points.
- If you do not block the data before error analysis but choose to use statistically dependent data (=data picked more frequently than τ_{int}), you probably have lots of data ⇒ use bootstrap.

---

[1] The number of bootstrap replicas is unlimited, so having $1/(N_{boot} - 1)$ as a factor would lead to ever-diminishing error, which cannot be correct.

How should one debug error-estimation subroutines? Many readers will have used canned error estimation routines, but here are some suggestions in case you are adventurous and want to code your own routines:

- Program all three: blocking, jackknife, bootstrap.
- Pick a simple set of five data points: 1,2,3,4,5.
- Compare your error estimates to the exact value $\sigma = 1/\sqrt{2} = 0.707\ 17$.
- If you get 'NaN' in bootstrap, your $\overline{A^2} - (\bar{A})^2$ is negative. Check that you have divided using the correct integer counts. $\bar{A}$ should be close to three.
- If your bootstrap error does not converge with increasing numbers of replicas, you have not divided by N_{boot} in computing the averages $\overline{A^2}$ or $\bar{A}$.
- If your bootstrap converges to $\sigma = \sqrt{2} = 1.414$, you forgot to multiply by $1/(N-1)$ under the square root. In this case, $N = 5$ (but do not hardwire N to five).
- After all seems to work, generate N data points using your old reliable Gaussian random number generator. Analyze the data: the average should go toward zero, and the error of the average should diminish as

$$\sigma = \sigma_{input}/\sqrt{N} = 1/\sqrt{N}$$

because the standard normal distribution has $\sigma_{input} = 1$.

IOP Publishing

A Practical Course on Quantum Monte Carlo

Vesa Apaja

Appendix G

Perturbation expansion

As an example of the continuous-time splitting of $e^{-\beta \hat{\mathcal{H}}}$, define $\hat{\mathcal{H}} = \hat{\mathcal{H}}_0 + \hat{\mathcal{H}}'$ and write a perturbation expansion for the soluble Hamiltonian $\hat{\mathcal{H}}_0$ and the perturbation $\hat{\mathcal{H}}'$:

$$e^{-\beta \hat{\mathcal{H}}} = e^{-\beta(\hat{\mathcal{H}}_0 + \hat{\mathcal{H}}')} := e^{-\beta \hat{\mathcal{H}}_0} \hat{\mathcal{U}}(\beta), \tag{G.1}$$

where $\hat{\mathcal{U}}(\beta)$ gives the desired expansion. Solve $\hat{\mathcal{U}}(\beta) = e^{\beta \hat{\mathcal{H}}_0} e^{-\beta(\hat{\mathcal{H}}_0 + \hat{\mathcal{H}}')}$, and take the derivative w.r.t. β,

$$\frac{\partial}{\partial \beta} U(\beta) = \hat{\mathcal{H}}_0 e^{\beta \hat{\mathcal{H}}_0} e^{-\beta(\hat{\mathcal{H}}_0 + \hat{\mathcal{H}}')} - e^{\beta \hat{\mathcal{H}}_0} (\hat{\mathcal{H}}_0 + \hat{\mathcal{H}}') e^{-\beta(\hat{\mathcal{H}}_0 + \hat{\mathcal{H}}')} \tag{G.2}$$

$$= -e^{\beta \hat{\mathcal{H}}_0} \hat{\mathcal{H}}' e^{-\beta(\hat{\mathcal{H}}_0 + \hat{\mathcal{H}}')} = -e^{\beta \hat{\mathcal{H}}_0} \hat{\mathcal{H}}' e^{-\beta \hat{\mathcal{H}}_0} e^{\beta \hat{\mathcal{H}}_0} e^{-\beta(\hat{\mathcal{H}}_0 + \hat{\mathcal{H}}')} \tag{G.3}$$

$$= -e^{\beta \hat{\mathcal{H}}_0} \hat{\mathcal{H}}' e^{-\beta \hat{\mathcal{H}}_0} \hat{\mathcal{U}}(\beta) := -\hat{\mathcal{H}}_0(\beta) \hat{\mathcal{U}}(\beta). \tag{G.4}$$

Integration gives $U(\beta) = 1 - \int_0^\beta dt_1 \hat{\mathcal{H}}_0(t_1) U(t_1)$, which can be solved iteratively. The final, exact 'operator splitting' is an imaginary-time version of the well-known Dyson series,

$$e^{-\beta \hat{\mathcal{H}}} = e^{-\beta \hat{\mathcal{H}}_0} \left[1 - \int_0^\beta dt_1 \hat{\mathcal{H}}_0(t_1) + \int_0^\beta dt_1 \int_0^{t_1} dt_2 \hat{\mathcal{H}}_0(t_1) \hat{\mathcal{H}}_0(t_2) - ... \right]. \tag{G.5}$$

As always with perturbation theory, there is no guarantee of convergence. From the quantum Monte Carlo (QMC) point of view, this also has its own 'sign problem.' For $\hat{\mathcal{H}}_0 = \hat{\mathcal{T}}$ and $\hat{\mathcal{H}}' = \hat{\mathcal{V}}$, the perturbation expansion with *purely attractive potentials* has only positive signs, and the integrals can be sampled. The number of integrals n varies, and their average number $\langle n \rangle$ turns out to be related to the total energy,

doi:10.1088/978-0-7503-6310-5ch13
G-1

$$\langle E \rangle = -\frac{\langle n \rangle}{\beta},$$ (G.6)

just as in stochastic series expansion (SSE) (see equation (6.1)). However, unlike SSE, this estimator is now valid only for purely attractive potentials. The perturbation expansion has not been used in QMC because, in most cases, there is little to gain.

www.ingramcontent.com/pod-product-compliance
Lightning Source LLC
Chambersburg PA
CBHW080519220326
41599CB00032B/6133